Lecture Notes in Mathematics

Edited by A. Dold and B. Eckmann

905

Differential Geometric Methods in Mathematical Physics

Clausthal 1980

Proceedings of an International Conference Held at the Technical University of Clausthal, FRG, July 23 – 25, 1980

Edited by H.-D. Doebner, S.I. Andersson, and H.R. Petry

Springer-Verlag
Berlin Heidelberg New York 1982

Editors
Heinz-Dietrich Doebner
Stig I. Andersson
Institut für Theoretische Physik, Technische Universität Clausthal
D-3392 Clausthal-Zellerfeld, FRG

Herbert Rainer Petry
Institut für Theoretische Kernphysik der Universität Bonn
Nußallee 14–16, D-5300 Bonn, FRG

AMS Subject Classifications (1980): 53-06, 53G05, 55R05, 58-06, 58G40, 81EXX, 81G30, 81G35, 83-06

ISBN 3-540-11197-2 Springer-Verlag Berlin Heidelberg New York
ISBN 0-387-11197-2 Springer-Verlag New York Heidelberg Berlin

2141/3140-543210

PREFACE

The 1980 conference on "Differential Geometric Methods in Mathematical Physics" at the Technical University of Clausthal, FRG, was part of the by now fairly long series of conferences on similar themes. Initiated by K. Bleuler in 1973 (Bonn) and continued in Bonn (1975, 1977), Aix-en-Provence (1974, 1979), Warsaw (1976), Clausthal (1978) and Salamanca (1979), these conferences have gathered a large number of prominent researchers in this special branch of mathematics/mathematical physics. No doubt, these conferences have become something of an institution.

As a tribute to one of the initiators of this series of conferences, one session was dedicated to K. Bleuler on the occasion of his retirement. Additionally, the 1980 conference also pursued a more local, Clausthal tradition of summer schools and meetings on special problems in mathematical physics.

The topics covered in this year's conference and in the attached workshop which are included in this volume could, roughly speaking, be described by the following keywords: symplectic category, differential operators on manifolds and vector bundles, mathematics of (non-abelian) gauge fields, geometric quantization and asymptotic expansions, all of which are, of course, central issues in the contemporary differential geometric-biased approach to a variety of mathematical questions in classical and quantum physics. Notable achievements were, more specifically, reports on the asymptotics for spherical functions, bifurcation theory, mathematical structure of gauge theories, space-time geometry and representation theory. The editors regret that due to a general editorial requirement of homogeneity in a lecture notes volume, which applies also for these proceedings , it was not possible to include contributions (invited talks as well as contributed papers) with a very strong bias towards physics or having definitely the form of a pure review paper or of a research announcement. In some other cases manucripts were not received in time.

Acknowledgments

We wish to express our gratitude to the following persons and organizations for financial support and for other assistance rendering the publication of this proceedings volume possible:

- Der Niedersächische Minster für Wissenschaft und Kunst,
- Alexander von Humboldt-Stiftung, Bonn.

We thank especially, for generous grants

- The Office for Foreign Studies and Activities at the Techn.Univ. of Clausthal (Prof.Dr. H.Quade and Dr. R.Pestel) and
- The Volkswagen Foundation.

We also want to thank Springer Verlag, Heidelberg, for their kind assistance in matters of publication.

Last but not least, we whish to thank Mrs. J. Gardiner, Institut für Theoretische Physik der TU Clausthal, for an exellent and speedy complete preparation of this volume, as well as the other members of the institute whose help made the organization smooth and efficient.

Clausthal, January 1982

The Editors.

TABLE OF CONTENTS

I. SESSION IN HONOUR OF KONRAD BLEULER

INTRODUCTION

This session is dedicated to Professor Konrad Bleuler, who is technically scheduled for retirement this year, but who we all know will continue as always to be in the forefront of the seekers for a true understanding of the physical world. As a man of broad scientific experience and deep vision, he knew that fundamental progress required coordination of a variety of scientific disciplines, and could see which strands needed pulling together. One of the means he chose towards this end was the sponsorship of the present series of conferences on Differential Geometry and Mathematical Physics. His scientific culture, communal dedication, and not least, personal interest and warmth have contributed enormously to the effectiveness of these conferences.

Today, theoretical physicists speak of bundles, group actions, operator algebras, and topology almost as easily as they did a decade or two ago of diagrams, form factors, Green's functions, and the like. From gauge theories of elementary particles to general relativity and cosmology, these notions are vital to the discussions of the theoretical community dedicated to their explication. At the same time, a new generation of mathematicians has found inspiration and motivation in central problems and ideas of theoretical physics. Konrad Bleuler has been one of the primary catalyzers of the interaction between disciplines that these changes represent. As a fundamental innovator and broad authority in his own right, and as one who has worked so effectively to keep the torch burning in other times and places, we salute him.

But we do so as he would want, by hearing now about new insights into challenging and fundamental problems.

I.E. Segal
Massachusetts Institute of Technology

MASSLESS LIMITS AND DUAL FIELD THEORIES

S. Deser *

Brandeis University,
Waltham, MA 02254, USA

We discuss some aspects of mass zero limits of massive gauge theories and gauge invariant representations of massless spin 0 and $\frac{1}{2}$ systems constructed from higher rank fields, which are not reachable from massive ones. A number of open problems of interest are mentioned.

I. Introduction

Two different but related topics will be discussed in this lecture. Neither is new, but both have found recent applications in areas of current interest such as supergravity, and the second especially is under active current investigation.

The first topic is the mass zero limit of massive gauge theories, which is well-understood in the abelian vector case. I will give an amusing application of this limit in a model involving supergravity, where just the opposite of the usual situation occurs, and then, in the spirit of this conference, state some open mathematical problems from a physicist's point of view in the non-abelian case. The second and main topic concerns alternative representations of lower spin by higher rank fields in terms of gauge invariant actions, which are sometimes called dual field theories. A considerable literature is accumulating on this, from a variety of motivations, and this subject too should have some interest for mathematicians. Rather than give details or extensive bibliography, I refer to the works on which this talk is based and where these may be found.

---

* Supported in part by NSF grant PHY 78-08644 A01

II. Massless Limit of Massive Gauge Theories

We begin with the abelian case. It is an old story (see e.g. [1] for some earlier literature) that the massless limit of Proca theory (massive photons) exists if and only if the current to which it couples is such that $m^{-1}\partial_\mu j^\mu$ stays finite in this limit. The latter can be analyzed in terms of the decomposition

$$A_\mu \equiv B_\mu^T + m^{-1}\partial_\mu \theta \quad , \quad \partial^\mu B_\mu^T \equiv 0, \tag{1}$$

applied to the action

$$I(m) = \int dx \left[-\frac{1}{4} F_{\mu\nu}^2 (A) - \frac{1}{2} m^2 A_\mu^2 + j^\mu A_\mu \right] \tag{2}$$

The point is that the kinetic term depends only on B_μ^T, i.e. $F_{\mu\nu}(A) \equiv F_{\mu\nu}(B)$, and orthogonality of the transverse and longitudinal parts of (1) in the mass term, $\int A^2 = \int B^2 + \int (\partial\theta)^2 m^{-2}$, decouples them there. Consequently, the action reads

$$\begin{aligned} I(m) = & \int dx \left[-\frac{1}{4} F_{\mu\nu}^2 (B) - \frac{1}{2} m^2 B^2 + j_\mu^T B_\mu^T \right] \\ & + \int dx \left[-\frac{1}{2} (\partial_\mu \theta)^2 - m^{-1}\theta\, \partial_\mu j^\mu \right] \end{aligned} \tag{3}$$

The first part of the action is no longer constrained by the Bianchi identity $\partial^2_{\mu\nu} F^{\mu\nu} \equiv 0$, since both $m^2 B_\mu^T$ and j_μ^T are automatically divergenceless, and the $m \longrightarrow 0$ limit there simply consists in dropping the B^2 term. This is precisely the usual Maxwell theory coupled to a conserved current j_μ^T, and although it appears to be expressed in terms of a vector potential in a specific (Lorentz) gauge ($\partial \cdot B = 0$), it is in fact gauge invariant. The second part represents a scalar field θ related to the original longitudinal photon, coupled to the quantity $m^{-1}\partial \cdot j$. Normal currents, such as $j^\mu = e\bar{\psi}\gamma^\mu\psi$, remain strictly conserved and the longitudinal field decouples from the charges. Consequently, walls become transparent to these photons, which removes the old equipartition paradox about the jump from three to two degrees of freedom. There are always three degrees of freedom, even at $m = 0$, but one is decoupled from charges (although it still couples to gravity, for example, where its energy can be measured in principle).

However, not all sources need be of this type. The first dynamical example, to my knowledge, occurs when a massive vector supermultiplet is coupled to supergravity. In this system, the current consists of an identically conserved part from a non-minimal coupling $\sim M^{\mu\nu} F_{\mu\nu}$ together with a nonconserved part explicitly proportional to m, i.e. a term of the form $m\, j^\mu A_\mu$. This leads to just the opposite situation from the usual one: *only* the longitudinal photons remain coupled to this finite term: $(\theta m^{-1})(m\, \partial \cdot \mathcal{J}) = \theta\, \partial \cdot \mathcal{J}$, while the transverse ones do not see this part of the current! The resulting limit is in fact that of an appropriate massless supermultiplet coupled to supergravity, which confirms the established wisdom in a novel setting. Details may be found in [1].

In the non-abelian situation, self-interaction complicates the corresponding analysis even in the absence of external sources, and I am not aware of any definite answer on the existence of the zero mass limit. One may still try to analyze the massive action

$$I(m) = \mathrm{tr} \int dx \left[-\tfrac{1}{4} F_{\mu\nu}^2 (A) - \tfrac{1}{2} m^2 A_\mu^2 \right]$$
$$F_{\mu\nu}(A) \equiv \partial_\mu A_\nu - \partial_\nu A_\mu + g \left[A_\mu, A_\nu \right] \tag{4}$$

in terms of a Stueckelberg-like decomposition [2]

$$A_\mu = \Omega^{-1} B_\mu^T \Omega + \Omega^{-1} \partial_\mu \Omega \tag{5}$$

which can be shown to exist (it has the form of a gauge transformation between arbitrary and Lorentz gauge). However, there are difficulties: the first is that although $\mathrm{tr}\, F^2(A) = \mathrm{tr}\, F^2(B)$ with (5), there is no longer orthogonality between the two parts of A_μ in the mass term; the second is that by dimensions there is no room for the m^{-1} factor in front of the longitudinal parts of (5), while its introduction through a factor $\varepsilon \equiv M/m$ spoils the gauge form of (5), and thereby also the $F^2(A) = F^2(B)$ relation. What seems difficult to achieve is the simultaneous satisfaction of the two conditions which made the limit possible in the abelian case, namely that the kinetic term depend only on B^T which is unconstrained while the mass term decomposes into separate transverse and longitudinal parts. Perhaps a better decomposition exists, or it may be necessary to investigate the question perturbatively in the coupling constant g, since the self-interaction is equivalent to a current $\mathcal{J}^\mu(A)$. What is known is that if A_μ is of the pure gauge form $\Omega^{-1}\partial_\mu\Omega$, then $F_{\mu\nu}(A)$ vanishes, while the mass term becomes $\sim \mathrm{tr}\, m^2 \int \partial_\mu \Omega^{-1} \partial_\mu \Omega$ which is the action of a scalar self-interacting multiplet, the σ-model, when Ω is written in terms of a scalar field as $\Omega(m^{-1}\varphi)$. Likewise, if A_μ is pure transverse, one obtains the limit of usual Yang-Mills theory, but the presence of both parts simultaneously leads to complications.

There are indications of discontinuities as $m \longrightarrow o$ in other aspects of this theory. For example, quantum corrections are well-known to differ in the two cases, and even at the classical level it has been established that there are no classical solutions for $m \neq o$ in the Euclidean domain, in contrast to the well-known instantons in four dimensions; conformal invariance seems to be crucial here and might preclude a smooth limit, unlike the abelian case. For spins greater than 1 (e.g., $\frac{3}{2}$ or 2), it is known that classical discontinuity is present at the abelian level already. It would be interesting to see a mathematician's evaluation of the relation between the geometrical situation where $m \equiv o$ and only the curvature (F) appears in the action and the $m \neq o$ one where also the connection forms (A) enter explicitly.

A final mathematical problem to which I draw your attention in the purely massless case is the absence of a direct proof from the field equations that there are no static solutions in the non-abelian theory in any but five space-time dimensions, whereas this is easy to show explicitly in the abelian situation in any number of dimensions. What I mean specifically is this: in abelian theory, the absence of well-behaved static solutions (all time-derivatives vanish in an appropriate gauge, or all gauge-invariants are time independent) follows immediately from Maxwell's equations. Dropping electric fields (which always vanish in static situations), it follows directly from $\nabla \cdot B = 0 = \nabla \times B$ that B vanishes in a topologically Euclidean manifold, say, and this is also the case in an arbitrary number of dimensions from the Maxwell equations $\partial_i F^{ij} = o$. In the non-abelian situation, the field equations read instead $D(A) \cdot B = o = D(A) \times B$, where D is the covariant derivative with respect to the connection A_i, $D_i \equiv \partial_i + g[A_i, \;]$. How can one infer from these equations directly the (correct) result that $B = o$, and why is it not true (from this point of view) for static solutions in (and only in) five space-time dimensions?

Let us restate the problem: We require a direct proof in Euclidean signature for an arbitrary dimension $D \neq 4$ (which is equivalent to the static case for D + 1 space-time dimensions) that the equations

$$D_i F^{ij}(A) = o, \qquad D_i \equiv \partial_i + g[A_i, \;] \qquad (6)$$
$$F_{ij} \equiv \partial_i A_j - \partial_j A_i + g[A_i, A_j]$$

have no well-behaved solutions. The indirect proof [3] is based on energy integrals, and as such is not applicable in a background or dynamical gravitational field, even if static and well-behaved (although a proof can be given for coupled Einstein-Maxwell fields [4]).

III. Lower Spins as Gauge Theories

There are a number of physical motivations for studying exotic representations of spin o and $\frac{1}{2}$ fields in terms of higher rank variables. A brief list would include: a uniform gauge invariant formulation of massless systems for any spin; such representations arise naturally in supergravity, e.g., from dimensional reduction; a theory in which a massless scalar field could not grow a mass would be a possible way of dealing with the hierarchy problem in grand unified models; possible quantum inequivalences of different formulations; the possible introduction of consistent couplings of higher spin fields to electromagnetism or gravity in an appropriate unorthodox representation. This is of some importance since the normal forms of higher ($\geqslant$ 2) spin fields are well-known not to admit consistent coupling, and this difficulty could a priori be representation-dependent rather than intrinsic.

We shall concentrate here on certain representations of spin o where there do indeed exist formulations which cannot grow a mass (and others which have no m = o limit!). We shall then turn to the spinor case, in which spin $\frac{1}{2}$ and/or $\frac{3}{2}$ systems are represented by higher rank spinors, and show that, for these, gravitational coupling is inconsistent (it is lucky that this was not the form first discovered by Dirac in 1927). Details may be found in [5].

Since this lecture was given, more general situations have been treated as well; we will not discuss them here, but refer to a recent preprint [6].

Consider first the most general quadratic first-order action involving a vector potential A_μ and independent field strength $F_{\mu\nu}$. [For simplicity, we deal here only with the abelian situation, although generalization to non-abelian cases is possible.] It is given by

$$I = \int dx \left[-\frac{1}{2} F^{\mu\nu} f_{\mu\nu}(A) - \frac{\lambda}{2} A_\mu^2 + \frac{\beta}{4} F_{\mu\nu}^2 + \alpha F_{\mu\nu} {}^*F^{\mu\nu} \right] \quad (7)$$

where $f_{\mu\nu}(A) \equiv \partial_\mu A_\nu - \partial_\nu A_\mu$, ${}^*F^{\mu\nu} \equiv \frac{1}{2} \varepsilon^{\mu\nu\alpha\beta} F_{\alpha\beta}$. We have omitted a term $\sim F\,{}^*f$ which may be removed by a change of variable; parity-violation is allowed if $\alpha \neq o$. One may analyze the dynamics of these models in terms of the parameters (λ, α, β). If only one fails to vanish, the results are:

$\beta \neq o$: Maxwell theory in first-order form
$\lambda \neq o$: massless scalar theory
$\alpha \neq o$: no dynamics.

If only one parameter vanishes, we find

$\alpha = o$: massive vector (Proca) theory
$\beta = o$: no dynamics
$\lambda = o$: Maxwell theory

When none of the parameters vanish, one has again massive vector theory. Note that there is never real parity violation even with $\alpha \neq o$.

Let us comment on these statements. The $\beta = \alpha = o$ form is the usual route for obtaining a gauge field description of massless spin o theory by antisymmetric tensors. That it describes spin o is clear from the field equations, which state that $f_{\mu\nu}(A) = o$, i.e. that $A_\mu = \partial_\mu \varphi$, while $\partial_\mu F^{\mu\nu} = \lambda A^\nu$ implies that $\partial_\mu A^\mu = o$ for consistency, and hence that $\Box \varphi = o$. Alternatively, inserting the solution $A_\mu = \partial_\mu \varphi$ of the $f_{\mu\nu} = o$ constraint into the action gives the usual scalar action $\int (\partial_\mu \varphi)^2$.

At this point, we mention another important problem in this field, which is the general mathematical one of when a change of variables involving explicit derivatives is permitted in a variational principle without altering the content of the Euler-Lagrange equations. A priori, it is not permitted, and many examples are known in which inequivalent results are obtained [6,7]. One toy example is $I = \int A_\mu^2$; decomposing $A_\mu = \partial^\nu H_{\mu\nu} + \partial_\mu \varphi$ with $H_{\mu\nu} = -H_{\nu\mu}$ keeps the correct number of variables but yields an action representing two scalar fields, rather than no dynamics if A_μ itself is varied. The only safe procedure is to insert only constraints involving spatial derivatives and use Hamiltonian decomposition at each step. All results given here have been checked that way, but it would clearly be useful to have general theorems in this domain.

What is relevant to us is that elimination of A_μ in terms of $\partial_\nu F^{\mu\nu}$ gives the antisymmetric tensor form

$$I \sim \int dx\, (\partial_\mu F^{\mu\nu})^2 \tag{8}$$

which is manifestly invariant under the three gauge transformation parameters ξ_β,

$$\delta F^{\mu\nu} = \varepsilon^{\mu\nu\alpha\beta} \partial_\alpha \xi_\beta \tag{9}$$

There is, however, no possible addition to (8) which will describe a massive scalar, as is seen by detailed Hamiltonian analysis of (8), or more generally of (7) which demonstrates the results stated earlier: There is no room for a massive scalar. On the other hand, the action obtained by starting from the first order form of a massive scalar field,

$$I = \int dx\, [\pi^\mu \partial_\mu \phi - \tfrac{1}{2} \pi_\mu^2 - \tfrac{1}{2} m^2 \phi^2] \tag{10}$$

can only describe massive excitations. If $m \neq o$, we may obtain the form

$I \sim \int dx\, [(\partial_\mu \pi^\mu)^2 - m^2 \pi_\mu{}^2]$ as a vector description [8], but if m = o, there is no way to write $I (\pi_\mu)$ only.

To return to (8), it provides one (but not the only one [6]) gauge-invariant formulation of a scalar theory which cannot grow a mass. Unfortunately, it cannot interact consistently with electromagnetism either, because the required gauge transformation $\delta F^{\mu\nu} = \varepsilon^{\mu\nu\alpha\beta} D_\alpha(A) \xi_\beta$ is no longer consistent when ∂_α is replaced by the gauge invariant derivative $D_\alpha(A)$. That is, the action $I = \int dx\, (\overline{D_\mu F^{\mu\nu}})(D_\alpha F^{\alpha\nu})$ violates the Bianchi identity since the curl of the field equations,

$$\varepsilon^{\mu\nu\alpha\beta} D_\beta (D_\mu D_\rho F^\rho{}_\nu) \equiv {}^*f_{\nu\alpha}(A)\; D_\rho F^{\rho\nu} \neq 0 \tag{11}$$

fails to vanish, and non-minimal coupling brings no improvement. Gravitational coupling is allowed, however, since ordinary and covariant derivatives coincide both in the gauge transformation and in the action (taking the contravariant tensor density $\mathcal{F}^{\mu\nu} \equiv - \mathcal{F}^{\nu\mu}$ as basic variable makes this manifest). We shall see in the next example that even gravitational coupling can become inconsistent in higher rank descriptions, so that these seem to be worse rather than better in this respect.

The first model of a massless fermionic system along these lines, via gauge-invariant action, was given in [9], whose action is equivalent to

$$I = -i \int (\partial_\mu \bar{\chi}^{\mu\nu}) \gamma^\lambda (\chi^{\lambda\nu} + 2 \gamma_5 {}^*\chi^{\lambda\nu}) \tag{12}$$

where $\chi^{\mu\nu} = - \chi^{\nu\mu}$ is an antisymmetric tensor-spinor. Although originally believed to represent pure spin $\frac{1}{2}$, it has since been shown [6,10] to include also spin $\frac{3}{2}$ modes. However, there is now a model which does only contain spin $\frac{1}{2}$, and for which the present conclusions also hold [6], so we will restrict ourselves to this model for simplicity and state the results, common to both cases. Here the gauge invariance is under

$$\delta \chi^{\mu\nu} = \varepsilon^{\mu\nu\alpha\beta} \gamma_\alpha \partial_\beta \alpha(x) \tag{13}$$

where $\alpha(x)$ is an arbitrary spinor function. It is obvious for the $(\partial_\mu \bar{\chi}^{\mu\nu})$ part; for the other it follows from Dirac matrix identities that $\delta [\gamma (\chi + 2 \gamma_5 {}^*\chi)] \sim \partial_\nu \alpha(x)$ which vanishes upon integration by parts onto $(\partial_\mu \bar{\chi}^{\mu\nu})$. Now, however, gravitational coupling is inconsistent because the derivative on a spinor must be made covariant. But then (13) with $\partial_\beta \longrightarrow D_\beta$ is no longer an invariance and correspondingly the Bianchi identities on the χ field equations are violated. In particular, the term in the variation which is proportional to

$$D_\nu D_\mu \bar{\chi}^{\mu\nu} \equiv \frac{1}{2} [D_\nu , D_\mu] \chi^{\mu\nu} \sim R_{\mu\nu ab} \sigma^{ab} \bar{\chi}^{\mu\nu} \tag{14}$$

involves the full curvature tensor. We emphasize this point because the Weyl tensor is not determined by the field equations, unlike the Einstein or Ricci tensor, so there is no hope of the supergravity-like miracle occurring here. We recall that for the traditional spin $\frac{3}{2}$ vector-spinor formulation, gravitational coupling led to the weaker dependence $\sim G_{\mu\nu} \gamma^\mu \psi^\nu$ of the corresponding "inconsistency", which was then removed by use of the field equations $G_{\mu\nu} = T_{\mu\nu} (\psi_\alpha)$, together with use of spinor identities which essentially implied that $T_{\mu\nu} (\psi) \gamma^\mu \psi^\nu \equiv 0$. That is, the original symmetry under $\delta \psi_\mu = \partial_\mu \alpha(x)$ of the Rarita-Schwinger action led to $\delta I_{3/2} \sim \int G_{\mu\nu} \bar{\alpha}(x) \gamma^\mu \psi^\nu$ when $\partial_\mu \rightarrow D_\mu$, which was cancelled, as a result of supersymmetry, by variation of the vierbein variables $\delta e_{\mu a} = \bar{\alpha}(x) \gamma_a \psi_\mu$ in $\delta \int R(x) \sim - \int G_{\mu\nu} \bar{\alpha} \gamma^\mu \psi^\nu$. Here there is no corresponding symmetry since spin $\frac{1}{2}$ and 2 are too far apart for a graded algebra to be possible. Indeed, what occurs is exactly similar to the problems of genuine higher spin fields [11] $s \geq \frac{5}{2}$, where the higher rank fields $\psi_{\mu\nu\ldots}$ led to terms proportional to the full curvature. This time we have been forced into inconsistency precisely because we have used higher rank representations (and again, non-minimal additions cannot help). The same problem occurs for fourth rank tensor representations of spin 0 or 1 [6], so it is not a purely fermionic phenomenon. Conversely, it seems unlikely from the above that reformulating high spin actions through still higher rank fields will overcome the difficulties there.

As mentioned earlier, the action (12) actually represents a mixture of spin $\frac{1}{2}$ and $\frac{3}{2}$, but exactly the same conclusions hold for the (actually more complicated) action which is pure spin $\frac{1}{2}$. Thus, the coupling difficulty really holds for both $\frac{1}{2}$ and $\frac{3}{2}$ when they are represented by higher rank fields, even though they are consistent in their traditional forms. It would clearly be useful to have a deeper mathematical analysis of different realizations of a given spin, which would certainly also clarify the basis for the interaction inconsistencies.

IV. Conclusions

We have considered a number of unusual problems in gauge theories. One area was that of the massless limit of traditional spin 1 theory, with an application in the abelian case to a system with a current that couples only to longitudinal photons, in contrast to the standard situation. We then emphasized the importance of understanding the massless limit of classical non-abelian theory where the mass term is proportional to the square of the connection and breaks the usual geometrical construction. We also remarked on the need to understand directly why the usual result, that vanishing curl and divergence of a vector implies that it vanishes, also holds when the derivatives become covariant ones on the non-abelian case.

The discussion of higher rank representations of massless lower spin fields as gauge invariant systems also led to a number of interesting problems. We mentioned in passing the general question in the calculus of variations of legality of changes of variable involving explicit derivatives, which empirically sometimes works and sometimes fails, and pointed out that these systems gave examples of discontinuous theories in the sense that there were no "nearby" massive theories of which they were the limit. Finally, we noted the need for a uniform understanding of the different representations allowed for a given spin and for the fact that the higher the rank of field used, the more difficult it is to couple the system to electromagnetism and most significantly to gravity, since all fields must couple universally to it if they couple to any other of its sources.

REFERENCES

[1] Deser, S., Phys. Rev. D21 2436 (1980).

[2] Boulware, D.G., Ann. Phys. 56, 140 (1970).

[3] Coleman, S., in Subnuclear Physics, A. Zichichi, ed., Plenum N.Y. 1977
Deser, S., Phys. Lett. 64B, 463 (1976).

[4] Lichnerowicz, A., Theories Relativistes de la Gravitation, Masson Paris 1955.
Thiry, Y., J. Math. Pure & Appl. 30, 275 (1951)

[5] Deser, S. and Witten, E., Nucl. Phys. B178, 491 (1981).

[6] Deser, S., Townsend, P.K., and Siegel, W., Nucl. Phys. B184, 333 (1981).

[7] Regge, T. and Teitelboim, C., Proc. M. Grossman Conf. Trieste (1975)
Deser, S., Pirani, F.A.E., and Robinson, D., Phys. Rev. D14, 3301 (1976).
Deser, S. and Townsend, P.K., Phys. Lett 98B, 188 (1981).

[8] Curtright, T.L. and Freund, P.G.O., Nucl. Phys. B172, 413 (1980).

[9] Townsend, P.K., Phys. Lett. 90B, 275 (1980).

[10] Aragone, C., (private communication).

[11] Aragone, C. and Deser, S., Phys. Lett 86B 161; Nucl. Phys. 170 FSI 329 (1980)
Berends, F.A., van Holten, J.W., van Nieuwenhuizen, P., and de Witt, B., J. Phys. A13, 1643 (1980).

POISSON COMMUTATIVITY AND THE GENERALIZED PERIODIC TODA LATTICE

Bertram Kostant*

Mathematics Department, M.I.T.,
Cambridge, Mass. 02139, U.S.A.

With affection and respect this paper is dedicated to
Professor Konrad Bleuler on the occasion of his retirement.

1. Introduction.

1. Let $\mathbb{R}^{2n}$ be classical phase space with canonical momentum and position coordinates $p_i, q_j \in C^\infty(\mathbb{R}^{2n})$, i, j = 1,...,n. Let $\mathcal{Q} \subset C^\infty(\mathbb{R}^{2n})$ be the finite dimensional (in fact n-dimensional) subspace spanned by the q_j. We consider here a mechanical system whose Hamiltonian $H \in C^\infty(\mathbb{R}^{2n})$ is of the form

$$H = \sum_{i=1}^{n} \frac{p_i^2}{2m_i} + U \tag{1.1.1}$$

where the $m_i > 0$ are masses, and U, the potential has the following form: For a number $\ell \leq n$ one has constants r_i and linear functions $\psi_i \in \mathcal{Q}$, $i = 0,\ldots,\ell$ where $r_0 \geq 0$ and $r_i > 0$ for $i \geq 1$ such that

$$U = \sum_{i=0}^{\ell} r_i e^{\psi_i}. \tag{1.1.2}$$

Now there is a natural positive definite bilinear form B_H on $\mathcal{Q}$, given in terms of Poisson bracket with H (see §7.1 in [3]). Since the q_i are a basis of $\mathcal{Q}$ the bilinear form may be given explicitly by the relation

$$B_H(q_i,q_j) = \delta_{ij}/m_i. \tag{1.1.3}$$

* Supported in part by the NSF under Grant MCS 7804007.

In [3], generalizing the finite non-periodic Toda lattice we considered the case where $r_0 = 0$ and the ψ_i, $i = 1,\ldots,\ell$ define a Dynkin diagram. With this assumption we explicitly solved the corresponding Hamilton's equations and determined, in terms of representation theory the state of the system for all values of time, given an arbitrary initial state.

The assumption above, that the ψ_i define a Dynkin diagram means the following: There exists a real split semi-simple Lie algebra $\mathfrak{g}$ of rank ℓ, a Cartan subalgebra $\mathfrak{h} \subseteq \mathfrak{g}$, a positive definite invariant bilinear form Q on $\mathfrak{g}$ (which extends naturally to $\mathfrak{h}'$, the dual to $\mathfrak{h}$) a set of simple roots, $\alpha_i \in \mathfrak{h}'$, $i = 1,\ldots,\ell$, such that

$$B_H(\psi_i,\psi_j) = Q(\alpha_i,\alpha_j) \tag{1.1.4}$$

$i, j = 1,\ldots,\ell$.

An important step in [3] is the proof of the complete integrability of the generalized non-periodic Toda lattice. Involved in this is the proof that certain functions Poisson commute with one another. In more detail let $S(\mathfrak{g})$ be the symmetric algebra over $\mathfrak{g}$. If we identify $\mathfrak{g}$ with its dual space using Q we may regard $S(\mathfrak{g})$ as the algebra of polynomial functions on $\mathfrak{g}$ itself. But $S(\mathfrak{g})$ is, in a natural way, a module for the adjoint group G of $\mathfrak{g}$ and hence the algebra of invariants, $S(\mathfrak{g})^G$, is the algebra of G-invariant polynomial functions on $\mathfrak{g}$. In effect in [3] we have defined a map, $\sigma \circ \delta$ (see (7.2.3) and (7.4.8)), now written:

$$\gamma_0 : \mathbb{R}^{2n} \longrightarrow \mathfrak{g}$$

such that if γ_0^* is the pull-back on functions,

$$\gamma_0^* : S(\mathfrak{g}) \longrightarrow C^\infty(\mathbb{R}^{2n}),$$

where $\gamma_0^*(\phi) = \phi \circ \gamma_0$ then

Theorem 1. *The Hamiltonian* H *(recall* $r_0 = 0$*) Poisson commutes with any function in* $\gamma_0^*(S(\mathfrak{g})^G)$. *Moreover any pair of functions in* $\gamma_0^*(S(\mathfrak{g})^G)$ *Poisson commute with each other.*

For the proof of Theorem 1 see Theorem 1.5, Proposition 6.5 and §7.1-7.4 in [3].

1.2. Some time ago we made the observation that the use of the extended Dynkin diagram (see below) in (1.1.2) generalizes the periodic Toda lattice. Furthermore as announced in §0.3 in [3] we were able to prove the analogue of Theorem 1 for the generalized periodic Toda lattice. Subsequently Adler and van Moerbeke went further, and in [1] and in their paper following [1] they, impressively, solved Hamilton's equations for the generalized periodic Toda lattice. In the course of doing so they also proved the analogue of Theorem 1. However we feel that our earlier proof of this result is clearer and conceptually simpler than theirs and it will be the main objective of this paper to present that proof. For one thing our proof avoids the introduction of an infinite dimensional Lie algebra and associated attendant difficulties. Furthermore it deals directly with the invariants rather than through the characteristic polynomial of some faithful matrix representation. The proof relies on some remarkable algebraic cancellations associated with the invariants and the highest root. See the lemmas in §3.

Now assume $\mathfrak{g}$ is simple where $\ell > 1$ and let α_0 be the negative of the highest root. The Dynkin diagram associated to $\alpha_1,\ldots,\alpha_\ell$ extends to the $\ell+1$ roots $\alpha_0,\ldots,\alpha_\ell$ and is generally referred to as the extended Dynkin diagram. It is a familiar tool in the structure theory of simple Lie algebras. Among other places it occurs (1) in the determination of the maximal subalgebras of $\mathfrak{g}$ (2) in the determination of the fundamental domain of the affine Weyl group and (3) in the structure of the Kac-Moody Lie algebras.

Recalling (1.1.2) we now consider the potential U for the case where $r_0 > 0$ so that $r_i > 0$ for $i = 0,\ldots,\ell$. We then say that the ψ_i, $i = 0,\ldots,\ell$, define an extended Dynkin diagram in case $\mathfrak{g}$, $\mathfrak{h}$, Q, $\alpha_i \in \mathfrak{h}'$, $i = 0,\ldots,\ell$ exists where $\alpha_1,\ldots,\alpha_\ell$ are simple roots and α_0 is the negative of the highest root such that

$$B_H(\psi_i,\psi_j) = Q(\alpha_i,\alpha_j) \tag{1.2.1}$$

for $i, j = 0,\ldots,\ell$ and ψ_0 is in the span of the ψ_i, $i = 1,\ldots,\ell$.

Remark 1.2. It is immediate from 1.2.1 that ψ_i, $i = 1,\ldots,\ell$, are linearly independent.

We note now that if the ψ_i define an extended Dynkin diagram then one, justifiably, may refer to the mechanical system in question as a generalized periodic Toda lattice. Indeed if $\ell = n-1$, $\psi_i = q_i - q_{i+1}$, $i = 1,\ldots,\ell$, $\psi_0 = q_n - q_0$, all $r_i = 1$, so that the potential U is of the form

$$U = e^{q_1-q_2} + \ldots + e^{q_{n-1}-q_n} + e^{q_n-q_0}$$

and all $m_i = 1$ then ψ_i define the extended Dynkin diagram of the simple Lie algebra

A_{n-1}. On the other hand one knows that this mechanical system is exactly the periodic Toda lattice.

We now give some other examples, by writing down the potential U, where the ψ_i define an extended Dynkin diagram. The Lie algebras in question in these examples are classical simple Lie algebras and also the exceptional Lie algebra F_4. In all these examples the r_i and the m_i are taken to be equal to unity. For a larger collection of examples and where all the simple Lie algebras occur see [1].

(a) $\ell = n, \quad \mathfrak{g}_{\mathbb{C}} = B_n, \quad \alpha_0 = -(q_1+q_2)$

$$U = e^{q_1-q_2} + \dots + e^{q_{n-1}-q_n} + e^{q_n} + e^{-(q_1+q_2)}$$

(b) $\ell = n, \quad \mathfrak{g}_{\mathbb{C}} = C_n, \quad \alpha_0 = -2q_1$

$$U = e^{q_1-q_2} + \dots + e^{q_{n-1}-q_n} + e^{2q_n} + e^{-2q_1}$$

(c) $\ell = n, \quad \mathfrak{g}_{\mathbb{C}} = D_n, \quad \alpha_0 = -(q_1+q_2)$

$$U = e^{q_1-q_2} + \dots + e^{q_{n-1}-q_n} + e^{q_{n-1}+q_n} + e^{-(q_1+q_2)}$$

(d) $\ell = n = 4, \quad \mathfrak{g}_{\mathbb{C}} = F_4, \quad \alpha_0 = -(q_1+q_4)$

$$U = e^{q_1-q_2} + e^{q_2-q_3} + e^{q_3} + e^{\frac{q_4-q_1-q_2-q_3}{2}} + e^{-(q_1+q_4)} .$$

It should be emphasized that the number of examples is uncountable since what is relevant is not the ψ_i themselves but the geometry (the lengths and angles) of the ψ_i.

1.3 As mentioned in §1.2 the objective here is to present our proof of the Poisson commutativity theorem (Theorem 3) for the generalized periodic Toda lattice. This theorem will be analogous to the Poisson commutativity theorem, Theorem 1, for the generalized non-periodic Toda lattice. The first problem is to modify the map γ_0 in Theorem 1 obtaining a map γ so that the conclusion of Theorem 1 holds where U in (1.1.1) and (1.1.2) is such that $r_0 > 0$ and the ψ_i, $i = 0,\dots,\ell$, define an extended Dynkin diagram.

2. The map γ.

1. In order to define the map γ we recall notation and results in [3]. Let $\Delta \subsetneq h'$ be the set of roots of (g, h) and let Δ_+ be the set of positive roots (the $\mathbb{Z}_+$ span of $\alpha_1, \ldots, \alpha_\ell$ in Δ). For any $\phi \in \Delta$ let $e_\phi \in g$ be a corresponding root vector. The choice is assumed normalized so that $Q(e_\phi, e_{-\phi}) = 1$. Let h_i, $i = 1, \ldots, \ell$ be the basis of h defined by putting

$$h_i = [e_{\alpha_i}, e_{-\alpha_i}]. \qquad (2.1.1)$$

Let $\bar{b}$ (resp. b) be the real Borel subalgebra spanned by the h_i and $e_{-\phi}$ (resp. e_ϕ) for $\phi \in \Delta_+$. The bilinear form Q sets up a non-singular pairing of b and $\bar{b}$ and we will identify the dual space $\bar{b}'$ with b (and also $\bar{b}$ with b'). That is

$$\bar{b}' = b . \qquad (2.1.2)$$

Now if $\tilde{b} \subsetneq b$ is any open set and $u \in C^\infty(\tilde{b})$ then the differential du defines a mapping of $\tilde{b}$ into b'. But then using the additional identification $b' = \bar{b}$ one therefore has a map

$$\delta u : \tilde{b} \longrightarrow \bar{b}. \qquad (2.1.3)$$

For any $y \in b$ let $\iota(y)$ be the partial directional derivative defined by the vector y. It is then immediate that for any $z \in \tilde{b}$

$$(\delta u)(z) = \sum_{\phi > 0} ((\iota(e_\phi)u)(z))e_{-\phi} + \sum_{j=1}^{\ell} ((\iota(y_j)u)(z))h_j \qquad (2.1.4)$$

where $\{y_j\}$ is the basis of h such that $Q(h_i, y_j) = \delta_{ij}$.

Now let $\bar{B} \subsetneq G$ be the Lie subgroup corresponding to $\bar{b}$. Using (2.1.2) the coadjoint action of $\bar{B}$ decomposes b into a union of $\bar{B}$-coadjoint orbits each of which is a symplectic $\bar{B}$-homogeneous space with respect to its coadjoint symplectic structure. See §6.3 in [3] where we take f = 0 and $a = \bar{b}$. One then defines a Poisson structure, see §1.1 in [2], on $C^\infty(\tilde{b})$ by defining $[u,v](z) = [u_1,v_1](z)$ where $z \in \tilde{b}$, $u, v \in C^\infty(\tilde{b})$ and $u_1 = u \mid O$, $v_1 = v \mid O$ where O is the $\bar{B}$-coadjoint orbit containing z.

Proposition 2.1. *If* $\tilde{b} \subsetneq b$ *is any open set,* $z \in \tilde{b}$, *and* $u, v \in C^\infty(\tilde{b})$ *one has*

$$[u,v](z) = Q(z, [\delta u(z), \delta v(z)]). \qquad (2.1.5)$$

Proof. Since $[u,v](z)$ depends only on the differentials of u and v at z it suffices to prove (2.1.5) for a coordinate system. But using Q elements of $\bar{\mathfrak{b}}$ provide such a coordinate system. But if $\psi^w \in C(\tilde{\mathfrak{b}})$, for $w \in \bar{\mathfrak{b}}$, is defined by $\psi^w(z) = Q(w,z)$ then $[\psi^w, \psi^y] = \psi^{[w,y]}$ (see Theorem 5.3.1 in [2]) for $w, y \in \bar{\mathfrak{b}}$. But clearly $\delta\psi^w = w$, and $\delta\psi^y = y$ (by (2.1.4). This establishes (2.1.5). Q.E.D.

2.2. Now let $\mathcal{O} \subseteq \mathfrak{b}$ be the 2ℓ-dimensional submanifold defined as the set of all $z \in \mathfrak{b}$ of the form

$$z = \sum_{i=1}^{\ell} c_i h_i + \sum_{i=1}^{\ell} a_i e_{\alpha_i} \tag{2.2.1}$$

where $c_i, a_i \in \mathbb{R}$ but that all $a_i > 0$.

Proposition 2.2.1. *$\mathcal{O}$ is a $\bar{B}$-coadjoint orbit in $\mathfrak{b}$.*

Proof. Ignoring the translation by f (in §6.4 of [3]) this is part of the statement of Proposition 6.4 in [3]. Q.E.D.

Now, recalling §1.1, let $\mathcal{Q}_1 \subseteq \mathcal{Q} \subseteq C^\infty(\mathbb{R}^{2n})$ be the ℓ-dimensional subspace spanned by ψ_i, $i = 1,\ldots,\ell$. Then as noted in §7.1 in [3] there exists uniquely $p_j' \in [\mathcal{Q}_1, H]$, $j = 1,\ldots,\ell$ such that in $C^\infty(\mathbb{R}^{2n})$

$$[\psi_i, p_j'] = \delta_{ij}. \tag{2.2.2}$$

Furthermore one has

$$[\psi_i, \psi_j] = [p_i', p_j'] = 0 \tag{2.2.3}$$

for $i, j = 1,\ldots,\ell$.

Now let $\tilde{\mathfrak{b}}$ be the open subset of $\mathfrak{b}$ defined by

$$\tilde{\mathfrak{b}} = \{ z \in \mathfrak{b} \mid Q(e_{-\alpha_i}, z) > 0,\ i = 1,\ldots,\ell \}. \tag{2.2.4}$$

It follows immediately from (2.2.1) that

$$\mathcal{O} \subseteq \tilde{\mathfrak{b}}. \tag{2.2.5}$$

We now define a map

$$\tau : \mathbb{R}^{2n} \longrightarrow \mathcal{O} \subseteq \tilde{\mathfrak{b}}$$

by putting

$$\tau(x) = \sum_{i=1}^{\ell} p_i'(x)h_i + \sum_{i=1}^{\ell} (r_i \exp \psi_i(x))e_{\alpha_i}. \tag{2.2.6}$$

Regarding τ as a map $\mathbb{R}^{2n} \longrightarrow \tilde{b}$ one defines the pull-back on functions

$$\tau^*: C^\infty(\tilde{b}) \longrightarrow C^\infty(\mathbb{R}^{2n}) \tag{2.2.7}$$

by putting $\tau^* u = u \circ \tau$ for any $u \in C^\infty(\tilde{b})$. Now both $C^\infty(\tilde{b})$ and $C^\infty(\mathbb{R}^{2n})$ have Poisson structures.

Proposition 2.2.2. τ^* *is a homomorphism of Poisson structures.*

Proof. Since $\tau(\mathbb{R}^{2n}) \subseteq O$ one may extend τ^* so that also

$$\tau^*: C^\infty(O) \longrightarrow C^\infty(\mathbb{R}^{2n}). \tag{2.2.8}$$

However since O is a $\bar{B}$-coadjoint orbit the map $C^\infty(\tilde{b}) \longrightarrow C^\infty(O)$ induced by restriction is a homomorphism of Poisson structures. Thus it suffices to show that (2.2.8) is a homomorphism of Poisson structures. But replacing Z by O in Proposition 6.5 of [3] (i.e. ignoring f) one has, in the notation of Proposition 6.5 in [3] that for $i = 1,\ldots,\ell$,

$$\tau^* g_i = p_i' \quad \text{and} \tag{2.2.9}$$

$$\tau^* \phi_i = \psi_i + \log r_i.$$

But then τ^*, by (2.2.2-3) and (6.5.4) in [3], preserves Poisson bracket for a coordinate system on O. Since O is symplectic this proves that (2.2.8) is a homomorphism of Poisson structures. Q.E.D.

2.3. We can now define the map γ. As in [3] let $f = \sum_{i=1}^{\ell} e_{-\alpha_i}$. Also e_{α_0} is the lowest root vector and $e_{-\alpha_0}$ is the highest root vector. One has (well-known) negative integers k_i so that

$$\alpha_0 = \sum_{i=1}^{\ell} k_i \alpha_i. \tag{2.3.1}$$

Since ψ_0 is in the span of the ψ_i, $i = 1,\ldots,\ell$, by assumption it then follows from (1.2.1) that

$$\psi_0 = \sum_{i=1}^{\ell} k_i \psi_i. \tag{2.3.2}$$

Let d be the positive constant defined so that $d \prod_{i=1}^{\ell} r_i^{k_i} = r_0$.
We now put for any $x \in \mathbb{R}^{2n}$

$$\gamma(x) = \tau(x) + f + de_{-\alpha_0} + \left(\prod_{i=1}^{\ell} (r_i \exp \psi_i(x))^{k_i}\right) e_{\alpha_0}. \quad (2.3.3)$$

This defines a map

$$\gamma : \mathbb{R}^{2n} \longrightarrow \mathfrak{g} \quad (2.3.4)$$

and hence by pull-back a map

$$\gamma^*: S(\mathfrak{g})^G \longrightarrow C^{\infty}(\mathbb{R}^{2n}) \quad (2.3.5)$$

where $\gamma^*(I) = I \circ \gamma$ for any invariant $I \in S(\mathfrak{g})^G$.

3. The Poisson Commutativity Theorem.

1. We now establish the following Poisson commutativity theorem for the generalized periodic Toda lattice.

Theorem 3. *Let the notation be as in* 1.1-2 *so that* H *is a Hamiltonian function on phase space* $\mathbb{R}^{2n}$ *of the form* (1.1.1) *where the potential* U *has the form* (1.1.2). *That is*

$$U = r_0 e^{\psi_0} + \ldots + r_\ell e^{\psi_\ell} \quad (3.1.1)$$

where $\ell \leq n$, *all* $r_i > 0$ *and the* ψ_i, $i = 0,\ldots,\ell$ *define an extended Dynkin diagram. Let* γ^* *be the map* $S(\mathfrak{g})^G \longrightarrow C^{\infty}(\mathbb{R}^{2n})$ *defined as in* (2.3.5). *Then* H *Poisson commutes with all elements in* $\gamma^*(S(\mathfrak{g})^G)$ *and any pair of functions in* $\gamma^*(S(\mathfrak{g})^G)$ *Poisson commute with each other.*

Proof. (The proof will occupy all the sections of chapter 3.) Let $z \in \mathfrak{b}$ so that we can write

$$z = \sum_{i=1}^{\ell} c_i h_i + \sum_{\phi > 0} a_\phi e_\phi$$

where $a_\phi \in \mathbb{R}$. Then by (2.2.4) $z \in \tilde{\mathfrak{b}}$ if and only if $a_{\alpha_i} > 0$ for $i = 1,\ldots,\ell$.
Write $a_{\alpha_i} = a_i(z)$. Now let

$$\nu : \tilde{\mathfrak{b}} \longrightarrow \mathfrak{g} \quad (3.1.2)$$

be the map defined by putting

$$\nu(z) = z + f + de_{-\alpha_0} + \left(\prod_{i=1}^{\ell} a_i(z)^{k_i}\right) e_{\alpha_0}. \quad (3.1.3)$$

Now for any invariant $I \in S(\mathfrak{g})^G$ let $u_I \in C^\infty(\tilde{\mathfrak{b}})$ be defined by putting $u_I(z) = I(\nu(z))$. But then recalling (2.2.3) one has $u_I(\tau(x)) = I(\gamma(x))$ for any $x \in \mathbb{R}^{2n}$ since clearly $\nu \circ \tau = \gamma$. That is,

$$\gamma^*(I) = \tau^*(u_I). \tag{3.1.4}$$

Thus to prove that the elements in $\gamma^*(S(\mathfrak{g})^G)$ Poisson commute with one another it suffices by Proposition 2.2.2 to show that the commutator $[u_I, u_J]$, for any I, $J \in S(\mathfrak{g})^G$, vanishes on the orbit O. That is, by (2.1.5) it suffices to show that

$$Q(z, [\delta u_I(z), \delta u_J(z)]) = 0 \tag{3.1.5}$$

for any $z \in O$ and I, $J \in S(\mathfrak{g})^G$.

Now let $I \in S(\mathfrak{g})^G$ and let $z \in O$. Then by (2.1.4)

$$\delta u_I(z) = \sum_{\phi>0} ((\iota(e_\phi)u_I)(z))e_{-\phi} + \sum_{j=1}^{\ell} ((\iota(y_j)u_I)(z))h_j. \tag{3.1.6}$$

But now of course the partial derivatives $\iota(e_\phi)$ and $\iota(y_j)$ are defined on $\mathfrak{g}$. But for $\phi > 0$ and not simple one has $\nu(z+te_\phi) = \nu(z) + te_\phi$. Also $\nu(z+ty_j) = \nu(z) + ty_j$, where $t \in \mathbb{R}$. Thus for $\phi > 0$, but not simple, one has

$$(\iota(e_\phi)u_I)(z) = (\iota(e_\phi)I)(\nu(z)). \tag{3.1.7}$$

Similarly

$$(\iota(y_j)u_I)(z) = (\iota(y_j)I)(\nu(z)). \tag{3.1.8}$$

On the other hand since

$$\nu(z+te_{\alpha_j}) = z + te_{\alpha_j} + de_{-\alpha_0} + \left(\prod_{i=1}^{\ell} a_i(z)^{k_i}\right)\left[\frac{a_j(z)+t}{a_j(z)}\right]^{k_j} e_{\alpha_0}$$

one has, upon differentiating $I(\nu(z+te_{\alpha_j}))$ with respect to t, and then setting $t = 0$

$$(\iota(e_{\alpha_j})u_I)(z) = (\iota(e_{\alpha_j})I)(\nu(z)) + g(I)(z)\,\frac{k_j}{a_j(z)} \tag{3.1.9}$$

where for any $J \in S(\mathfrak{g})^G$ we let $g(J) \in C^\infty(\tilde{\mathfrak{b}})$ be the function defined by putting

$$g(J)(z) = (\iota(e_{\alpha_0})J)(\nu(z)) \prod_{i=1}^{\ell} a_i(z)^{k_i}. \tag{3.1.10}$$

Now let $\mathfrak{g}_{-1} \subseteq \mathfrak{g}$ be the span of the $e_{-\alpha_i}$, $i = 1, \ldots, \ell$. For any $J \in S(\mathfrak{g})^G$ let

$$\beta(J) \;:\; \tilde{b} \longrightarrow g_{-1}$$

be the map defined by putting

$$\beta(J)\,(z) \;=\; g\,(J)\,(z) \sum_{j=1}^{\ell} \frac{k_j}{a_j(z)}\, e_{-\alpha_j} \;. \tag{3.1.11}$$

Also let $n = [b,b]$ and $\bar{n} = [\bar{b},\bar{b}]$. For any $J \in S(g)^G$ let

$$\delta_n(J) \;:\; g \longrightarrow n$$

be defined by

$$\delta_n(J)\,(w) \;=\; \sum_{\phi>0} ((\iota\,(e_{-\phi})J)\,(w))e_{\phi} \;. \tag{3.1.12}$$

Similarly one defines a map

$$\delta_{\bar{n}}(J) \;:\; g \longrightarrow \bar{n}$$

by interchanging ϕ and $-\phi$ in (3.1.12). Also one defines the map

$$\delta_h(J) \;:\; g \longrightarrow h$$

by putting

$$\delta_h(J)\;\;(w) \;=\; \sum_{j=1}^{\ell} ((\iota\,(y_j)I)\;(w))h_j \;. \tag{3.1.13}$$

Finally put

$$\delta_{\bar{b}}(J) \;=\; \delta_h(J) + \delta_{\bar{n}}(J) \;. \tag{3.1.14}$$

3.2. We need a series of lemmas to prove Theorem 3. The first is the observation

Lemma 3.2.1. *For any* $I \in S(g)^G$ *and any* $z \in 0$ *one has*

$$(\delta\, u_I)\,(z) \;=\; (\,\delta_{\bar{b}}(I))\;(\nu(z)) + \beta\,(I)\,(z) \;. \tag{3.2.1}$$

The proof of Lemma 3.2.1 follows directly from the formula (3.1.6) together with the relations (3.1.7), (3.1.8) and (3.1.9) together with the definitions (3.1.11) and (3.1.14).

Lemma 3.2.2. *Let* I, J $\in S(\mathfrak{g})^G$ *then for any* $z \in O$

$$Q([z,\beta(I)(z)], \delta_{\bar{\mathfrak{b}}}(J)(\nu(z))) = (g(I)(z)) \langle \alpha_0, \delta_{\mathfrak{h}}(J)(\nu(x))\rangle . \quad (3.2.2)$$

Proof of Lemma 3.2.2. Since $\beta(I)(z) \in \mathfrak{g}_{-1}$ one has $[z,\beta(I)(z)] \in \bar{\mathfrak{b}}$. Thus only the component, x, of $[z,\beta(I)(z)]$ in $\mathfrak{h}$ can contribute to the left side of (3.2.2). But then clearly by (2.2.1), recalling that $a_i = a_i(z)$, and (3.1.11)

$$x = g(I)(z) \sum_{j=1}^{\ell} \frac{k_j}{a_j(z)} a_j(z)\left[e_{\alpha_j}, e_{-\alpha_j}\right].$$

Thus

$$x = g(I)(z) \sum_{j=1}^{\ell} k_j h_j . \quad (3.2.3)$$

But also then only the component, $\delta_{\mathfrak{h}}(J)(\nu(z))$, of $\delta_{\bar{\mathfrak{b}}}(J)(\nu(z))$ in $\mathfrak{h}$ can contribute to the left side of (3.2.2). However since $Q(e_{\alpha_j}, e_{-\alpha_j}) = 1$ one has $Q(h_j,y) = \langle\alpha_j,y\rangle$ for any $y \in \mathfrak{h}$. Thus $\langle \sum_j k_j h_j, y\rangle = \langle\alpha_0,y\rangle$ by (2.3.1). Thus if $y = \delta_{\mathfrak{h}}(J)(\nu(z))$ the lemma follows from (3.2.3). Q.E.D.

3.3. Unless otherwise stated the notation will remain as in Lemma 3.2.2 so that $I, J \in S(\mathfrak{g})^G$ and $z \in O$. Let $b_{I,J}(z)$ be the scalar given by the left side (and hence the right side) of (3.2.2). By the invariance of Q one has

$$b_{I,J}(z) = Q(z, [\beta(I)(z), \delta_{\bar{\mathfrak{b}}}(J)(\nu(z))]). \quad (3.3.1)$$

Also let

$$w = \left[\delta_{\bar{\mathfrak{b}}}(I)(\nu(z)), \delta_{\bar{\mathfrak{b}}}(J)\nu(z)\right] .$$

Lemma 3.3.1. *One has*

$$Q(z, [\delta u_I(z), \delta u_J(z)]) = Q(z,w) + b_{I,J}(z) - b_{J,I}(z) . \quad (3.3.2)$$

The proof of Lemma 3.3.1 follows immediately from (3.2.1) and (3.3.1) using the fact that, since $\beta(I)(z)$ and $\beta(J)(z)$ are scalar multiples of the same element, $[\beta(I)(z),\beta(J)(z)] = 0$.

Now let $\delta_{\mathfrak{g}}(K) = \delta_{\mathfrak{n}}(K) + \delta_{\bar{\mathfrak{b}}}(K)$ for any $K \in S(\mathfrak{g})^G$. By Proposition 1.3 in [3] one has that $\delta_{\mathfrak{g}}(K)(y)$ is in the center of the centralizer of y for any $y \in \mathfrak{g}$. Thus

Lemma 3.3.2. *One has*

$$0 = [\delta_{\mathfrak{g}}(I)(\nu(z)), \delta_{\mathfrak{g}}(J)(\nu(z))] \tag{3.3.3}$$

and also

$$0 = [\nu(z), \delta_{\mathfrak{g}}(I)(\nu(z))] = [\nu(z), \delta_{\mathfrak{g}}(J)(\nu(z))]. \tag{3.3.4}$$

Next we note

Lemma 3.3.3. *One has*

$$Q(e_{-\alpha_0}, w) = 0.$$

Proof of Lemma 3.3.3. Since $e_{-\alpha_0}$ is the highest root vector it is in the center of $\mathfrak{n}$. Thus

$$[e_{-\alpha_0}, \delta_{\mathfrak{n}}(I)(\nu(z))] = [e_{-\alpha_0}, \delta_{\mathfrak{n}}(J)(\nu(z))] = 0.$$

But then, by the invariance of Q, one has $Q(e_{-\alpha_0}, w) = Q(e_{-\alpha_0}, w')$ where $w' = [\delta_{\mathfrak{g}}(I)(\nu(z)), \delta_{\mathfrak{g}}(J)\nu(z)]$. However $w' = 0$ by Lemma 3.3.2. Thus $Q(e_{-\alpha_0}, w') = Q(e_{-\alpha_0}, w) = 0$. Q.E.D.

Lemma 3.3.4. *One has*

$$Q(z,w) = Q(\nu(z),w).$$

Proof of Lemma 3.3.4. It suffices to show $\nu(z) - z$ is Q-orthogonal to w. But $\nu(z) - z$ is in the span of f, e_{α_0} and $e_{-\alpha_0}$. But $w \in \bar{\mathfrak{n}}$ and f, $e_{\alpha_0} \in \bar{\mathfrak{n}}$. Thus w is Q-orthogonal to f and e_{α_0}. But it is also Q-orthogonal to $e_{-\alpha_0}$ by Lemma 3.3.3. Q.E.D.

Let

$$y = [\delta_{\mathfrak{n}}(I)(\nu(z)), \delta_{\mathfrak{n}}(J)(\nu(z))].$$

Lemma 3.3.5. *One has*

$$Q(\nu(z),w) = Q(\nu(z),y).$$

Proof of Lemma 3.3.5. Let $v_1 = [\delta_{\mathfrak{g}}(I)(\nu(z)), \delta_{\mathfrak{n}}(J)(\nu(z))]$ and $v_2 = [\delta_{\mathfrak{b}}(I)(\nu(z)), \delta_{\mathfrak{g}}(J)(\nu(z))]$. It is then immediate that $y - w = v_1 - v_2$. But by (3.3.4) and the invariance of Q it follows that v_1 and v_2 are both Q-orthogonal to $\nu(z)$. Thus $y - w$ is Q-orthogonal to $\nu(z)$. Q.E.D.

Lemma 3.3.6. *One has*

$$Q(\nu(z),y) = \left(\prod_{j=1}^{\ell} a_j(z)^{k_j}\right) Q(e_{\alpha_0},y) . \qquad (3.3.5)$$

Proof of Lemma 3.3.6. One has $y \in [\mathfrak{n},\mathfrak{n}]$ and hence y is Q-orthogonal to f. But since z and $e_{-\alpha_0} \in \mathfrak{n}$ it is also true that y is Q-orthogonal to z and $e_{-\alpha_0}$. The result then follows from (3.1.3), the definition of $\nu(z)$. Q.E.D.

For any $K \in S(\mathfrak{g})^G$ let $\delta_{\mathfrak{b}}(K) = \delta_{\mathfrak{h}}(K) + \delta_{\mathfrak{n}}(K)$. Let

$$w' = [\delta_{\mathfrak{b}}(I)(\nu(z)), \delta_{\mathfrak{b}}(J)(\nu(z))] . \qquad (3.3.6)$$

Lemma 3.3.7. *One has*

$$Q(e_{\alpha_0},w') = 0 .$$

Proof of Lemma 3.3.7. Let $v_1' = [\delta_{\bar{\mathfrak{n}}}(I)(\nu(z)), \delta_{\mathfrak{g}}(J)(\nu(z))]$ and $v_2' = [\delta_{\mathfrak{b}}(I)(\nu(z)), \delta_{\bar{\mathfrak{n}}}(J)(\nu(z))]$. But now by (3.3.3) one has $w' + v_1' + v_2' = 0$. But since e_{α_0} is in the center of $\bar{\mathfrak{n}}$ one has

$$[e_{\alpha_0}, \delta_{\bar{\mathfrak{n}}}(I)(\nu(z))] = [e_{\alpha_0}, \delta_{\bar{\mathfrak{n}}}(J)(\nu(z))] .$$

Thus v_1' and v_2' are Q-orthogonal to e_{α_0} by the invariance of Q. This proves the lemma since $w' = -(v_1' + v_2')$. Q.E.D.

3.4. Let

$$c_{I,J}(z) = [\delta_{\mathfrak{h}}(I)(\nu(z)), \delta_{\mathfrak{n}}(J)(\nu(z))] .$$

But then since $\mathfrak{h}$ is commutative one clearly has

$$w' = y + c_{I,J}(z) - c_{J,I}(z) \qquad (3.4.1)$$

and hence by Lemma 3.3.7. one has

$$Q(e_{\alpha_0},y) = Q(e_{\alpha_0},c_{J,I}(z)) - Q(e_{\alpha_0},c_{I,J}(z)) . \qquad (3.4.2)$$

Lemma 3.4.1. *One has*

$$Q(e_{\alpha_0}, c_{I,J}(z)) = -\langle \alpha_0, \delta_h(I)(\nu(z))\rangle (\iota(e_{\alpha_0})J)(\nu(z)). \quad (3.4.3)$$

Proof of Lemma 3.4.1. Put $\langle \alpha_0, \delta_h(I)(\nu(z))\rangle = c$. Since e_{α_0} is a root vector clearly

$$[e_{\alpha_0}, \delta_h(I)(\nu(z))] = -ce_{\alpha_0}.$$

But then by the invariance of Q one has

$$Q(e_{\alpha_0}, c_{I,J}(z)) = -cQ(e_{\alpha_0}, \delta_n(J)(\nu(z))).$$

But then by (3.1.12) one has (since $Q(e_{\alpha_0}, e_{-\alpha_0}) = 1$ and $Q(e_{\alpha_0}, e_\phi) = 0$ for all $\phi \in \Delta_+$ where $\phi \neq -\alpha_0$)

$$Q(e_{\alpha_0}, \delta_n(J)(\nu(z))) = (\iota(e_{\alpha_0})J)(\nu(z)). \quad (3.4.4)$$

But then (3.4.3) follows from (3.4.4). Q.E.D.

Now recall (3.3.1) and the paragraph preceding it.

Lemma 3.4.2. *One has*

$$Q(\nu(z), y) = b_{J,I}(z) - b_{I,J}(z).$$

Proof of Lemma 3.4.2. By Lemma 3.4.1 and (3.1.10) one has

$$\prod_{i=1}^{\ell} a_i(z)^{k_i} Q(e_{\alpha_0}, c_{I,J}(z)) = -g(J)(\nu(z))\langle \alpha_0, \delta_h(I)(\nu(z))\rangle$$

But then by Lemma 3.2.2 one has

$$\prod_{i=1}^{\ell} a_i(z)^{k_i} Q(e_{\alpha_0}, c_{I,J}(z)) = -b_{J,I}(z).$$

But then by Lemma 3.3.6 and (3.4.2)

$$Q(\nu(z), y) = b_{J,I}(z) - b_{I,J}(z).$$ Q.E.D.

We can now prove

Lemma 3.4.3. *The elements* $\gamma^*(I)$ *and* $\gamma^*(J)$ *in* $C^\infty(\mathbb{R}^{2n})$ *Poisson commute.*

Proof of Lemma 3.4.3. Let s be the left side of (3.3.2). As noted in (3.1.5) it suffices to show that s = 0. But by Lemma 3.3.1 one has $b = Q(z,w) + b_{I,J}(z) - b_{J,I}(z)$.

Thus by Lemma 3.4.2 it suffices to show that $Q(z,w) = Q(\nu(z),y)$. But this follows from Lemmas 3.3.4 and 3.3.5. Q.E.D.

Since I and J are arbitrary in $S(\mathfrak{g})^G$ Lemma 3.4.3 implies that the elements in $\gamma^*(S(\mathfrak{g})^G)$ Poisson commute with one another. We have only to show that H Poisson commutes with the elements of $\gamma^*(S(\mathfrak{g})^G)$. If $n = \ell$ we could do this immediately since in that case as we will see $H = \gamma^*(Q/2)$.

3.5. We recall that $\mathcal{Q}_1$ is the ℓ-dimensional subspace of $\mathcal{Q}$ spanned by ψ_i, $i = 1, \ldots, \ell$. By assumption $\psi_0 \in \mathcal{Q}_1$. Let $\mathcal{Q}_2$ be the B_H-orthogonal complement of $\mathcal{Q}_1$ in $\mathcal{Q}$. (See (1.1.3)). Let q_i', $i = \ell+1,\ldots,n$ be a B_H-orthonormal basis of $\mathcal{Q}_2$.

Proposition 3.5.1. *There exists uniquely elements* $p_i' \in [\mathcal{Q}_2,H]$, $i = \ell+1,\ldots,n$ *such that*

$$[q_i',p_j'] = \delta_{ij}$$

for $i, j = \ell+1,\ldots,n$. *Moreover the elements* p_j', $j = \ell+1,\ldots,n$ *not only Poisson commute among themselves but they all Poisson commute with all the elements in* $\gamma^*(S(\mathfrak{g})^G)$.

Proof of Proposition 3.5.1. We use the arguments and notation of §7.1 in [3]. If $\mathcal{P}_2 = [\mathcal{Q}_2,H]$ it is clear that $\mathcal{P}_2$ is a subspace of the space $\mathcal{P} \subseteq C^\infty(\mathbb{R}^{2n})$ spanned by the p_i, $i = 1,\ldots,n$. It follows therefore that the elements in $\mathcal{P}_2$ Poisson commute with one another. On the other hand since B_H is positive definite it follows from (7.1.12) in [3] that $\mathcal{P}_2$ in the notation of §7.1 in [3] is non-singularly paired to $\mathcal{Q}_2$ by B_ω. This proves the existence and uniqueness of the p_j', $j = \ell+1,\ldots,n$. Furthermore since $\mathcal{Q}_1$ is B_H-orthogonal to $\mathcal{Q}_2$ it also follows from (7.1.12) in [3] that $\mathcal{P}_2$ is B_ω orthogonal to $\mathcal{Q}_1$. Thus $[p_j', \psi_i] = 0$ for $j = \ell+1,\ldots,n$ and $i = 1,\ldots,\ell$. But of course $[p_j',p_i'] = 0$ for the same values of j and i. On the other hand if $\phi \in \gamma^*(S(\mathfrak{g}))$ then, recalling the argument in the proof of Proposition 2.2.2, the differential $d\phi$ is spanned by the $d\psi_i$ and dp_k', $i, k = 1,\ldots,\ell$. This proves that ϕ Poisson commutes with p_j' for $j = \ell+1,\ldots,n$. Q.E.D.

Put

$$H_2 = \sum_{j=\ell+1}^{n} \frac{(p_j')^2}{2}. \tag{3.5.1}$$

Let $b_{ij}' = B_H(\psi_i,\psi_j)$ for $i, j = 1,\ldots,\ell$. For the kinetic energy one has

Lemma 3.5. *One has*

$$\sum_{i=1}^{n} \frac{p_i^2}{2m_i} = H_2 + \frac{1}{2} \sum_{i,j=1}^{\ell} b'_{ij} p'_i p'_j .$$

The proof of Lemma 3.5 is given in the proof of Proposition 7.1 in [3].

The Hamiltonian H is related to Q by

Proposition 3.5.2. *One has*

$$H = \gamma^*(Q/2) + H_2 .$$

Proof of Proposition 3.5.2. Let $H_1 = \gamma^*(Q/2)$ so that for any $x \in \mathbb{R}^{2n}$

$$H_1(x) = \frac{1}{2} Q(\gamma(x), \gamma(x)).$$

But if $b_{ij} = Q(h_i, h_j)$ for $i, j = 1, \ldots, \ell$ it follows from (2.2.6) and (2.3.3) that

$$H_1(x) = \left(\sum_{i,j=1}^{\ell} b_{ij} p'_i p'_j \right)(x) + \sum_{i=1}^{\ell} r_i e^{\psi_i(x)} Q(e_{\alpha_i}, f)$$

$$+ \left(d \prod_{i=1}^{\ell} r_i^{k_i}\right) \left(\prod_{i=1}^{\ell} e^{k_i \psi_i(x)} \right) Q(e_{\alpha_0}, e_{-\alpha_0}) .$$

But $Q(e_{\alpha_i}, f) = Q(e_{\alpha_0}, e_{-\alpha_0}) = 1$. Also recall the relation (2.3.2) and the definition of d (see sentence following (2.3.2)). One thus has

$$H_1 = \sum_{i,j=1}^{\ell} b_{ij} p'_i p'_j + \sum_{i=0}^{\ell} r_i e^{\psi_i} .$$

But since $h_i = [e_{\alpha_i}, e_{-\alpha_i}]$ and $Q(e_{\alpha_i}, e_{-\alpha_i}) = 1$ it follows that h_i corresponds to α_i under the isomorphism $\mathfrak{h} \longrightarrow \mathfrak{h}'$ induced by Q. Thus $b_{ij} = Q(\alpha_i, \alpha_j)$. Hence $b_{ij} = b'_{ij}$ by (1.2.1). Thus $H = H_1 + H_2$ by Lemma 3.5, (1.1.1) and (3.1.1).

Q.E.D.

It follows from Proposition 3.5.1 that H_2 Poisson commutes with all the elements in $\gamma^*(S(\mathfrak{g})^G)$. But of course $Q \in S(\mathfrak{g})^G$. Thus by Proposition 3.5.2 and Lemma 3.4.3 it follows that H Poisson commutes with all the elements in $\gamma^*(S(\mathfrak{g})^G)$. This finishes the proof of Theorem 3.

Q.E.D.

Let I_j, $j = 1, \ldots, \ell$ be fundamental invariants in $S(\mathfrak{g})^G$ so that $S(\mathfrak{g})^G$ is isomorphic

(Chevalley) to the polynomial algebra $\mathbb{C}[I_1,\ldots, I_\ell]$. As a consequence of Theorem 3 and Propositions 3.5.1 and 3.5.2, one has

Corollary 3. *The* n *functions* $\gamma^*(I_1),\ldots, \gamma^*(I_\ell)$, $p'_{\ell+1},\ldots,p'_n$ *are a system in involution in* $C^\infty(\mathbb{R}^{2n})$. *Furthermore the Hamiltonian* H *of the generalized periodic Toda lattice is in the algebra generated by these functions.*

Remark 3.5. By noting the homogeneous terms of highest degree in the variables p'_i, $i = 1,\ldots,\ell$ of the functions $\gamma^*(I_j)$, $j = 1,\ldots,\ell$ it is easy to see that the differentials of the n functions in Corollary 3 are linearly independent on a non-empty open subset of phase space $\mathbb{R}^{2n}$. (These homogeneous terms are directly related to the generators of the algebra of Weyl group invariants in the symmetric algebra $S(\mathfrak{h})$).

REFERENCES

[1] Adler, M. and van Moerbeke, P., Completely integrable systems, Euclidean Lie algebras and curves, Advances in Math., vol. 38 (1980), 267-316.

[2] Kostant, B., Quantization and unitary representations, in "Lecture Notes in Mathematics", No. 170, pp. 87-208, Springer-Verlag, New York/Berlin 1970.

[3] Kostant, B., The solution to a generalized Toda lattice and representation theory, Advances in Math., vol. 34 (1979), 195-338.

SPACES OF SOLUTIONS OF RELATIVISTIC FIELD THEORIES WITH CONSTRAINTS

Jerrold E. Marsden +

Department of Mathematics, University of California,
Berkeley, California 94720, U.S.A.

Dedicated to Professor Bleuler on the occasion of his retirement.

1. Introduction

In this paper I shall explain how the reduction results of Marsden and Weinstein [38] can be used to study the space of solutions of relativistic field theories. Two of the main examples that will be discussed are the Einstein equations and the Yang-Mills equations.

The basic paper on spaces of solutions is that of Segal [49]. That paper deals with unconstrained systems and is primarily motivated by semilinear wave equations. We are mainly concerned here with systems with constraints in the sense of Dirac. Roughly speaking, these are systems whose four dimensional Euler-Lagrange equations are not all hyperbolic but rather split into hyperbolic evolution equations and elliptic constraint equations.

The methods that have been used to study these problems are of two types. First, there have been direct four dimensional attacks which, for example, put symplectic and multi-symplectic structures on the space of all solutions. These procedures are geometrically appealing since they are manifestly covariant. Since so many people have worked in this area, we merely refer the reader to [27,29,34,52,53] and references therein. Secondly, people have used the 3 + 1 or "geometrodynamic" approach. For the latter, one selects appropriate projections of the four dimensional fields on each spacelike hypersurface and imposes Hamiltonian evolution equations together with constraints. This procedure is generally called the "Dirac theory of constraints". Two

+ Research partially supported by the National Science Foundation.

good references are [30] and [31]. For vacuum relativity the procedure is sometimes called the "ADM formalism" after Arnowitt, Deser, Misner and Dirac (see [39]). From an analytical point of view, this second method is more powerful. It enables one to prove that spaces of solutions are a manifold at most points and to precisely investigate their symplectic structure. This paper will discuss this second method in the context of [38].

2. Some Additional Background and History.

Before embarking on a discussion of the mathematics we shall continue to review some of the background and history. This review does not pretend to be exhaustive and does omit a number of basic papers. However our intent is only to highlight some of the papers that are basic to the point of view we wish to develop.

As we have already mentioned, Segal's paper [49] gives a framework for the unconstrained theory. This leads naturally to an abstract theory of infinite dimensional Hamiltonian systems, as in [36] and [11].

The first example with constraints whose solution space was seriously studied was general relativity. In retrospect, general relativity is a harder example than Yang-Mills fields. However, developments in perturbation theory and historical circumstances dictated that relativity be done first.

2a. General Relativity

The first thing to do is to set up an infinite dimensional symplectic manifold and to realize the Einstein equations as Hamiltonian evolution equations together with constraints. An important point is that the constraints are the zero set of the conserved quantity generated by the gauge group of general relativity, i.e. the group of diffeomorphisms of spacetime. This is a fairly routine procedure given the existing ADM formalism and was carried out in [22]. (There were associated advances in the existence and uniqueness theorems; cf. [32,21,33,15] etc.).

The notation we shall use for this formalism is as follows. Let $(V, {}^{(4)}g)$ be a given spacetime. Let a slicing be given that is based on a fixed 3 manifold M. By restricting ${}^{(4)}g$ to each hypersurface in the slicing, we get a curve $g(\lambda)$ of Riemannian metrics on M. The basic symplectic space is $T^*\mathcal{M}$, the L^2-cotangent bundle of the space of Riemannian metrics on M. The conjugate variables π are symmetric tensor densities related to the extrinsic curvature (second fundamental form) of the hypersurface. The tangents to the parameter lines of the slicing decompose into normal and tangential parts determining the lapse function N and the shift vector field X of the slicing. The choice of a particular slicing is basically a choice of gauge.

Einstein's vacuum equations $\mathrm{Ein}({}^{(4)}g) = 0$ (the Einstein tensor formed from ${}^{(4)}g$) are equivalent to the evolution equations in adjoint form

$$\frac{\partial}{\partial\lambda}\begin{bmatrix} g \\ \pi \end{bmatrix} = -\mathbb{J}\circ D\Phi(g,\pi)^{*}\begin{pmatrix} N \\ X \end{pmatrix}; \quad \mathbb{J} = \begin{bmatrix} 0 & I \\ -I & 0 \end{bmatrix}$$

together with the constraints

$$\Phi(g,\pi) = 0$$

where $\Phi(g,\pi) = (\mathcal{H}(g,\pi), \mathcal{J}(g,\pi))$ is the super energy-momentum. This quantity Φ is the Noether conserved quantity generated by the group $\mathcal{D}$ of diffeomorphisms of spacetime. (For asymptotically flat spacetimes, only diffeomorphisms that are spatially asymptotic to the identity are needed to generate Φ. As in [48], the Lorentz group at infinity generates the total energy momentum tensor for the spacetime; see [14]).

This machinery may now be used as a tool to investigate the structure of the space of solutions of Einstein's equations.

Let V be a fixed four manifold and let $\mathcal{E}$ be the set of all globally hyperbolic Lorentz metrics $g = {}^{(4)}g$ that satisfy the vacuum Einstein equations $\mathrm{Ein}(g) = 0$ on V (plus some additional technical smoothness conditions). Let $g_0 \in \mathcal{E}$ be a given solution. We ask: what is the structure of $\mathcal{E}$ in the neighbourhood of g_0?

There are two basic reasons why this question is asked. First of all, it is relevant to the problem of finding solutions to the Einstein equations in the form of a perturbation series:

$$g(\lambda) = g_0 + \lambda h_1 + \frac{\lambda^2}{2} h_2 + \ldots$$

where λ is a small parameter. If $g(\lambda)$ is to solve $\mathrm{Ein}(g(\lambda)) = 0$ identically in λ then clearly h_1 must satisfy the *linearized Einstein equations:*

$$\mathrm{DEin}(g) \cdot h_1 = 0$$

where $\mathrm{DEin}(g)$ is the derivative of the mapping $g \longrightarrow \mathrm{Ein}(g)$. For such a perturbation series to be possible, is it sufficient that h_1 satisfy the linearized Einstein equations? i.e. is h_1 necessarily a direction of *linearization stability?* We shall see that in general the answer is no, unless additional conditions hold. The second reason why the structure of $\mathcal{E}$ is of interest is in the problem of quantization of the Einstein equations. Whether one quantizes by means of direct phase space techniques (due to Dirac, Segal, Souriau and Kostant in various forms) or by Feynman path integrals, there will be difficulties near places where the space of classical solutions is such that the linearized theory is *not* a good approximation to the nonlinear theory.

The dynamical formulation mentioned above is crucial to the analysis of this problem. Indeed, the essence of the problem reduces to the study of structure of the space of solutions of the constraint equations $\Phi(g,\pi) = 0$.

The final answer to these questions is this: $\mathcal{E}$ has a conical or quadratic singularity at g_0 if and only if there is a non-trivial Killing field for g_0 that belongs to the gauge group generating $\Phi = 0$ (thus, the flat metric on $T^3 \times \mathbb{R}$ has such Killing fields, but the Minkowski metric has none.) When $\mathcal{E}$ has such a singularity, we speak of a bifurcation in the space of solutions. When $\mathcal{E}$ has no singularity, the symplectic form induced on $\mathcal{E}$ has a kernel that equals the orbits of the gauge group, so $\mathcal{E}/\mathcal{D}$ is a smooth symplectic manifold. (See Theorem 3 below).

2b. Yang-Mills Equations

There is a similar situation for gauge field theories of Yang-Mills type, possibly coupled to gravity. The final situation here is as follows. The space of solutions is a smooth manifold near solutions with no gauge symmetries and this space, modulo the gauge group, is a smooth symplectic manifold. Near solutions with a symmetry, the space of solutions has a conical singularity.

2c. History and References

The historical circumstances leading up to statements of this type are as follows:

(a) Brill and Deser [10] considered perturbations of the flat metric on $T^3 \times \mathbb{R}$ and discovered the first example of trouble in perturbation theory. They found, by going to a second order perturbation analysis, that they had to readjust the first order perturbations in order to avoid inconsistencies at second order. This was the first hint of a conical structure for $\mathcal{E}$ near solutions with symmetry.

(b) Fischer and Marsden [23] found general sufficient conditions for $\mathcal{E}$ to be a manifold in terms of the Cauchy data for vacuum spacetimes and coined the term "linearization stability". Related results were proved by O'Murchadha and York [46].

(c) Choquet-Bruhat and Deser [12,13] proved that $\mathcal{E}$ is a manifold near Minkowski space. (This was later improved by Choquet-Bruhat, Fischer and Marsden [14]).

(d) An abstract theory for systems with constraints was developed (and applied to a number of examples, including relativity) by Marsden and Weinstein [38]. They studied the general problem of the structure of the level sets (and in particular the zero sets) of conserved quantities, i.e. momentum maps, associated with a gauge or symmetry group and proved that the quotient of these level sets by the gauge group is a symplectic manifold near nonsingular (i.e. non-symmetric) points. This theory will be briefly described below with further indications of how it fits into the general scheme of relativistic field theories with constraints.

(e) Moncrief [40] showed that the sufficient conditions derived by Fischer and Marsden for the compact case where equivalent to the requirement that (V, g_0) have no Killing fields. This then led to the link between symmetries and bifurcations.

(f) Moncrief [41] discovered the general splitting of gravitational perturbations generalizing Deser's decomposition [19]. The further generalization to momentum maps (general Noether currents) was found by Arms, Fischer and Marsden [4] in the context of [38]. This then applies to other examples such as gauge theory and also gives York's decomposition [56] as special cases.

(g) D'Eath [18] obtained the basic linearization stability results for Robertson-Walker universes.

(h) Moncrief [42] discovered the spacetime significance of the second order conditions that arise when one has a Killing field and identified them with conserved quantities of Taub [54]. (Arms and Marsden [5] showed that the second order conditions for compact spacelike hypersurfaces are nontrivial conditions.)

(i) A Hamiltonian formalism for pure gauge theories of Yang-Mills type was well-known by about 1975; see [31,17,43] and references therein. This implied that the abstract results in [38] on the space of solutions can be applied directly as is explained in §3 below (once the ellipticity of the adjoint of the derivative of the constraint map is known; this simple calculation was noted in [43]). Similar facts for the pure Yang-Mills case were obtained independently by Segal [50,51] and Garcia [28].

(j) Case (i) deals with points where the space of solutions is nonsingular. The singular case was studied by Moncrief [43]. A complete proof that the singularities are conical was given by Arms [3].

(k) Coll [16] and Arms [1] carried out a study of both the singular and nonsingular points for the Einstein-Maxwell equations. In [2] the general coupled Einstein-Yang-Mills system was studied.

(l) Moncrief [44] investigated the quantum analogues of linearization stabilities. Using $T^3 \times \mathbb{R}$, he shows that unless such conditions are imposed, the correspondence principle is violated.

(m) For general relativity a detailed description of the conical singularity in $\mathcal{E}$ near a spacetime with symmetries is due to Fischer, Marsden and Moncrief [26] for one Killing field and to Arms, Marsden and Moncrief [7] in the general case.

(n) An abstraction of the results in the singular case to the general context of [38] was obtained by Arms, Marsden and Moncrief [6]. They showed quite generally that zero sets of momentum maps have conical singularities near a point with symmetry.

(o) Pilati [47] developed a Hamiltonian formalism for supergravity. This is used by Bao [9] to study the space of solutions.

3. Spaces of Solutions Near Regular Points.

To study the space of solutions of a relativistic field theory and its symplectic structure, one can carry out the following steps:

1. A "3+1 procedure" of Dirac is carried out. A symplectic manifold for the dynamics is found and the constraint equations Φ (fields = ϕ, conjugate momentum = π_ϕ) = 0 are isolated.

2. The constraints Φ are identified with the momentum map J for the action of an appropriate gauge group; i.e. one proves that $\Phi = J$.

3. One checks that $DJ^* = D\Phi^*$ is elliptic (in the sense of Douglis and Nirenberg for mixed systems).

4. Invoke [38] near generic (regular) points.

5. Invoke [6] near singular points; i.e. solutions with gauge symmetry.

Let us comment a little further on points 1 to 4. Step 1 is the classical Dirac procedure; we have already referred to [31] and [30] for it.

So far, Step 2 has been checked by hand for each example. The general philosophy that the constraint set can be identified with the zero set of a momentum mapping seems to be true in a remarkably large number of cases. Another example is the Einstein-Dirac equations; see [45]. Several people (Gotay, Isenberg, Marsden, Sniatycki and Yasskin) are currently investigating general contexts in which this can be proved.

Step 3 is generally a simple calculation. However, it is essential so one can justify the splitting theorems of Moncrief. This abstract theorem (see [4]) generalizes the usual decompositions of gravitational perturbations and decompositions of the Maxwell field etc. It is analogous to a Hodge-type decomposition in a symplectic context and is stated below.

Next we recall a few of the features of Step 4. To do this, we first need a bit of notation. Let M be a given manifold (possibly infinite dimensional) and let a Lie

group G act on M. In examples, G will be infinite dimensional, such as the group of diffeomorphisms of a manifold or bundle automorphisms. (The proper sense in which these are Lie groups is discussed in [20]). Associated to each element ξ in the Lie algebra $\mathfrak{g}$ of G, we have a vector field ξ_M naturally induced on M. We shall denote the action by $\Phi : G \times M \longrightarrow M$ and we shall write $\Phi_g : M \longrightarrow M$ for the transformation of M associated with the group element $g \in G$. Thus
$\xi_M(x) = \frac{d}{dt} \Phi_{\exp(t\xi)}(x)\big|_{t=0}$.

Now let (P,ω) be a symplectic manifold, so ω is a closed (weakly) non-degenerate two-form on P and let Φ be an action of a Lie group G on P. Assume the action is symplectic: i.e. $\Phi_g^*\omega = \omega$ for all $g \in G$. A *momentum mapping* is a smooth mapping $J : P \longrightarrow \mathfrak{g}^*$ such that

$$\langle dJ(x) \cdot v_x , \xi \rangle = \omega_x (\xi_p(x), v_x)$$

for all $\xi \in \mathfrak{g}$, $v_x \in T_xP$ where $dJ(x)$ is the derivative of J at x, regarded as a linear map of T_xP to $\mathfrak{g}^*$ and $\langle \, , \rangle$ is the natural pairing between $\mathfrak{g}$ and $\mathfrak{g}^*$.

A momentum map is Ad^*-*equivariant* when the following diagram commutes for each $g \in G$:

$$\begin{array}{ccc} P & \xrightarrow{\ g\ } & P \\ {\scriptstyle J}\downarrow & & \downarrow{\scriptstyle J} \\ \mathfrak{g}^* & \xrightarrow[Ad^*_{g^{-1}}]{} & \mathfrak{g}^* \end{array}$$

where $Ad^*_{g^{-1}}$ denotes the co-adjoint action of G on $\mathfrak{g}^*$. If J is Ad^* equivariant, we call (P,ω,G,J) a *Hamiltonian G-space*.

Momentum maps represent the (Noether) conserved quantities associated with symmetry groups acting on phase space. This topic is of course a very old one, but it is only with more recent work of Souriau and Kostant that a deeper understanding has been achieved.

Let S_{x_0} = (the component of the identity of) $\{ g \in G | gx_0 = x_0 \}$, called the *symmetry group* of x_0. Its Lie algebra is denoted $\mathfrak{s}_x$, so

$$\mathfrak{s}_{x_0} = \{ \xi \in \mathfrak{g} \mid \xi_p(x_0) = 0 \} .$$

Let (P,ω,G,J) be a Hamiltonian G-space. If $x_0 \in P$, $\mu_0 = J(x_0)$ and if

$$dJ(x_0) \;:\; T_xP \longrightarrow \mathfrak{g}^*$$

is surjective (with split kernel), then locally $J^{-1}(\mu_0)$ is a manifold and $\{J^{-1}(\mu) \mid \mu \in \mathfrak{g}^*\}$ forms a regular local foliation of a neighbourhood of x_0. Thus, when $dJ(x_0)$ fails to be surjective, the set of solutions of $J(x) = 0$ could fail to be a manifold.

Theorem 1. $dJ(x_0)$ *is surjective if and only if* $\dim S_{x_0} = 0$; *i.e.* $s_{x_0} = \{0\}$.

Proof. $dJ(x_0)$ fails to be surjective if there is a $\xi \neq 0$ such that $\langle dJ(x_0)\cdot v_{x_0}, \xi\rangle = 0$ for all $v_{x_0} \in T_{x_0}P$. From the definition of momentum map, this is equivalent to $\omega_{x_0}(\xi_P(x_0), v_{x_0}) = 0$ for all v_{x_0}. Since ω_{x_0} is non-degenerate, this is, in turn equivalent to $\xi_P(x_0) = 0$; i.e. $s_{x_0} \neq 0$. ∎

This theorem assumes implicitly that there is a splitting

$$\mathfrak{g} = \text{Range } dJ(x_0) \oplus \text{Kernel } dJ(x_0)^* .$$

In the finite dimensional case this is automatic. In the infinite dimensional case it holds if $dJ(x_0)^*$ is an elliptic operator. In this case one also has the splitting

$$T_{x_0}P = \text{Ker } dJ(x_0) \oplus \text{Range } dJ(x_0)^*$$

These splittings are usually called the Fredholm alternative.

A corollary of theorem 1 is that $J^{-1}(\mu)$ is a smooth manifold near points x_0 with no symmetries.

Theorem 2. *The kernel of the symplectic form restricted to* $\ker dJ(x_0)$ *equals the tangent space to the* G_μ *orbit of* x_0 *at* x_0. *Here* $G_\mu = \{g \in G \mid Ad^*_{g^{-1}}\mu = \mu\}$ *and* $\mu = J(x_0)$.
Thus, near those points with no symmetry, $P_\mu = J^{-1}(\mu) / G_\mu$ *is a smooth symplectic manifold.*

We call P_μ the *reduced* symplectic manifold.
This result is proved in [38], but it also follows from Moncrief's decomposition which, for $\mu = 0$ reads

$$T_{x_0} P = \text{Range } dJ(x_0)^* \oplus \text{Ker } dJ(x_0) \cap \text{Ker}(dJ(x_0) \circ \mathbb{J})$$
$$\oplus \text{ Range } \mathbb{J} \circ dJ(x_0)^*$$

where $\mathbb{J}$ is a complex structure associated with the symplectic form. The middle summand represents the tangent space to P_μ. The proof of Moncrief's decomposition is conveniently available in a number of places, such as [24,25] and [37].

As has already been indicated, for relativistic field theories, the four dimensional equations usually split into hyperbolic evolution equations and the constraint equations $J = 0$. If the gauge group includes time translations, the evolution equations take the abstract form

$$\frac{d}{d\lambda} x(\lambda) = - \mathbb{J} \circ DJ(x(\lambda))^* \xi$$

where $\xi(\lambda) \in \mathfrak{g}$ represents a gauge choice. The space of solutions is thus represented by the set $J = 0$ in Cauchy-data space.

The symplectic structure one gets on the space of solutions by this procedure coincides with the one obtained by direct four dimensional methods (although this is not established in complete generality, it can be checked directly for a class of examples that includes all of those of interest to us).

4. A Simple Example: Electromagnetism.*

We now give a simple example of how the reduction procedure in the previous section works. We give it for electromagnetism for simplicity; the construction easily generalizes to Yang-Mills fields.

The four dimensional set-up consists of the usual Kaluza-Klein formalism. One has a circle bundle over spacetime whose connections represent electromagnetic potentials. A 3+1 analysis gives a circle bundle $\pi : B \longrightarrow M$ over a three manifold M representing a spacelike hypersurface.

Let CM be the bundle over M whose sections A are connections for the bundle B. The bundle CM is, in a canonical way, a symplectic manifold which is constructed via

---

* The point of view developed in this example was obtained jointly with Alan Weinstein.

reduction as follows (cf. [55]): the group S^1 acts on B and hence on T^*B. It produces a momentum map $J : T^*B \longrightarrow \mathbb{R}$. The reduced manifold $J^{-1}(1) / S^1$ is then CM . (The choice of $1 \in \mathbb{R}$ represents a normalization for a unit charge).

Let $\mathfrak{A}$ denote all sections of CM and let $\mathcal{G}$ denote the group of all automorphisms of the bundle B. Then via reduction, $\mathcal{G}$ acts on $\mathfrak{A}$.

Elements of $T^*\mathfrak{A}$ represent pairs (A-E), where A is the potential and E is the electric field. We put on $T^*\mathfrak{A}$ the canonical symplectic structure.

Maxwell's vacuum equations in terms of A and E may be summarized as

1. Hamiltonian evolution equations in $T^*\mathfrak{A}$ for the Hamiltonian

$$H = \frac{1}{2} \int_M \left[E^2 + (dA)^2\right] dx$$

and 2. the constraint equation $J = \operatorname{div} E = 0$.

Here J is the momentum map for the action of $\mathcal{G}$ on $T^*\mathfrak{A}$. This is a straightforward calculation. (For sources, use $J = \rho$ or, better, couple Maxwell's equations to a source and the full momentum map will be $\operatorname{div} E - \rho$.)

How are Maxwell's vacuum equations in terms of E and B obtained? One merely reduces $T^*\mathfrak{A}$ by the gauge group $\mathcal{G}$ at the value 0; i.e. we form the symplectic manifold $J^{-1}(0) / \mathcal{G}$. If M is simply connected, say $\mathbb{R}^3$, the reduced space is isomorphic to the space of pairs (B,E) where B and E are divergence free. By Theorem 2 above, this reduced space is naturally a symplectic manifold. The Poisson bracket on it may be computed to be

$$\{F,G\} = \int \left[\frac{\delta F}{\delta E} \operatorname{curl} \frac{\delta G}{\delta B} - \frac{\delta G}{\delta E} \operatorname{curl} \frac{\delta F}{\delta B}\right] dx$$

where F and G are real valued functions of E and B and the functional derivatives are defined in terms of the Frechet derivative by

$$DF(E,B) \cdot (E',B') = \int \left[\frac{\delta F}{\delta E} \cdot E' + \frac{\delta F}{\delta B} \cdot B'\right] dx .$$

The usual decompositions of electromagnetic fields are seen to be a special case of Moncrief's decomposition.

This example is linear so the spaces of solutions are always manifolds. However it does demonstrate nicely how the constraint equations are the zero set of a momentum map.

For other Yang-Mills fields however, the space of solutions is not a manifold, as was already pointed out in [43]. We will briefly discuss the singular case next.

5. Spaces of Solutions near Singular Points.

In general the space of solutions of a nonlinear relativistic field theory with constraints will have singularities at solutions with symmetries. As we have pointed out already, this was first hinted at in relativity by Brill and Deser [10]. For both relativity and Yang-Mills fields these singularities are known to be conical. (See the references in §2). This is especially surprising for relativity in view of the complexity of the field equations. However, from [6] there is good reason to think that this is fairly general, independent of how badly nonlinear the field theory is. On the other hand, it requires a somewhat special and complex argument for relativity.

For vacuum gravity, let us state one of the main results in the cosmological case: suppose (V, g_0) is a vacuum spacetime that has a *compact* spacelike hypersurface $M \subset V$. (Actually we also require the existence of at least one of constant mean curvature for technical reasons). Let S_{g_0} be the Lie group of isometries of g_0 and let k be its dimension.

Theorem 3.

1. ([23,40]) $k = 0$, then $\mathcal{E}$ is a smooth manifold in a neighbourhood of g_0 with tangent space at g_0 given by the solutions of the linearized Einstein equations. The symplectic form inherited naturally from $T^*\mathcal{M}$ has kernel on $\mathcal{E}$ equal to the infinitesimal gauge transformations, so the space $\mathcal{E}/\mathcal{D}$ is a symplectic manifold near such points.+

2. ([42,26,6,7]) If $k > 0$ then $\mathcal{E}$ is *not* a smooth manifold at g_0. A solution h_1 of the linearized equations is tangent to a curve in $\mathcal{E}$ if and only if h_1 is such that Taub conserved quantities vanish; i.e. for every Killing field X for g_0,

$$\int_M X\cdot\left[D^2\mathrm{Ein}(g_0)\cdot(h_1, h_1)\right]\cdot Z\,dx = 0$$

where Z is the unit normal to the hypersurface M and " • " denotes contraction with respect to the metric g_0.

+ The proof of a technical, but important item, namely that near points with no symmetry, $\mathcal{E}/\mathcal{D}$ is a manifold has not yet appeared in the literature. It will appear in a forthcoming publication of Isenberg and Marsden.

All explicitly known solutions possess symmetries, so while 1. is "generic", 2. is what occurs in examples. This theorem gives a complete answer to the perturbation question: a perturbation series is possible if and only if all the Taub quantities vanish. Thus, the second order conditions of Taub tell us the tangents to the conical singularity. There is a similar theorem for Yang-Mills fields [6,3].

Let us give a brief abstract indication of why such second order conditions should come in. Suppose X and Y are Banach spaces and $F : X \longrightarrow Y$ is a smooth map. In our examples, F will be a momentum map. Suppose $F(X_0) = 0$ and $x(\lambda)$ is a curve with $x(0) = x_0$ and $F(x(\lambda)) \equiv 0$. Let $h_1 = x'(0)$ so by the chain rule $DF(x_0) \cdot h_1 = 0$. Now suppose $DF(x_0)$ is not surjective and in fact suppose there is a linear functional $\ell \in Y^*$ orthogonal to its range: $\langle \ell , DF(x_0) \cdot u \rangle = 0$ for all $u \in X$. (Recall from Theorem 1 that dJ fails to be surjective at points with symmetry.) By differentiating $F(x(\lambda)) = 0$ twice at $\lambda = 0$, we get

$$D^2F(x_0) \cdot (h_1, h_1) + DF(x_0) \cdot x''(0) = 0 .$$

Applying ℓ gives

$$\langle \ell , D^2F(x_0) \cdot (h_1, h_1) \rangle = 0$$

which are necessary second order conditions that must be satisfied by h_1.

It is by this general method that one arrives at the Taub conditions. The issue of whether or not these conditions are sufficient is much deeper requiring extensive analysis and bifurcation theory (for $k = 1$, the Morse lemma is used, while for $k > 1$ the Kuranishi deformation theory is needed; see [35,8,6]).

REFERENCES

[1] Arms, J., Linearization stability of the Einstein-Maxwell system, J. Math. Phys. 18, 830-3 (1977).

[2] Arms, J., Linearization stability of gravitational and gauge fields, J. Math. Phys. 20, 443-453 (1979).

[3] Arms, J., The structure of the solution set for the Yang-Mills equations (preprint) (1980).

[4] Arms, J., Fischer, A., and Marsden, J., Une approche symplectique pour des théorèmes de decomposition en geométrie ou relativité générale, C.R. Acad. Sci. Paris 281, 517-520 (1975).

[5] Arms, J., and Marsden, J., The absence of Killing fields is necessary for linearization stability of Einstein's equations, Ind. Univ. Math. J. 28, 119-125 (1979).

[6] Arms, J., Marsden, J., and Moncrief, V., Bifurcations of momentum mappings, Comm. Math. Phys. 78, 455-478 (1981).

[7] Arms, J., Marsden, J., and Moncrief, V., The structure of the space of solutions of Einstein's equations; II Many Killing fields, (in preparation) (1981).

[8] Atiyah, M.F., Hitchin, N.J., and Singer, I.M., Self-duality in four-dimensional Riemannian geometry, Proc. Roy. Soc. London, A 362, 425-461 (1978).

[9] Bao, D., Linearization Stability of Supergravity, Thesis, Berkeley (1982).

[10] Brill, D. and Deser, S., Instability of closed spaces in general relativity, Comm. Math. Phys. 32, 291-304 (1973).

[11] Chernoff, P. and Marsden, J., Properties of Infinite Dimensional Hamiltonian Systems, Springer Lecture Notes in Math. No. 425 (1974).

[12] Choquet-Bruhat, Y. and Deser, S., Stabilité initiale de l'espace temps de Minkowski, C.R. Acad. Sci. Paris 275, 1019-1027 (1972).

[13] Choquet-Bruhat, Y. and Deser, S., On the stability of flat space. Ann. of Phys. 81, 165-168 (1973).

[14] Choquet-Bruhat, Y., Fischer, A, and Marsden, J., Maximal Hypersurfaces and Positivity of Mass, in *Isolated Gravitating Systems and General Relativity*, J. Ehlers ed., Italian Physical Society, 322-395 (1979).

[15] Christodoulou, D. and O'Murchadha, N., The boost problem in general relativity, (preprint) (1980).

[16] Coll, B. Sur la détermination, par des donnés de Cauchy, des champs de Killing admis par un espace-time, d'Einstein-Maxwell, C.R. Acad. Sci. (Paris), 281, 1109-12, 282, 247-250, J. Math. Phys. 18, 1918-22 (1975)

[17] Cordero, P. and Teitelboim, C., Hamiltonian treatment of the spherically symmetric Einstein-Yang-Mills system, Ann. Phys. (N.Y.), 100, 607-31 ((1976).

[18] D'Eath, P.D., On the existence of perturbed Robertson-Walker universes, Ann. Phys. (N.Y.) 98, 237-63 (1976).

[19] Deser, S., Covariant decomposition of symmetric tensors and the gravitational Cauchy problem, Ann. Inst. H. Poincaré VII, 149-188 (1967).

[20] Ebin, D. and Marsden, J., Groups of Diffeomorphisms and the motion of an incompressible fluid, Ann. of Math. 92, 102-163 (1970).

[21] Fischer, A. and Marsden, J., The Einstein Evolution equations as a first-order symmetric hyperbolic quasilinear system, Commun. Math. Phys. 28, 1-38 (1972).

[22] Fischer, A. and Marsden, J., The Einstein equations of evolution -- a geometric approach, J. Math. Phys. 13, 546-68 (1972).

[23] Fischer, A. and Marsden, J., Linearization stability of the Einstein equations, Bull. Am. Math. Soc. 79, 995-1001 (1973).

[24] Fischer, A. and Marsden, J., The initial value problem and the dynamical formulation of general relativity, in "General Relativiy, An Einstein Centenary Survey, ed. S. Hawking and W. Israel, Cambridge, Ch. 4 (1979).

[25] Fischer, A. and Marsden, J., Topics in the dynamics of general relativity, in *Isolated gravitating systems in General Relativity*, J. Ehlers, ed., Italian Physical Society, 322-395 (1979).

[26] Fischer, A., Marsden, J., and Moncrief, V., The structure of the space of solutions of Einstein's equations, I: One Killing Field, Ann. Inst. H. Poincaré 33, 147-194 (1980).

[27] Garcia, P., The Poincaré-Cartan invariant in the calculus of variations, Symp. Math. 14, 219-246 (1974).

[28] Garcia, P., (See his article in this volume) (1981).

[29] Goldschmidt, H. and Sternberg, S., The Hamilton-Jacobi formalism in the calculus of variations, Ann. Inst. Fourier 23, 203-267 (1973).

[30] Gotay, M., Nester, J., and Hinds, G., Pre-symplectic manifolds and the Dirac-Bergman theory of constraints, J. Math. Phys. 19, 2388-99 (1978).

[31] Hansen, A., Regge, T., and Teitelboim, C., Constrained Hamiltonian Systems, Accademia Nazionale dei Lincei, Scuola Normale Superlore, Pisa (1976).

[32] Hawking, S. and Ellis, G., *The Large Scale Structure of Spacetime,* Cambridge University Press, (1973).

[33] Hughes, T., Kato, T., and Marsden, J., Well-posed quasi-linear hyperbolic systems with applications to nonlinear elastodynamics and general relativity, Arch. Rat. Mech. An. 63, 273-294 (1977).

[34] Kijowski, J. and Tulczyjew, W., *A symplectic framework for field theories,* Springer Lecture Notes in Physics No. 107 (1979).

[35] Kuranishi, M., New proof for the existence of locally complete families of complex structures. Proc. of Conf. on Complex Analysis, (A. Aeppli et al., eds.), Springer (1975).

[36] Marsden, J., Hamiltonian One Parameter Groups, Arch. Rat. Mech. An. 28, 361-396 (1968).

[37] Marsden, J., Lectures on Geometric Methods in Mathematical Physics, SIAM (1981).

[38] Marsden, J. and Weinstein, A., Reduction of symplectic manifolds with symmetry, Rep. Math. Phys. 5, 121-130 (1974).

[39] Misner, C., Thorne, K., and Wheeler, J., Gravitation, Freeman, San Francisco, (1973).

[40] Moncrief, V., Spacetime symmetries and linearization stability of the Einstein equations, J. Math. Phys. 16, 493-498 (1975).

[41] Moncrief, V., Decompositions of gravitational perturbations, J. Math. Phys. 16, 1556-1560 (1975).

[42] Moncrief, V., Spacetime symmetries and linearization stability of the Einstein equations II, J. Math. Phys. 17, 1893-1902 (1976).

[43] Moncrief, V., Gauge symmetries of Yang-Mills fields, Ann. of Phys. 108, 387-400 (1977).

[44] Moncrief, V., Invariant states and quantized gravitational perturbations, Phys. Rev. D. 18, 983-989 (1978).

[45] Nelson, J. and Teitelboim, C., Hamiltonian for the Einstein-Dirac field, Phys. Lett. 69B, 81-84 (1977), and Ann. of Phys. (1977).

[46] O'Murchadha, N. and York, J., Existence and uniqueness of solutions of the Hamiltonian constraint of general relativity on compact manifolds, J. Math. Phys. 14, 1551-1557 (1973).

[47] Pilati, M., The canonical formulation of supergravity, Nuclear Physics B 132, 138-154 (1978).

[48] Regge, T. and Teitelboim, C., Role of surface integrals in the Hamiltonian formulation of general relativity, Ann. Phys. (N.Y.), 88, 286-318 (1974).

[49] Segal, I., Differential operators in the manifold of solutions of nonlinear differential equations, J. Math. Pures. et Appl. XLIV, 71-132 (1965).

[50] Segal, I., General properties of Yang-Mills fields on physical space, Nat. Acad. Sci. 75, 4638-4639 (1978).

[51] Segal, I., The Cauchy problem for the Yang-Mills equations, J. Funct. An. 33, 175-194 (1979).

[52] Sniatycki, J., On the canonical formulation of general relativity, Journées Relativistes de Caen, Soc. Math. de France and Proc. Camb. Phil. Soc. 68, 475-484 (1970).

[53] Szczyryba, W., A symplectic structure of the set of Einstein metrics: a canonical formalism for general relativity, Comm. Math. Phys. 51, 163-182 (1976).

[54] Taub, A., Variational Principles in General Relativity, C.I.M.E. Bressanone, 206-300 (1970).

[55] Weinstein, A., A Universal Phase Space for Particles in Yang-Mills fields, Lett. Math. Phys. 2, 417-420 (1979).

[56] York, J.W., Covariant decompositions of symmetric tensors in the theory of gravitation, Ann. Inst. H. Poincaré 21, 319-32 (1974).

II. SYMPLECTIC GEOMETRY

THE SYMPLECTIC "CATEGORY"

Alan Weinstein

Dept. of Mathematics, University of California
Berkeley, CA 94720, U.S.A.

In this lecture, I would like to present an approach to symplectic geometry and geometric quantization, with the hope that it will contribute to progress in those areas by unifying many known constructions and by suggesting new ones. The underlying idea was forseen by many people, including Van Vleck, Maslov, Kostant, Slawianowski, Sniatycki, and Tulczejew, and was made nearly explicit by Hörmander in his work on Fourier integral operators. Succinctly, it may be stated in the form of a "symplectic creed":

EVERYTHING IS A LAGRANGIAN SUBMANIFOLD

More formally, we think of symplectic geometry as taking place in a "category" $\mathcal{S}$ in which the objects are symplectic manifolds and the morphisms are canonical relations. The composition of canonical relations is again canonical if a transversality or clean intersection hypothesis is satisfied, but not always; hence, $\mathcal{S}$ is not a true category.

The symplectic "category" $\mathcal{S}$ was introduced in a survey article on symplectic geometry [10]. The aim of the present lecture is to supplement the earlier article by adding some further examples, especially as $\mathcal{S}$ relates to geometric quantization. As was already mentioned in [10], the subcategory $\mathcal{S}_L$ of $\mathcal{S}$ consisting of linear symplectic spaces and linear canonical relations was studied by Guillemin and Sternberg in [3]. (The clean intersection property is always satisfied in $\mathcal{S}_L$, so it is a true category.) Guillemin and Sternberg give a complete quantization of this category which extends the metaplectic representation; a related construction is also discussed in [9].

Analogies with linear spaces

An object in $\mathcal{S}$ is a symplectic manifold $\mathcal{P} = (P, \Omega)$. The "dual" $\mathcal{P}^*$ is defined to be $(P,-\Omega)$, and the product of $\mathcal{P}_j = (P_j, \Omega_j)$ for j = 1,2 is defined to be $\mathcal{P}_1 \times \mathcal{P}_2 = (P_1, \Omega_1) \times (P_2, \Omega_2)$. It is useful to keep in mind the following basic analogies between $\mathcal{S}$ and categories of hermitian vector spaces.

symplectic manifold $\mathcal{P} = (P, \Omega)$	hermitian vector space $\mathcal{H}$
$\mathcal{P}^*$	dual space $\bar{\mathcal{H}}$
$\mathcal{P}_1 \times \mathcal{P}_2$	tensor product $\mathcal{H}_1 \hat{\otimes} \mathcal{H}_2$
$\mathcal{P}_1 \times \mathcal{P}_2^*$	mapping space $\mathrm{Hom}(\mathcal{H}_2, \mathcal{H}_1)$
(a point, 0)	$\mathbb{C}$

An element of $\mathcal{H}$ may be thought of as a morphism from $\mathbb{C}$ to $\mathcal{H}$; by analogy, we may think of the "elements" of $\mathcal{P}$ in $\mathcal{S}$ to be the morphisms from (a point, 0) to $\mathcal{P}$; these are just the lagrangian submanifolds of $\mathcal{P}$. In particular, the "elements" of the "Hom object" $\mathcal{P}_1 \times \mathcal{P}_2^*$ are just the canonical relations from $\mathcal{P}_2$ to $\mathcal{P}_1$, i.e. the $\mathcal{S}$ - morphisms from $\mathcal{P}_2$ to $\mathcal{P}_1$, so our sequence of definitions and analogies has come full circle.

Starting with the basic analogies in the box above, we can define other notions in $\mathcal{S}$ which correspond to linear space concepts. For instance, a "conjugate linear" mapping from $\mathcal{P}_1$ to $\mathcal{P}_2$ is a morphism from $\mathcal{P}_1$ to $\mathcal{P}_2^*$; the dual of a morphism from $\mathcal{P}_1$ to $\mathcal{P}_2$ is the lagrangian submanifold of $(P_1, \Omega_1) \times (P_2, -\Omega_2)$ obtained from the given one in $(P_2, \Omega_2) \times (P_1, -\Omega_1)$ by exchanging the factors, and a "hermitian form" is a morphism from $\mathcal{P}$ to itself which is equal to its dual. The diagonal (i.e. identity), is the natural hermitian form on each object. A morphism is "unitary" if its dual is its inverse; it is easy to see that the unitary morphisms are just the graphs of canonical transformations.

Algebras

A $*$-algebra in $\mathcal{S}$ is a symplectic manifold $\mathcal{Q}$ with a multiplication morphism from $\mathcal{Q} \times \mathcal{Q}$ to $\mathcal{Q}$, a conjugation morphism from $\mathcal{Q}$ to $\mathcal{Q}^*$, and an identity "element" satisfying the appropriate composition identities. The three most important examples of algebras in $\mathcal{S}$ which I know are:

(i)	$\mathcal{P} \times \mathcal{P}^*$	(algebras of endomorphisms)
(ii)	$(T^* G, \Omega_G) = \mathcal{G}$	(group algebras of Lie groups G)
(iii)	$(T^* X, \Omega_X) = \mathcal{X}$	(function algebras on manifolds X)

The details of the algebra structures are given in [10].

An $\mathcal{S}$-morphism which is compatible with algebra structures may be called a homomorphism. A homomorphism from $\mathcal{L}$ to $\mathcal{P} \times \mathcal{P}^*$ plays the role in $\mathcal{S}$ of a representation. It turns out (see [9] or [10]) that such a homomorphism can be constructed in a natural way from a symplectic action of G on P together with an equivariant momentum mapping from P to $\mathfrak{g}^*$; thus, the representation in $\mathcal{S}$ contains both the global action of G on P and the infinitesimal homomorphism from $\mathfrak{g}$ to $C^\infty(P)$ given by the momentum mapping.

Given a representation of the type just described, we can apply it to the diagonal in $\mathcal{P} \times \mathcal{P}^*$ to get a lagrangian submanifold of (T^*G, Ω_G) which is the "character" of the representation. Although a clean intersection assumption may be necessary for the character to be defined, Kostant [6] has observed that the character is always defined if $\mathcal{P}$ is a *homogeneous* hamiltonian G-space; in fact, the transversality assumption sufficient for composibility of canonical relations is satisfied in this case. Kostant has connected this geometric description of the character with the character formula conjectured by Kirillov [5].

I can give the following geometric picture of Kostant's construction. Given a coadjoint orbit Θ in $\mathfrak{g}^*$; consider the character lagrangian submanifold $L \subseteq T^*G$ associated with the action of G on Θ. Pull back L to $T^*\mathfrak{g}$ by using the exponential map from G to $\mathfrak{g}$ to get a lagrangian submanifold $L_1 \subseteq T^*\mathfrak{g}$. Finally, use the symplectomorphism of $T^*\mathfrak{g} = \mathfrak{g} \times \mathfrak{g}^*$ with $T^*\mathfrak{g}^* = \mathfrak{g}^* \times \mathfrak{g}$ given by $(\xi, \mu) \longrightarrow (\mu, -\xi)$ (the "Fourier transform") to transform L_1 into $L_2 \subseteq T^*\mathfrak{g}^*$. Then one may verify that L_2 is just the conormal bundle of $\Theta \subseteq \mathfrak{g}^*$. This is the symplectic version of the statement "the pullback of the character is the Fourier transform of the orbit".

Geometric quantization [3] [8] [11]

Suppose that we have quantized some objects in $\mathcal{S}$ by equipping them with line bundles and polarizations. (These structures can themselves be interpreted in $\mathcal{S}$ [10], but we will not develop this point here.) The quantizations extend in a natural way to duals and products, and hence to objects of the form $\mathcal{P}_1 \times \mathcal{P}_2^*$. By this extension, we can reduce the problem of quantizing $\mathcal{S}$-morphisms, be they unitary or not, to the following single problem:

FUNDAMENTAL QUANTIZATION PROBLEM. Suppose that the symplectic manifold (P, Ω) is quantized by $\mathbf{H}$ (E, ∇, J), where E is a line bundle with connection ∇, J is a polarization, and $H = H(E, \nabla, J)$ is the space of J-parallel sections of E $\otimes$ (half-forms + normal to J). Given a triple (L, γ, Θ), where $L \subseteq P$ is lagrangian,

+ For most purposes, one may just as well use half-densities

γ is a parallel lift of L to E, and θ is a half-form on L, how should one construct an element (or class of elements) $\psi(L,\gamma,\theta)$ in H (or some extension of H) in a "natural" way? As much as possible, the elements $\psi(L,\gamma,\theta)$ associated with canonical relations should satisfy the same composition laws as the underlying relations; i.e. quantization should be "functorial".

The problem just posed can be solved under various special assumptions on the data given. The rest of this lecture will consist of descriptions of some of these solutions. I hope that it will not be too long until the general situation is well understood.

Ⓐ If L is globally transverse to J, the half-forms on L can be identified with half-forms normal to J, and the section γ can simply be extended from L to P, so that we get a well-defined element of H.* Special cases of this are:

1) Let $\mathcal{P} = (T^*X, \Omega_X)$, with the usual quantization by half-forms on X. If L is globally transverse to the foliation by fibres, and if there is a parallel lift of L, then L is the image of dS for some function S on X with values in $\mathbb{R}/2\pi\mathbb{Z}$. (The arbitrary additive constant in S is determined by the choice of parallel lift.) Any half-form θ on L may be pulled back to X by dS; the quantization of this data is then the half-form $e^{iS}\,[dS^*(\theta)]$ on X.

2) Given a fixed prequantization (E,∇) and two polarizations J_1 and J_2 on $\mathcal{P}$, we can equip the product $\mathcal{P} \times \mathcal{P}^*$ with the prequantization whose line bundle is Hom (E, E) and the polarization $J_1 \times J_2$. The diagonal Δ is lagrangian, it carries a natural half-density (from the Liouville form on P), and it has a natural parallel lift to Hom (E, E) (the identity bundle map). Δ is globally transverse to $J_1 \times J_2$ if and only if J_1 is globally transverse to J_2. In this case, the operator from H_1 to H_2 obtained by quantizing Δ is just the operator coming from the Blattner-Kostant-Sternberg pairing, but our $\mathcal{J}$-viewpoint sheds some new light on this pairing. First of all, it makes it clear that the pairing arises from essentially the same construction as the $dS \longrightarrow e^{iS}$ correspondence. Second, it shows that the question of unitarity is a special case of the functoriality problem, since the pairing gives operators $A : H_1 \longrightarrow H_2$ and $B : H_2 \longrightarrow H_1$ which are adjoints of each other, and functoriality would imply that AB and BA are identity operators. (We assume that the identity relation with a fixed polarization is quantized by the identity operator; see Ⓒ below.) Third, our construction may suggest approaches to the pairing of non-transverse polarizations.

---

* Here and elsewhere in this lecture, we are assuming that J is a real polarization. If J is Kählerian, the problem is the much-studied one of holomorphic extension from totally real submanifolds.

But, finally, if the pairing idea is to be used to quantize group actions which do not preserve a polarization, we will see below that it may be conceptually simpler not to think in terms of pairings at all.

Ⓑ Let $\mathcal{P} = (T^*X, \Omega_X)$ as in Ⓐ 1), and suppose that L is a union of rays in T^*X minus the zero section. Such a homogeneous L has a natural parallel lift. (When the usual line bundle over T^*X is given the usual trivialization, the connection 1-form is zero on L, so we may take the constant section equal to 1.) If the half-density γ on L has suitable growth properties along the rays, the theory of Fourier integral operators [4] gives a class of distributions on X associated with (L, γ). The difference between two elements of the class is of a "lower order of singularity", and functoriality is satisfied as far as the entire classes are concerned. The classes can be narrowed down if one replaces the half-density on L by a "total symbol" which includes information at all levels of singularity, but the calculus of these total symbols seems to be extremely complicated; nevertheless, the theory of Fourier integral operators does suggest that higher order symbols may play a useful role in geometric quantization.

Ⓒ A special case of Ⓑ in which the quantization is unambiguous is that where L is a single cotangent space $T^*_x X$ and γ is a constant half-density. The natural quantization of L and γ is a delta-distribution at x. More generally, if L is the conormal bundle to a submanifold $Y \subseteq X$, then certain homogeneous half-densities on L quantize unambiguously to delta-distributions concentrated along Y. (See [2] for a discussion of distributions from this geometric point of view.)

Suppose now that we have any quantization (E, ∇, J) of (P, Ω). If L is a leaf of J, it has a natural flat connection, so we can speak of a constant half-density γ on L. By analogy with the cotangent bundle case, we should quantize (L, γ) by a delta-section supported at the point L of the leaf space L/J. More generally, if L has clean intersection with the leaves of J, then L projects onto a submanifold Y in L/J, and we should quantize suitable homogeneous half-densities on L by delta-sections along Y. This construction applies to the situation described in Ⓐ 2) above, if the polarizations J_1 and J_2 have constant-dimensional intersection. Here, homogeneity of the half-density should correspond to the vanishing of the obstruction described by Blattner [1]. In particular, if Δ is the diagonal in $\mathcal{P} \times \mathcal{P}^*$ with the natural half-density and parallel lift, and $J_1 = J_2 = J$, then Δ is quantized by a delta-distribution along the diagonal in $P/J \times P/J$: namely, the kernel of the identity operator.

Ⓓ If $\mathcal{P}$ and (E, ∇, J) are given, and $\psi : P \longrightarrow P$ is a symplectomorphism, we may try to quantize the graph $\Gamma_\psi \subseteq \mathcal{P} \times \mathcal{P}^*$ with its natural half-density induced from P. A parallel lift of Γ_ψ corresponds to a lift of ψ to a connection-preserving automorphism of (E, ∇). If ψ preserves the polarization J, then Γ_ψ may be quantized by the method of Ⓒ, since Γ_ψ will have clean intersection with the leaves of J; the quantization in this case is just the pullback operator.

If ψ^*J is transverse to J, then Γ_ψ is transverse to $J \times J$, and we can apply the method of Ⓐ. The result is the same as that obtained by the usual pairing method, except that here we quantize in one step instead of two (pullback plus pairing). The difference between the two approaches is conceptual rather than mathematical, but it seems to me that the new point of view is a little bit neater. A more natural use of the pairing may be found in Rawnsley [7].

Ⓔ Suppose that we are given a hamiltonian action of G on $\mathcal{P}$, and that G is quantized by H. Then $\mathcal{P} \times \mathcal{P}^* \times \mathcal{G}^*$ is quantized by Hom $(L^2(G), \mathrm{Hom}(H, H))$, and quantizing the $\mathcal{S}$-homomorphism from $\mathcal{G}$ to $\mathcal{P} \times \mathcal{P}^*$ associated with the action amounts to constructing a representation of G on H. If the process of quantizing lagrangian submanifolds is sufficiently well understood, this should lead to new insights into quantizing hamiltonian group actions and computing the characters of the representations thus obtained. Some progress in this direction has recently been made by Boutet de Monvel, Guillemin, Kostant, and Sternberg, but much remains to be done.

REFERENCES

[1] Blattner, R.J., The metalinear geometry of non-real polarization, *Lecture Notes in Mathematics* 570 (1977), 11-45.

[2] Guillemin, V.W., and Sternberg, S., *Geometric Asymptotics*, American Math. Soc., Providence, 1977.

[3] Guillemin, V.W., and Sternberg, S., Some problems in integral geometry and some related problems in micro-local analysis, *Amer. J. Math.* 101 (1979), 915-955.

[4] Hörmander, L., Fourier integral operators, I, *Acta Math.* 127 (1971), 79-183.

[5] Kirillov, A.A., The characters of unitary representations of Lie groups, Funct. Anal. Appl. 2 (1968), 133-146.

[6] Kostant, B., Geometric quantization and character theory, lecture given in special session at Amer. Math. Soc. annual meeting, San Francisco, January 1981.

[7] Rawnsley, J., A nonunitary pairing of polarizations for the Kepler problem, Trans. Amer. Math. Soc. 250 (1979), 167-180.

[8] Sniatycki, J., *Geometric Quantization and Quantum Mechanics*, Springer-Verlag, New York-Heidelberg-Berlin, 1980.

[9] Weinstein, A., Lectures on symplectic manifolds, *Regional Conferences in Mathematics*, vol. 29, Amer. Math. Soc., Providence, 1977.

[10] Weinstein, A., Symplectic geometry, *Bull. Amer. Math. Soc.* (1981) (to appear).
[11] Woodhouse, N.M.H., *Geometric Quantization*, Oxford U. Press, Oxford,1981.

MOMENTS AND REDUCTIONS

Victor Guillemin

Dept. of Mathematics, M.I.T.
Cambridge, MA 02139, USA.

and

Shlomo Sternberg

Dept. of Maths., Harvard University,
Cambridge, MA 02139, USA.

In this note we gather together some diverse facts about the moment map for symplectic group actions and the corresponding reductions of the symplectic manifold on which the group acts. We also relate moment maps with the "character Lagrangian" so as to obtain "intertwining Lagrangians" for Hamiltonian group actions.

1.

Let G be a Lie group. For any $a \in G$ we let ℓ_a denote left multiplication by a, so

$$\ell_a c = ac .$$

We let r_b denote right multiplication by b^{-1} for any $b \in G$ so that

$$r_b c = cb^{-1} .$$

A differential form ω on G is called left invariant if

$$\ell_a^* \omega = \omega$$

for all $a \in G$. Similarly, a vector field ξ on G is called left invariant if it satisfies

$$\ell_a \xi = \xi$$

for all $a \in G$. We may identify the set of left invariant vector fields with the Lie algebra, g, of G. (Indeed, this is one possible definition of the Lie algebra.) Each left invariant vector field generates a one parameter group of right multiplications, since it is the right multiplications which commute with all left multiplications. Any subalgebra, h, of g defines a foliation on G whose leaves consist of cosets aH where H is the connected subgroup generated by h. Each left invariant differential form is completely determined by its values on the left invariant vector fields, we may thus identify the space of left invariant q-forms with $\wedge^q(g^*)$. If ω is left invariant, so is $d\omega$ and hence d induces a linear map from $\wedge^q(g^*)$ to $\wedge^{q+1}(g^*)$ which we shall denote by δ. The formula for δ is standard, and can be deduced inductively from the fundamental formula for the Lie derivative:

$$D_\xi \omega = i(\xi)d\omega + di(\xi)\omega$$

when applied to left invariant vector fields and forms. Here D_ξ denotes Lie derivative with respect to ξ and $i(\xi)$ interior product.

For example, suppose that ω is a linear differential form. Then $i(\xi)\omega$ is a constant and so the second term in the above expression vanishes. If η is a second left invariant differential form, then $\omega(\eta) = i(\eta)\omega$ is again a constant and so

$$\begin{aligned} 0 = D_\xi(\omega(\eta)) &= (D_\xi\omega)(\eta) + \omega([\xi,\eta]) \\ &= i(\xi)\, d\omega(\eta) + \omega([\xi,\eta]) \\ &= d\omega(\xi \wedge \eta) + \omega([\xi,\eta]) \end{aligned}$$

and so

$$\delta\omega(\xi\wedge\eta) = -\omega([\xi,\eta])$$

and so on.

Since right and left multiplication commute, if ω is left invariant so is $r_b^*\omega$ and furthermore

$$r_b^*\omega = r_b^*\ell_b^*\omega \quad .$$

Now $r_b\ell_b(c) = bcb^{-1}$ and the corresponding action on left invariant vector fields is usually denoted by Ad_b. Hence $b \rightsquigarrow r_{b^{-1}}^*$ defines a representation of G on each $\wedge^q(g^*)$ which we denote by $\mathrm{Ad}^\#$. Thus

$$\mathrm{Ad}^\#_a\omega = r_{a^{-1}}^*\omega .$$

2.

Let M be a differentiable manifold and let $G \times M \longrightarrow M$ be a left action of G on M. Thus for each $a \in G$ we get a diffeomorphism $\varphi_a : M \longrightarrow M$

$$\varphi_a(m) = am .$$

Similarly, for each point m of M we get a map $\psi : G \longrightarrow M$ given by

$$\psi_m(a) = am .$$

Since $b(am) = (ba)m$ we see that

$$\varphi_b \psi_m = \varphi_m \ell_b \tag{2.1}$$

and

$$\psi_{am} = \varphi_m r_{a^{-1}} . \tag{2.2}$$

Now let Ω be an invariant q-form on M, so that

$$\varphi_a^* \Omega = \Omega$$

for all $a \in G$. For each point m, the form $\psi_m^* \Omega$ on G is left invariant. Indeed, by (2.1)

$$\ell_b^* \psi_m^* \Omega = \psi_m^* \varphi_b^* \Omega = \psi_m^* \Omega .$$

Furthermore, by (2.2),

$$\psi_{am}^* \Omega = r_{a^{-1}}^* \Omega = Ad_a^\# \Omega .$$

We have thus proved:

PROPOSITION 2.1. *An invariant* q*-form* Ω *on* M *defines map* $\Psi : M \longrightarrow \Lambda^q(g^*)$ *given by*

$$\Psi(m) = \psi_m^* \Omega .$$

This map is a G *morphism, i.e.,*

$$\Psi(am) = Ad_a^\# \Psi(m) .$$

Each $\xi \in G$ defines a one parameter subgroup of G and hence a one parameter group of diffeomorphisms of M and hence a vector field on M which we shall denote by ξ_M. Letting $\xi_M(m)$ denote the value of the vector field ξ_M at the point m, we see that

$$\xi_M(m) = d(\psi_m)_e \xi$$

where, on the right-hand side of this equation we consider ξ as a tangent vector at the identity, e of the group and $d(\psi_m)_e : TG_e \longrightarrow TM_m$. Letting $\langle\ ,\ \rangle$ denote the pairing between $\wedge^q(g^*)$ and $\wedge^q(g)$, it follows that

$$\langle \Psi , \xi_1 \wedge \ldots \wedge \xi_q \rangle = \Omega(\xi_{1M} \wedge \ldots \wedge \xi_{qM}) \qquad \xi_1, \ldots, \xi_q \in g \tag{2.3}$$

as functions on M.

If Ω is closed then so is $\Psi(m) = \psi_m^* \Omega$ for each $m \in M$. In particular, if M is a symplectic manifold and Ω is the symplectic form, then

$$\Psi : M \longrightarrow Z^2(g)$$

where $Z^2(g) \subset \wedge^2(g^*)$ denotes the space of two cocycles, i.e. the space of those ω with $\delta\omega = 0$. Since Ψ is equivarient, the image of Ψ is a union of G orbits in $Z^2(g)$. In particular, if G acts transitively on M then the image of Ψ is a single orbit.

Let ω be an element of $Z^2(g)$. Define

$$h_\omega = \{\xi \in g \mid i(\xi)\omega = 0\}.$$

Notice that $h_\omega \subset g_\omega$, the isotropy algebra of ω, since g_ω consists of those ξ with $D_\xi\omega = 0$, and $D_\xi\omega = d(i(\xi)\omega)$ since $d\omega = 0$. If $\xi \in h_\omega$ and $\eta \in g_\omega$ then

$$0 = D_\eta(i(\xi)\omega) = i([\eta,\xi])\omega + i(\xi)D_\eta\omega = i([\eta,\xi])\omega$$

so $[\eta,\xi] \in h_\omega$. Thus h_ω is an ideal in g_ω, and in particular h_ω is a subalgebra of g.

Let $\mathcal{O}$ be a G orbit in $Z^2(g)$ and suppose that Ψ intersects $\mathcal{O}$ cleanly, so that $\Psi^{-1}(\mathcal{O})$ is a submanifold of M, and, at each $m \in \Psi^{-1}(\mathcal{O})$ the tangent space

$$V = T(\Psi^{-1}(\mathcal{O}))_m$$

is given by

$$V = d\Psi_m^{-1}(T\mathcal{O}_{\Psi(m)}).$$

We claim that

$$V = g_M(m) + [g,g]_M(m)^\perp \tag{2.4}$$

where the $\perp$ is relative to the symplectic form on TM_m. Indeed, since Ψ is equivariant, it is clear that the $\xi_M(m)$ all lie in V and that their image under $d\Psi_m$ span $T\mathcal{O}_{\Psi(m)}$. We thus have

$$V = g_M(m) + \ker d\Psi_m .$$

Now for any pair $\tilde{v}$, $\tilde{w}$ of symplectic vector fields on M we have

$$d(i(\tilde{v})\Omega) = 0$$

so

$$\begin{aligned} D_{\tilde{w}}(i(\tilde{v})\Omega) &= i([\tilde{w}, \tilde{v}]) \\ &= i(\tilde{w})\, d(i(\tilde{v})\Omega) + d(i(\tilde{w})\, i(\tilde{v})\Omega) \\ &= d\Omega(\tilde{v} \wedge \tilde{w}) . \end{aligned}$$

Applied to vector fields of the form ζ_M and η_M we conclude that

$$i([\eta_M, \zeta_M])\Omega = \langle \Psi , \eta \wedge \zeta \rangle$$

so, for any vector v in TM_m we have

$$\langle d\Psi_m(v), \eta \wedge \zeta \rangle = (v, [\eta , \zeta]_M(m))_m \tag{2.5}$$

where $(\, , \,)_m$ denotes the symplectic form, Ω_m, on TM_m. This proves (2.4). Notice that (2.4) implies that V is coisotropic with

$$V = [g,g]_M(m) \cap g_M(m)^{\perp} .$$

But for any $\xi \in g$, we have $(\xi_M, \eta_M)_m = \langle \Psi(m), \xi \wedge \eta \rangle$ so

$$\xi_M(m) \in g_M(m)^{\perp} \quad \text{if and only if} \quad \xi \in h_{\Psi(m)} . \tag{2.7}$$

For any $\omega \in Z^2(g)$ let

$$t_\omega = h_\omega \cap [g,g] . \tag{2.8}$$

Then t_ω is a subalgebra of g, and we let T_ω denote the connected subgroup it generates. Notice that T_ω is contained in the isotropy subgroup of ω. It follows from the above discussion that

$$V^{\perp} = (t_{\Psi(m)})_M(m) \tag{2.9}$$

which is just the tangent space to the orbit of the group $T_{\Psi(m)}$ through the point m. To summarize, we have proved

THEOREM 2.1. *Let* $G \times M \longrightarrow M$ *be a symplectic* G *action whose associated map* $\Psi : M \longrightarrow Z^2(g)$ *intersects cleanly with a* G *orbit* $\mathcal{O} \subset Z^2(g)$. *Then* $\Psi^{-1}(\mathcal{O})$ *is a coisotropic submanifold of* M. *The null foliation passing through any point* m *in this submanifold is the orbit through* m *of the group* $T_{\Psi(m)}$.

By a slight modification of the preceding argument we can prove the following:

THEOREM 2.2. *Let* $G \times M \longrightarrow M$ *and* $G \times N \longrightarrow N$ *be two symplectic actions whose associated maps* $\Psi_M : M \longrightarrow Z^2(g)$ *and* $\Psi_N : N \longrightarrow Z^2(g)$ *intersect cleanly. Then* $\Psi_M^{-1}(\Psi_N(N))$ *is a coisotropic submanifold of* M. *Let* m *be a point of this submanifold and* n *a point of* N *with* $\Psi_M(m) = \Psi_N(n)$. *Let* g_n *denote the isotropy subalgebra of* n, *let* $t_n = g_n \cap [g,g]$ *and let* T_n *denote the connected subgroup generated by the subalgebra* t_n. *Then the leaf of the null foliation through* m *is the orbit of* m *under the subgroup* T_n.

Proof. The fact that $\Psi_M^{-1}(\Psi_N(N))$ is coisotropic follows from Theorem 2.1 since it is a union of $\Psi^{-1}(\mathcal{O})$ which are coisotropic, provided that the relevant intersection are clean. But we can see this directly by a double application of (2.5). Let W denote the tangent space to $\Psi_M^{-1}(\Psi_N(N))$ at m. Then W consists of all $v \in TM_m$ such that there exists a $w \in TN_n$ with

$$(v,[\xi,\eta]_M)_m = (w,[\xi,\eta]_N)_n$$

for all ξ,η in g. Taking w = 0 in this equation shows that $W \supset [g,g]_M(m)^{\perp}$ and clearly $W \supset g_M(m)$ so $W \supset V$ where V is given by (2.4) and hence W is coisotropic. Also, $W^{\perp} \subset [g,g]_M(m)$, so we must determine which elements of $[g,g]$ give rise to an element of $W^{\perp}$. Let γ be an element of $\wedge^2 g$ and let (γ) denote its image in g under the Lie bracket mapping $\wedge^2 g \longrightarrow g$. Then (2.5) says that

$$(v,(\gamma)_M)_m = \langle d\Psi_M(v),\gamma\rangle$$

and v lies in W if and only if $d\Psi_M(v)$ lies in $d\Psi_M(TM_m) \cap d\Psi_N(TN_n)$ so $(\gamma)_M(m)$ lies in $W^{\perp}$ if and only if γ is orthogonal to the above intersection, i.e. if and only if

$$\gamma \in (d\Psi_M(TM_m))^0 + (d\Psi_N(TN_n))^0.$$

Decompose $\gamma = \gamma_1 + \gamma_2$ by the above, where $\gamma_1 \in (d\Psi_M(TM_m))^0$ and $\gamma_2 \in (d\Psi_N(TN_n))^0$. Then another two applications of (2.5) show that

$$(\gamma_1)_M(m) = 0$$

and

$$(\gamma_2)_N(n) = 0$$

since $(\ ,\ )_m$ and $(\ ,\ )_n$ are non-degenerate. Thus $(\gamma)_M(m) = (\gamma_2)_M(m)$ with $(\gamma_2)_N(n) = 0$, i.e. $(\gamma_2) \in t_n$. This completes the proof of Theorem 2.2.

3.

Suppose the group G has the property that $H^1(g) = H^2(g) = 0$. This means that $[g,g] = g$ and each $\omega \in Z^2(g)$ can be written uniquely as $\omega = \delta\beta$ with $\beta \in g$. Orbits in $Z^2(g)$ become identified with orbits in g^* and the algebra t_ω is just the isotropy algebra, g_β. For any symplectic group action $G \times M \longrightarrow M$ there is a unique map

$$\Phi : M \longrightarrow g^*$$

with

$$i(\xi_M)\Omega = -d\langle\Phi, \xi\rangle \quad . \tag{3.1}$$

The results of the preceding paragraph take on a simpler form when expressed in terms of the map Φ and will be formulated below. Many interesting groups have the above property. For instance, if g is semi-simple we know that the above property holds. Also, if g is the semi-direct product of a semi-simple algebra k with a commutative algebra, p, such that p has no trivial components under the action of k, and p does not admit an antisymmetric bilinear form invariant under k. This would include, for example, all "Euclidean" algebras $o(V) + V$ where V is a vector space with a non-degenerate scalar product, o(v) its orthogonal algebra and $\dim V > 2$, cf. [3] or [5].

For groups which do not necessarily have the above property, we say that we have a Hamiltonian action of G on M provided that we can find an equivariant map Φ satisfying (3.1) and have chosen one such Φ ; the map Φ is called the moment map of the Hamiltonian action. The crucial formula replacing (2.5) is

$$\langle d\Phi_m(v), \xi\rangle = (v, \xi_M(m))_m \, , \tag{3.2}$$

from which it follows that

$$\ker d\Phi_m = g_M(m)^{\perp} \quad . \tag{3.3}$$

Let $\mathcal{O}$ be an orbit in g^* and suppose that Φ intersects $\mathcal{O}$ transversally. Let m be a point of $\Phi^{-1}(\mathcal{O})$ and let W denote the tangent space to $\Phi^{-1}(\mathcal{O})$ at m. It then follows that

$$W = g_M(m) + g_M(m)^{\perp} \tag{3.4}$$

so that W is coisotropic and

$$W^{\perp} = g_M(m) \cap g_M(m)^{\perp} .$$

It is not hard to see that $\xi \in g$ has the property that $\xi_M(m) \in g_M(m)^{\perp}$ if and only if ξ lies in the isotropy algebra $g_{\Phi(m)}$. We shall in fact prove a generalization of this fact below. But this already is the content of the main theorem in [2] which asserts that

THEOREM 3.1. *If* $\Phi : M \longrightarrow g^*$ *intersects an orbit* $\mathcal{O}$ *cleanly, then* $\Phi^{-1}(\mathcal{O})$ *is coisotropic and the null foliation through a point* m *in* $\Phi^{-1}(\mathcal{O})$ *is the orbit of* M *under* $G^0_{\Phi(m)}$, *the connected component of the isotropy subgroup of* $\Phi(m)$.

More generally, suppose that we are given two Hamiltonian G actions with moment maps

$$\Phi_M : M \longrightarrow g^* \quad \text{and} \quad \Phi_N : N \longrightarrow g^*$$

which intersect cleanly. Let m be a point of $\Phi_M^{-1}(\Phi_N(N))$ and W the tangent space to this submanifold at m. Let n be a point of N with $\Phi_M(m) = \Phi_N(n)$. Then W consists of all $v \in TM_m$ such that there is a $w \in TN_n$ with

$$(v, \xi_M(m))_m = (w, \xi_N(n))_n$$

for all $\xi \in g$. Taking w = 0 shows that $W \supset g_M(m)^{\perp}$ and clearly $W \supset g_M(m)$ so that W is coisotropic. Also $W^{\perp} \subset g_M^{\perp\perp} = g_M(m)$ so our problem is to determine which $\xi \in g$ are such that $\xi_M(m) \in W^{\perp}$. Now $v \in W$ if and only if $d\Phi_m(v) \in d\Phi_M(TM_m) \cap d\Phi_N(TN_n)$ and thus, by (3.2) $\xi_M(m) \in W^{\perp}$ if and only if

$$\xi \in (d\Phi_M(TM_m) \cap d\Phi_N(TN_n))^0 = (d\Phi_M(TM_m))^0 + (d\Phi_N(TM_n))^0 .$$

Now by (3.2) and the non-singularity of the symplectic form on M, we know that $\zeta \in d\Phi_M(TM_m)^0$ if and only if $\xi_M(m) = 0$, and similarly for N. Hence, writing

$$\xi = \xi_1 + \xi_2$$

with $\xi_1 \in d\Phi_M(TM_m)^0$ and $\xi_2 \in d\Phi_N(TN_n)^0$ we see that $\xi_M(m) = \xi_{2M}(m)$ where $\xi_{2N}(n) = 0$. We have thus proved

THEOREM 3.2. *Let* G *have two Hamiltonian actions with moment maps* $\Phi_M : M \longrightarrow g^*$ *and* $\Phi_N : N \longrightarrow g^*$ *which intersect cleanly. Then* $\Phi_M^{-1}(\Phi_N(N))$ *is a coisotropic submanifold of* M. *If* $m \in M$ *and* $n \in N$ *are such that* $\Phi_M(m) = \Phi_N(n)$ *then the leaf of the null foliation through* m *is the orbit of* m *under* G_n^0, *the connected component of the isotropy group of* n.

If we take $N = \mathcal{O}$ and Φ the inclusion map of $\mathcal{O}$ in g^* we get Theorem 3.1 as a special case of Theorem 3.2.

Suppose that we are given a Hamiltonian action of G on M whose moment map, Φ, is a submersion, and let Z be a slice for the G action on g^*. Thus, Z is a submanifold of g, and at each point $\beta \in Z$, $TZ_\beta \cap T\mathcal{O}_\beta = 0$ and $TZ_\beta + T\mathcal{O}_\beta = g^*$ where $\mathcal{O}$ is the orbit through β. Let m be a point of N with $\Phi(m) = \beta$ and set $U = d\Phi_m^{-1}(TZ_\beta)$. Then $U \supset \ker d\Phi_m = g_M(m)^\perp$. Hence $U^\perp \subset g_M(m)$. We claim that $U \cap U^\perp = 0$, i.e. that U is a symplectic subspace. Indeed, if $\xi \in g$ is such that $\xi_M(m) \in U$, then, by the equivariance of Φ, we know that $d\Phi_m(\xi_M(m))$ must be tangent to $\mathcal{O}$ and also lie in TZ_β and hence = 0. This implies that $\xi \in T\mathcal{O}_\beta^0$. But, by (3.2), if $\xi_M(m) \in U^\perp$, then $\xi \in TZ_\beta^0$. Since $TZ_\beta + T\mathcal{O}_\beta = g^*$ this implies that $\xi = 0$. We have thus proved

THEOREM 3.3. *If the moment map* $\Phi : M \longrightarrow g^*$ *is a submersion, and* $Z \subset g^*$ *is a slice for the* G *action, then* $\Phi^{-1}(Z)$ *is a symplectic submanifold of* M.

4.

Let $Q \subset g^*$ be an invariant submanifold so that Q is a union of G orbits. At each $w \in g^*$ we have a standard identification of $T^*g^*_w$ into g, and hence, for each $w \in Q$ we can identify the normal space to Q, NQ_w as a subspace of g. Explicitly

$$\xi \in NQ_w \text{ if and only if } \langle v, \xi \rangle = 0 \text{ for all } v \text{ tangent to } Q. \tag{4.1}$$

If $\eta \in g$, then $\eta \cdot w$ is tangent to Q since W is invariant. Hence the above condition implies that $0 = \langle \eta \cdot w, \xi \rangle = \langle w, [\xi, \eta] \rangle = -\langle \xi \cdot w, \eta \rangle$ for all η, i.e. that $\xi \cdot w = 0$, in other words that ξ is in the isotropy algebra of w. Thus

$$NQ_w \subset g_w .$$

Also, G_w preserves Q and w and hence TQ_w and hence NQ_w. Thus NQ_w is invariant under the adjoint action of G_w on g_w. In particular NW_w is an ideal in g_w and hence a subalgebra. We will let $\mathcal{N}_w$ denote the connected subgroup generated by NQ_w.

THEOREM 4.1. *Let* M *be a Hamiltonian* G *manifold with moment map* $\Phi : M \longrightarrow g^*$ *which intersects* Q *cleanly, where* Q *is an invariant submanifold of* g^*. *Then* $\Phi^{-1}Q$ *is a coisotropic submanifold of* M *and the leaf of the null foliation through* $m \in \Phi^{-1}Q$ *is the orbit of* m *under the group* $n_{\Phi(m)}$.

Proof. Let $W = d\Phi_m^{-1}(TQ_{\Phi(m)})$. Then as before, $W \supset g_M(n) + g_M(m)^{\perp}$ and hence is coisotropic and thus $W^{\perp} \subset g_M(m) \cap g_M(m)^{\perp}$ and we must determine which ξ are such that $\xi_M(m) \in g_M(m)$. By (3.2) and (4.1) we see that this is precisely when $\xi \in n_{\Phi(m)}$, QED.

One way of producing the situation of Theorem 4.1 is as follows: Suppose that we have a symplectic or a Hamiltonian action of G on M, and we have another group, H, of diffeomorphisms of M which do not necessarily preserve the symplectic form, Ω, of M but which do commute with the action of G. The principal case that we have in mind for applications is where $H = \mathbb{R}^+$ so that M is a "homogeneous" symplectic manifold so each positive real number acts by "multiplication" on M and $t^*\Omega = t\Omega$, i.e. Ω is carried into some multiple of itself by the action of elements of H. (Actually, the assumptions that we make are so strong that this is essentially the only example for the case where G is simple, but we formulate the theorem in general since there may be other applications.) We thus assume that we are given an action of H on M which commutes with the action of G on M:

$$\tau a M = a \tau m \quad \text{for all} \quad a \in G,\ \tau \in H \ \text{and} \ m \in M$$

and a representation, γ, of H on g^* so that

$$\Phi(\tau m) = \gamma(\tau)\, \Phi(m). \tag{4.2}$$

This implies that $\gamma(\tau)$ commutes with $Ad_a^{\#}$, at least on $\operatorname{Im}\Phi$. For any orbit $\mathcal{O} \subset g^*$ we can consider its saturation under the action of H, i.e. set $Q = \gamma(H)\cdot\mathcal{O}$. We have thus proved

THEOREM 4.2. *Suppose that* H *is a group of diffeomorphisms commuting with the Hamiltonian action of* G *on* M, *and suppose that* γ *is a representation of* H *on* g^* *satisfying* (4.2). *Then if* Φ *intersects* $\gamma(H)\cdot\mathcal{O}$ *cleanly, then* $\Phi^{-1}(\gamma(H)\cdot\mathcal{O})$ *is a coisotropic submanifold of* M. *The leaf of the null foliation through any point* m *of this submanifold is the orbit through* m *of the connected group generated by the subalgebra of* g_β *given by all* ξ *satisfying* (4.1) *where* $\beta = \Phi(m)$.

We leave to the reader the formulation and proof of the analogue of Theorems 3.2 and 4.2.

Notice that in the important case where $H = \mathbb{R}^+$ so that $\gamma(t)$ consists of multiplication by t, the space of $\zeta(\beta)$ consists of all multiples of β and (4.2) becomes

$$\langle \beta, \eta \rangle = 0 . \qquad (4.3)$$

5.

We return to the situation studied in Section 3: We are given two Hamiltonian actions of the group G on symplectic manifolds M and N with moment maps $\Phi_M : M \longrightarrow g^*$ and $\Phi_N : N \longrightarrow g^*$. However let us now make the stronger assumption that these maps are transversal to each other. By definition, these two maps are transversal if and only if for every $(m,n) \in MN$ with $\Phi_N(m) = \Phi_N(n)$ we have

$$d\Phi_M(TM_m) + d\Phi_N(TN_n) = g^* , \qquad (5.1)$$

which is equivalent to the condition that

$$(d\Phi_M(TM_m))^0 \cap (d\Phi_N(TN_n))^0 = 0 . \qquad (5.2)$$

By (3.2) the subspace $(d\Phi_M(TM_m))^0$ of g consists precisely of the isotropy subalgebra of m, i.e. of those ξ for which $\xi_M(m) = 0$, with a similar assertion for N. Thus (5.2) is equivalent to the assertion that the map

$$g \longrightarrow TM_m \times TN_n \qquad \xi \leadsto (\xi_M(m), \xi_N(n))$$

is injective. The vector field (ξ_M, ξ_N) on M x N is just the vector field corresponding to the (diagonal)symplectic action of G on M x N. We have thus proved

THEOREM 5.1. *Equation* (5.1) *holds if and only if the isotropy subgroup* $G_{(m,n)}$ *of* (m,n) *in* G *is discrete.*

Thus, for example, if G is compact, the transversality condition guarantees that the isotropy $G_{(m,n)}$ is finite. A slightly stronger assumption would be that this isotropy group be trivial. This would mean that the group G_n acts freely on $\Phi_M^{-1}(\Phi(n))$, i.e. the leaves of the full foliation are orbits of the free action of a compact Lie group. Under mild conditions this would imply that we can make the space of leaves of the null foliation into a Hausdorff manifold, P, together with a smooth projection, $\pi : \Phi_M^{-1}(\Phi_N(N)) \longrightarrow P$. For the rest of this section we will take the existence of P and π as a standing hypothesis. That is, we assume that there is a smooth manifold, Z, a smooth projection π such that $\pi(m_1) = \pi(m_2)$ if and only if m_1 and m_2 lie on the same leaf of the null foliation. Letting ι denote the injection of $\Phi_M^{-1}(\Phi_N(N))$

into M and letting Ω_M denote the symplectic form on M, then $\iota^* \Phi_N$ is a presymplectic form on $\Phi_M^{-1}(\Phi_N(N))$ whose null foliation are the fibers of the map π. Hence there is a unique symplectic form, Ω_P on P with

$$\pi^* \Omega_P = \iota^* \Omega_M . \tag{5.3}$$

If $\Phi_M(m) = \Phi_N(n)$, then all the points on the leaf of the null foliation through m map, under Φ_M into the same point, $\Phi_N(n)$ in g^*. Thus Φ_M induces a map, $\Phi_P : P \longrightarrow g^*$ such that the diagram

$$\begin{array}{ccc} \Phi_M^{-1}(\Phi_N(N)) & & \\ \pi \downarrow & \searrow \Phi_M & \\ Z & \xrightarrow{\ \Phi_N\ } & g^* \end{array} \tag{5.4}$$

commutes. It is immediate that the group G permutes the fibers of the map π and hence induces an action of G on P. If $m \in \Phi_M^{-1}(\Phi_N(N))$ then, with the notations of Section 2, the maps $\psi_m : G \longrightarrow M$ and $\psi_{\pi m} : G \longrightarrow P$ are related by

$$\psi_{\pi m} = \pi \psi_m . \tag{5.5}$$

It follows from (5.3) that the group action on P preserves Ω_P and from (5.2) and (5.5) that the maps

$$\Psi^{\iota^*\Omega_M} : \Phi_M^{-1}(\Phi_N(N)) \longrightarrow Z^2(g) \quad \text{and} \quad \Psi^{\Omega_P} : P \longrightarrow Z^2(g)$$

are related by

$$\Psi^{\iota^*\Omega_M} = \Psi^{\Omega_P} \pi . \tag{5.6}$$

A little more argument that we leave to the reader shows that

THEOREM 5.2. *The action of* G *on* P *is Hamiltonian with moment map given by* Φ_P.

6.

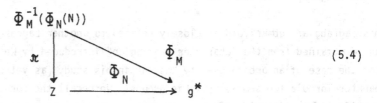

Given a coisotropic submanifold, $\mathcal{V}$, of a symplectic manifold, M, there is a standard way of associating to it a Lagrangian submanifold, $\Lambda = \Lambda(\mathcal{V})$ of $M \times M^-$ (where M^- denotes M with the symplectic form $-\Omega_M$ and where we put the obvious product symplectic structure on $M \times M^-$). The Lagrangian submanifold Λ consists of all pairs (m_1, m_2) where m_1 and m_2 lie on the same leaf of the null foliation of $\mathcal{V}$, cf. [4] for a proof that Λ is Lagrangian and a study of its geometric and analytic significance. Thus, if we are in the situation of Theorem 3.2, we get a Lagrangian submanifold, $\Lambda_{M,N}$ of $M \times M^-$ given by

$$\Lambda_{M,N} = \{ (m_1, m_2) \text{ with } \Phi_M(m_1) = \Phi_N(m_2) = \Phi_N(n) \\ \text{for some } n \in N \text{ and } m_1 = a\, m_2 \text{ with } a \in G_n^0 \} . \tag{6.1}$$

The condition that $a \in G_n^0$ means, of course, that a is in the connected component of the subgroup satisfying $an = n$.

This Lagrangian submanifold is closely related to another Lagrangian submanifold that can be obtained from the "character Lagrangian" introduced by Weinstein [1], [6] and (for the case of an orbit) used by Kostant in his study (as yet unpublished) of the charater formula for semi-simple Lie groups. We recall the construction. Let us identify T^*G with $G \times g^*$ by left multiplication. At a point $(c, \alpha) \in G \times g^* = T^*G$ the tangent space becomes identified with $g \times g^*$. If (η_1, β_1) and (η_2, β_2) are two such tangent vectors, then their symplectic scalar product is given by,

$$((\eta_1, \beta_1), (\eta_2, \beta_2))_{(c,\alpha)} = \langle \alpha , \eta_1, \eta_2 \rangle + \langle \beta_2, \eta_1 \rangle - \langle \beta_2, \eta_1 \rangle . \tag{6.2}$$

If we are given a Hamiltonian action of G on M with moment map Φ_M then it follows by a computation involving (3.2) and (6.2) that the submanifold $R_M \subset T^*G \times M \times M^-$ given by

$$R_M = \{ ((a, \Phi_N(m)), m, a\, m) \}$$

is Lagrangian. Let $\Delta_M \subset M \times M^-$ denote the diagonal, so that Δ_M is Lagrangian. Assuming the necessary clean intersection hypotheses, the composition of R_M with Δ_M will be a Lagrangian submanifold of T G which is called the character Lagrangian and which we shall denote by ch(M) . Thus

$$\mathrm{ch}(M) = R_M \circ \Delta_M = \{ (a, \Phi_M(m))\ a\, m = m \} . \tag{6.3}$$

Similarly, if N is a second Hamiltonian G manifold with moment map Φ_N we can construct its character Lagrangian, ch(m). Finally, we can compose ch(N) with R_M to obtain a Lagrangian submanifold of $M \times M^-$, i.e. form

$$\mathrm{ch}(N)\ R_M = (m, am) \quad {}_M(m) = {}_N(n) , \text{ with } an = n . \tag{6.4}$$

If we compare (6.4) and (6.1) we see that the only difference is that in (6.4) the condition on a is that $a \in G_n$ while in (6.1) the condition is that $a \in G_n^0$. Thus,

THEOREM 6.1. *If the isotropy groups* G_n *are connected, then the Lagrangian submanifold of* $M \times M^-$ *associated with the coisotropic* $\Phi_M^{-1}(\Phi_N(N))$ *is* $ch(N) \circ R_M$.

REFERENCES

[1] Abraham, R. and Marsden, J., *Foundations of Mechanics* 2nd Edit. Benjamin/Cummings Pub. Co. Reading, Mass. 1978 Ex. 5. 31, page 422.

[2] Kazhdan, D., Kostant, B., and Sternberg, S., "Hamiltonian group actions and dynamical systems of Calogero type", Comm. Pure and App. Math. 31 (1978) 481-507.

[3] Guillemin, V. and Sternberg, S., *Geometric Asymptotics,* Amer. Math. Soc., Providence, Rhode Island, (1977).

[4] Guillemin, V. and Sternberg, S., "Some problems in integral geometry and some related problems in micro-local analysis", Amer. Jour. of Math. 101 (1979) 915-955.

[5] Sternberg, S., "Symplectic homogeneous spaces", Trans. Am. Math. Soc. 212, (1975) 113-130.

[6] Weinstein, A., "Symplectic geometry", lecture to appear in the symposium on the mathematical heritage of Poincaré.

ELEMENTARY SYSTEMS FOR LIE ALGEBRA BUNDLE ACTIONS *

Tomas Ungar

Universität Bonn, West Germany

Introduction

The concept of a moment map for a Lie group action on a symplectic manifold proved to be a very useful tool in the study of a finite and infinite dimensional dynamical systems. In this paper we would like to study this concept in a situation where instead of a Lie group action we have its infinitesimal counterpart, namely an action of a Lie algebra vector bundle.

The basic idea is as follows. Let us assume that we are given a space of states of some dynamical system W together with a Lie algebra vector bundle E(W) over W and a strong, surjective vector bundle map $\lambda : E(W) \longrightarrow T(W)$, the tangent bundle over W. In other words, we have attached to each point $w \in W$ a Lie algebra E_w and to each vector $e_w \in E_w$ an infinitesimal symmetry $\lambda(e_w) \in T_w W$. Now, suppose that we are given a smooth section μ of $E^*(W)$, the dual of E(W) and linear connection ∇ on E(W). Then to any element e_w we can assign two vectors in $T_{\mu(w)} E^*_w$

a) $(ad^{\#} e_w) \cdot \mu(w)$

b) $(\lambda(e_w) \lrcorner \nabla\mu)_{\mu(w)}$

where $ad^{\#}$ is the coadjoint action i.e. $ad^{\#} = -ad^*$.

Now, if the vectors a) and b) agree for any $e_w \in E_w$, then we can use the canonical structure on the image of E_w in $T_{\mu(w)} E^*$ to construct a two form σ on W at w. On the other hand, if we are given a two form σ, we can try to find λ, μ, ∇ so that $\lambda(e_w) \lrcorner \sigma = (\nabla \mu e_w)$ for any $w \in W$. Now, let us assume that any motion of the system $\gamma : (\tau, \tau_c) \longrightarrow W$ belongs to a leaf of the characteristic foliation of σ

* This work was done under the program Sonderforschungsbereich "Theoretische Mathematik" (SFB 40) at the University of Bonn.

i.e. $(\gamma_* (\frac{\partial}{\partial \tau}) \lrcorner \sigma)_{\gamma(\tau)} = 0$ then $(\gamma_* (\frac{\partial}{\partial \tau}) \lrcorner \nabla \mu)_{\gamma(\tau)} = 0$ which means that μ is a covariant constant of motion. For many dynamical systems we can find a surjective Lie algebra vector bundle action and the corresponding moment μ with a straightforward physical interpretation. For example, the components of the moment at w can be energy, electric charge and other physical quantities. In this case, we can regard the equation (1) as a part of a more general conservation law. Such a law can be for example written as $\nabla J = 0$ where J is some p-form with values in $E^*(M)$. Now the correspondence $\mu \sim J$ has two analogies. The first analogy is that of an electric charge and electric current. The second analogy is that of energy momentum and the stress energy momentum tensor.

In order to understand the $\mu \sim J$ correspondence, we shall use an infinite dimensional version of a Lie algebra vector bundle action instead of the variational principle and the principle of general covariance. This will enable us to construct an abstract theory of relativity modelled both on the Einstein theory and the Yang-Mills field theory.

The plan of this paper is as follows. In the rest of section 0 we shall try to provide a bridge between the standard Hamiltonian formalism on a cotangent bundle and the Hamiltonianless presymplectic formalism. Here the material is almost standard and textbooks like [1, 15] will provide any necessary background material. See also [8]. This section also provides a more concrete material which serves to illustrate rather abstract formalism of section 1, where we shall reformulate some basic results concerning the moment map of a Lie group action for a Lie algebra vector bundle action. Here the basic idea is very strongly motivated by [4, 8, 9, 10, 15, 16, 17].

In section 2 we shall construct an abstract theory of relativity. This theory should be regarded as a basic common denominator of the following theories:

a) Einstein theory of relativity
b) Its Newtonian limit in a curved space-time
c) Extension of the Einstein theory by Yang-Mills type of interactions
d) Conformal and graded Lie algebra extensions.

The basic idea of section 2 was motivated by [7, 13, 14, 17, 18].

In section 3 we illustrate the abstract approach of section 2 on two concrete situations, namely, the Einstein theory of relativity and its Newtonian approximation. The Einstein theory is treated in a more or less standard fashion. The Newtonian approximation is based on the work of E. Cartan [3, 6].

Nevertheless, we have deviated from the Cartan approach by taking into account the nontrivial central extention of the Galilei group as a Newtonian limit of the Poincaré group [2,5,15]. It is exactly this additional consideration which enables us to write down the Newtonian counterpart of the Einstein equations.

I would like to thank Dr. Min Oo, Dr. H. Petry, Prof. E. Ruh and Prof. D. Simms for helpful discussions, and the Sonderforschungsbereich Theoretische Mathematik, Universität Bonn for their hospitality.

0.1.

Usually, a Hamiltonian dynamical system is a differential manifold W, given together with a closed nondegenerate two form σ and a smooth vector field X on W so that $X \lrcorner \sigma = dH$. Here $X \lrcorner \sigma$ denotes the value of the two form σ on X and H is a smooth function on W.

Now, let us assume that we are given a manifold W' together with a two form σ', vector field X' on W' and a smooth projection $\pi : W' \longrightarrow W$ so that $\pi_* X' = X$ and $\pi^* \sigma = \sigma'$. So we have $X' \lrcorner \pi^* \sigma = d(H \circ \pi)$ where $H \circ \pi$ is the pull back of H onto W' via π. Then we shall consider (W', σ', X') to be equivalent to (W, σ, X). Let us consider two simple examples of such a situation:

Example 1:

Let (W, σ, X) be a Hamiltonian system. $W' = W \times R$, $X' = X + \frac{\partial}{\partial t}$, $\sigma = \sigma'$, and π is the projection on the first factor. Then the integral curves of X' are of the form $\gamma'(\tau) = (\gamma(\tau), \tau + \tau_0)$. So to any integral curve of X there corresponds an equivalence class of integral curves of X'. The difference between two curves within the same equivalence class rests in the different identification of the dynamical time τ along the curve and the kinematical time t as a coordinate on W x R.

Example 2:

Let (W, σ, X) be as in Example 1. Furthermore, let us assume that X is completely integrable, i.e. any integral curve of X can be prolonged from $-\infty$ to $+\infty$. Now, let us suppose that through any $w_0 \in W$ we have such a curve i.e. $\gamma_{w_0}(o) = w_0$, $\gamma : (-\infty, +\infty) \longrightarrow W$. So we can define $\pi : W \times R \longrightarrow W$ by setting $\pi(w, t) = w_0$ if $\gamma_{w_0}(\tau) = w$ and $\tau = \tau_0$. Then the system $(W', \sigma - dH \wedge dt, X' = X + \frac{\partial}{\partial t})$ is equivalent (via π) to the system (W, σ, 0) i.e. to any integral curve $\gamma(\tau) = w_0$ of the trivial vector field there exists an equivalence class of integral curves of X' on W. They differ again by an additive constant entering into the identification of τ and t. A little more complicated example of the above described situation is obtained if we try to compare a family of distinct dynamical systems (W_i, σ_i, X_i) via realisation as subsystems of a single system $(\hat{W}, \hat{\sigma}, \hat{X})$. For example, let (W_i, σ_i, X_i) be elementary dynamical systems of a Lie group G, i.e. W_i are homogeneous G-spaces and σ and X are G-invariant. Furthermore, let us assume that $\sigma_i = d\alpha_i$ for some G-invariant one form α_i on W. Then, as we shall see below, we can realise (W_i, σ_i, X_i) (up to the above defined equivalence) as a subsystem of (T^*G, ω, X) where ω is the canonical two form on $T^*(G)$, the cotangent bundle of G and X is some vector field on $T^*(G)$. Moreover, we can introduce interaction as a deformation of this subsystem within the large system. This in turn will allow us to introduce "universal interaction" as deformations of different subsystems caused by the same geometrical mechanism.

0.2.

Let G be a connected Lie group and $T^*(G)$ its cotangent bundle. Then any nondegenerate one form θ on G with values in g, the Lie algebra of G, will induce a map $\phi : T^*(G) \longrightarrow g^*$ the dual of g, by setting: $\langle \phi(p), v \lrcorner \theta \rangle = p(v)$ for any $p \in T_a(G)$, $v \in T_a(G)$ and $a \in G$. Here $\langle\ ,\ \rangle$ denotes the pairing between g and g^*. For any choice of basis e_i of g and its dual e^i of g^* we obtain a frame of lineary independent one forms $d\phi_i, \pi^*\theta$ on $T^*(G)$. Here $\pi : T^*(G) \longrightarrow G$ is just the canonical projection. So any form α on $T^*(G)$ can be written as

$$(1) \qquad \alpha = \langle \alpha^\phi, d\phi \rangle + \langle \alpha^\theta, \pi^*\theta \rangle$$

Here α^ϕ is a g valued function on $T^*(G)$ and α^θ is a g^* valued function on $T^*(G)$. In particular, from (1) it follows that the canonical one form β on $T^*(G)$ can be written as

$$(2) \qquad \beta = \langle \beta^\theta, \pi^*\theta \rangle = \langle \phi, \pi^*\theta \rangle \quad \text{or} \quad \beta^\phi = 0, \quad \beta^\theta = \phi$$

Similarly, to any vector field Y on $T^*(G)$, one can assign a g valued function $Y^\theta = Y \lrcorner \pi^*\theta$ and g^* valued function $Y^\phi = Y \lrcorner d\phi$ on $T^*(G)$.

Using $d\beta = \omega$, the canonical symplectic form on $T^*(G)$, we can assign to any smooth function H on $T^*(G)$ a smooth vector field X so that

$$(3) \qquad X \lrcorner \omega = dH$$

We can rewrite (3) as

a) $\langle (X \lrcorner d\phi) \wedge \pi^* \theta \rangle + X \lrcorner \langle \phi \wedge \pi^* d\theta \rangle = \langle (dH)^\theta , \pi^* \theta \rangle$

b) $\langle d\phi \wedge X \lrcorner \pi^* \theta \rangle = + \langle (dH)^\phi \cdot d\phi \rangle$

So if (4) $X \lrcorner \pi^* d\phi = 0$ then we obtain

$$(5) \qquad \begin{aligned} X^\phi &= dH^\theta \\ X^\theta &= dH^\phi \end{aligned}$$

0.3.

A particular example of such a situation is obtained if $\theta = \theta_L$ is a left invariant one form on G and $H = \hat{H} \circ \phi$ where $\hat{H}$ is a smooth function on g^*, invariant under the coadjoint action of G on g^*. In this case $(dH)^\theta = 0$ and from the invariance of H and from (3b) it follows that

$$(6) \qquad \langle Y \lrcorner d\phi, X \lrcorner \pi^* \theta \rangle = - Y \lrcorner \langle (dH)^\phi \wedge d\phi \rangle_p = 0$$

for any Y such that $(Y \lrcorner d\phi)_p = ad^* A \cdot \phi(p)$ for some $A \in g$. This means that $ad^\#(X \lrcorner \pi\theta_L) \cdot \phi(p) = 0$. So the above condition (4) is satisfied and $X^\phi = 0$. In other words, X is tangent to the level surfaces of ϕ and our original Hamiltonian system on $T^*(G)$ decomposes into a family of more elementary systems. As each level surface of ϕ is diffeomorphic to G, we can pull back any such elementary system onto G. In this way we will obtain for each $f \in g^*$ a dynamical system $(G, i_f^*\omega, X_f)$ on G. Here i_f is the submanifold map which injects G into $T^*(G)$ as the level surface $\phi(\ ) = f$. So $i_f^*\omega = -\frac{1}{2}\langle f\ [\theta_L, \theta_L] \rangle$ is an exact left invariant two form of constant rank on G.

X_f is obtained by the projection via π_* of the restriction of X to the level surface $\phi(\ ) = f$. Clearly $(\pi_* X_f \lrcorner \theta) = d\hat{H}^\phi(f)$ where the differential $d\hat{H}$ of the smooth function $\hat{H}$ on g^* at f is identified with a g valued function on g^* at f.

Let G_f be the isotropy subgroup of f. Then the right action of G_f on G preserves $i_f\omega$ and X_f. This means that $X_f \lrcorner \theta_L$ is a function on G with values in the Lie algebra z_f of the centrum Z_f of G_f. More generally, let us assume that $f = Ad\ a.f'$ for some $a \in G$. Then the right translation R_a of G induces a diffeomorphism of $T^*(G)$ which moves the level set $\phi(.) = f$ into the level set $\phi(.) = f'$. This means that the elementary systems constructed via i_f and $i_{f'}$ are equivalent.

0.4.
Let us return to the generic situation where θ is not necessarily a left invariant form. Nevertheless, we shall retain H of the form $H = \hat{H} \circ \Phi$ where H is again an Ad^* invariant function on g^*. Let us define a g valued two form $D\theta$ by setting $D\theta = \frac{1}{2}[\theta,\theta] + d\theta$. Here

$$\frac{1}{2}[\theta,\theta] = \sum_{\substack{i,\,j\\ i<j}} C^i_{jK}\ \theta^j \wedge \theta^K e_i$$

where C^i_{jK} are the structure constants of g with respect to the basis e_i. If for some $p \in T^*_a(G)$ we will have $\langle \Phi(p) \wedge \pi^* D\theta \rangle_p = 0$ then (4) and (6) are again valid at p. This means that if for some $f \in g^*$ we have $\langle f\, D\theta \rangle = 0$ identically on G, then X is again tangent to the level surface $\Phi(\ ) = f$. So we can use the above described procedure to obtain an elementary system. $i_f^*\omega = -\frac{1}{2}\langle f\,[\theta,\theta]\rangle$ is again an exact two form of constant rank (as follows from the nondegeneracy of θ). Note that $(X_f \lrcorner\ \theta)$ and $d\hat{H}(f)$ are independent of the choice of θ.

0.5.
We can interpret the construction of 0.3 as follows. The g valued one form θ on G can be regarded as a co-frame (i.e. a dual to a moving frame of linearly independent vector fields on G). With respect fo this co-frame, we can express any two form on G (a smooth section of $\Lambda^2 T^*(G)$) as a function on G with values in $\Lambda^2(g^*)$. In particular, if this form is $\sigma_f = -\frac{1}{2}\langle f, [\theta,\theta]\rangle$ then it corresponds to a constant function on G whose value is the canonical symplectic structure on O_f, the coadjoint orbit of G through f, at the point $f \in g^*$. On the other hand, let us consider the commutive diagram

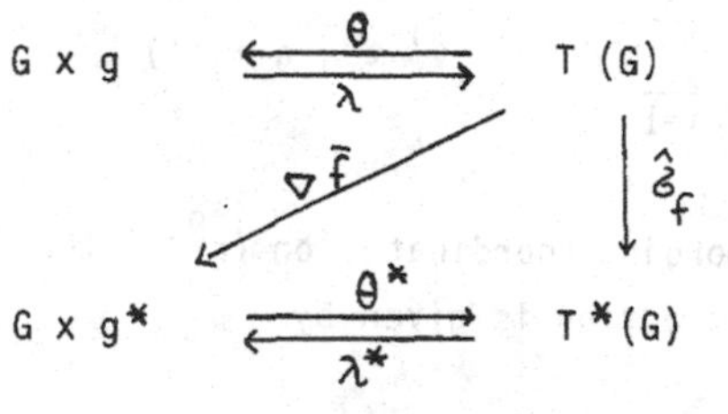

$\lambda = \theta^{-1}$

Here $\hat{\sigma}_f$ is the $T^*(G)$ valued one form on G determined by σ and $\nabla\bar{f} = df + ad\theta^*.f$ is an exterior covariant derivative of a smooth section $\bar{f} : G \longrightarrow Gxg^*$ given by $\bar{f}(a) = (a, f)$ for any $a \in G$.

Now, let φ be any smooth function on G with values in G, then we can use φ and $\bar{f}$ to define a new nondegenerate one form θ' on G with values in g by setting:
$\theta'(a) = (Ad\,\varphi(a))\ \theta(a)$.

We can write:

$$\omega_f = -\frac{1}{2}\langle \mu_f [\theta', \theta'] \rangle$$

where $\mu_f(a) = \mathrm{Ad}^{*}\varphi(a)\cdot\bar{f}$. In other words, with respect to the new co-frame ω_f corresponds to a function on G with values in $\Lambda^2 g^{*}$ whose value at the point a can be identified with the canonical two form on O_f at the point $\mu_f(a) = \mathrm{Ad}^{*}\varphi(a)f$. Again, we can find a connection ∇' on G x g^* so that $\lambda^{*}\omega = \nabla'\mu_f$ but now ∇' will not be associated with θ' in such a straightforward fashion. For example, let $\theta = \theta_L$ be the left invariant one form on G, then if φ is given by $\varphi(a) = a$, we will obtain as a right invariant one form θ' and $\nabla' = d$ is the flat exterior derivative, because μ_f is the Kostant-Souriau moment map.

0.6.
As an example of the above given construction, let us consider a dynamical system corresponding to the massive spinless charged particle moving under the influence of inertial, gravitational and Coulomb forces within the framework of the Newtonian mechanics. Let M be the flat Newtonian space-time given together with a preferred inertial coordinate system (x^i, t) on M, $x^i, t \in R$, $i = 1,2,3$. Then we shall consider as a space of states of our particle a 9 dimensional manifold G, diffeomorphic to R^9, and equipped with a preferred set of coordinates (v^i, x^i, t, s, d). Here $v^i, x^i, t, s, d \in R$, $i = 1,2,3$. Using those preferred coordinates, we have a fibering $\pi : G \longrightarrow M$ which sends a point with coordinates (v^i, x^i, t, s, d) into a point m on M with coordinates (x^i, t). Now, we shall give G a structure of a Lie group. With respect to the above given coordinate system the multiplication law is given by:

$$(v^i,x^i,t,s,a)\,(v^{i'},x^{i'},t',s',a') = (v^i + v^{i'},\ x^i + x^{i'} + v^i t',\ t + t',$$
$$s + s' + \sum_{i=1}^{3} v^i x^{i'} + \sum_{i=1}^{3} \frac{1}{2} v^i\ v^i t',\ a + a')$$

Let (k^i, p^i, E, m, e) be the corresponding coordinates on $(r^9)^{*}$ the dual of the Lie algebra $\underline{r}^9$ of $\underline{R}^9$. Then the coadjoint action is given by

$$\mathrm{Ad}^{*}(v^i,x^i,t,s,a)\cdot(k^i,p^i,E,m,e) =$$
$$(k^i + p^i t + mx^i,\ p^i - mv^i,\ E + \sum_{i=1}^{3} p^i v^i + m\sum_{i=1}^{3} v^i v^i t,\ m,\ e)$$

So the ring of invariant polynomials is generated by

$$H_m(k,p,E,m,e) = m,\quad H_e(k,p,E,d,e) = e$$

and

$$H_2(k,p,E,m,e) = \sum_{i=1}^{3} p^i p^i - 2\,mE$$

Let $\theta = \theta_L + \varrho\,dt$, where θ_L is a left invariant $\underline{r}^9$ valued one form on $\underline{R}^9$, i.e.

$$\theta_L = (dv^i,\ dx^i - v^i dt,\ dt,\ ds - \sum_{i=1}^{3} x^i dv^i + \sum_{i=1}^{3} \langle v^i, v^i\rangle \tfrac{1}{2}\,dt,\ da)$$

and ϱ is a g valued function on $\underline{R}^9$. Let us assume that $R_a \varrho\,dt = (Ad^{*} b)\,\varrho\ dt$ for any $b \in \underline{R}^9$ of the form $b = (v,0,0,s,a)$. Then ϱ must be of the form $(\varrho_v^i, 0, 0, \varrho_s, \varrho_a)$ where $\varrho_v^i, \varrho_s, \varrho_a$, $i = 1,2,3$ are functions of x and t only.

We can restrict the choice of ϱ still further by requiring that $\langle f\ D\,\theta\rangle = 0$ identically on $\underline{R}^9$ for some fixed choice of $f \in g^{*}$. For example, let $f = (o,o,o,m,e)$, then this choice of f together with the choice $H = H_2$ leads to the elementary dynamical system

$$(\underline{R}^9, \sum_{i=1}^{3} m\,(dv^i - dt) \wedge (dx^i - v^i dt), X_{m,e})\ .$$

Here

$$(\varrho_v)^i = (-\frac{\partial \varrho_s}{\partial x^i} + \frac{e}{m}\,\frac{\partial \varrho_a}{\partial x^i}) \quad \text{for} \quad i = 1,2,3$$

and

$$X_{m,e} = \sum_{i=1}^{3} ((\varrho_v)^i\,\frac{\partial}{\partial v^i} + v^i \frac{\partial}{\partial x^i}) + \frac{\partial}{\partial t}$$

This dynamical system can be interpreted as a Newtonian particle of mass m and charge e moving under the influence of the inertial, gravitational and electric forces. All the forces are treated on the same footing.

0.7.

More generally, we shall understand under a Hamiltonian system a triplet $(W, \eth, X)$ where X is a vector field on W and $\eth$ is a two form on W, so that $X \lrcorner\ \eth = dH$ for some smooth function H on W.

We shall say that a system (W,X) admits a differential manifold M as its configuration space, if there exists a projection $\pi : W \longrightarrow M$ so that $(\pi_{*} X)_{\pi(W)} = 0$ only if $X_W = 0$.

Section 1:

In this section we would like to show that the standard notion of a moment map for a Lie group action can be modified to a Lie algebra vector bundle action as well. With small modifications, we can retain some of the basic results concerning the usual moment map. This will enable us to introduce minimal systems as a counterpart to elementary dynamical systems for a Lie group:

1.1.
Let E(W) be a vector bundle over a differential manifold W, then E(W) will be called a Lie algebra vector bundle (l.a.v.b for short), if we have a smooth assignment of a Lie algebra structure on each fiber E_w of E(W). Note that a Lie algebra vector bundle is not necessarily a Lie algebra bundle. We do not have to have a local trivialization with locally constant structure constants.

1.2.
Given l.a.v.b E(W) we shall call any strong vector bundle map $\lambda : E(W) \longrightarrow T(W)$ an action of E(W) on W. Here T(W) denotes the tangent bundle and by a strong vector bundle map we understand a vector bundle map which maps the fiber E_w at w into the fiber $T_w(W)$. In particular, if E(W) is a trivial Lie algebra bundle $E(W) = W \times E$, then we shall refer to λ as a Lie algebra action of E on W. Let $\wedge E^*$ be the exterior algebra over the dual of a Lie algebra E. Then we can define an operator $\delta_E : \wedge E^* \longrightarrow \wedge E^*$ by setting

$$\delta_E \, \Phi(h_o, h_1, \ldots h_p) =$$

$$= \sum_{i,j=1,\ i<j}^{n} (-1)^{i+j} \Phi([h_i, h_j], h_0, h_1 .. \hat{h}_i .. \hat{h}_j .. h_p)$$

$\Phi \in \wedge E^*$, $h_i \in E$ and $\hat{h}_i$, $\hat{h}_j$ means that h_i and h_j are deleted. Clearly $\delta_E^2 = 0$ and $(\wedge E^*, \delta_E)$ is a graded differential algebra. The corresponding cohomology algebra is called the cohomology algebra of E.

1.3.
Let F(W) be a vector bundle over W, then we shall denote by $A^p(F(W))$ the space of all smooth p-forms on W with values in F(W). Let E(W) be a Lie algebra vector bundle given together with an l.a.v.b. action λ of E(W) on W. Then λ induces a map $\lambda^*_k : A^p(F(W)) \longrightarrow A^{p-k}((F \otimes \wedge^k E^*)(W))$ given by: $\lambda^*_k \alpha(y) = \langle \lambda(y) \lrcorner \alpha \rangle$ for any $y \in \wedge^k E_w(W)$ $w \in W$ and $\alpha \in A^p(F(W))$. The operator δ_E of 1.2. defines a map $\delta_E : A^p(\wedge E^*(W)) \longrightarrow A^p(\wedge E^*(W))$ and any exterior covariant derivative induces a map $\nabla : A^p(F(W)) \longrightarrow A^{p+1}(F(W))$.

Using the above given notation, we shall make the following definitions:

1.4.

a) $\nu \in A^p(\wedge E^*(W))$ is a moment of λ_k at $w \in W$, if $\langle \nu(w) \cdot s \rangle$ whenever $\lambda(s) = 0$ for $s \in \wedge^k E_w(W)$

b) $\mu \in A^p(\wedge^k E^*(W))$ will be called a moment of λ and $\sigma \in A^{p+1}(\wedge^{k-1} E^*(W))$ at $w \in W$ if:

$$(1) \quad \begin{array}{ll} a) & \lambda_2^* \sigma_w = \delta_E \mu_w \\ b) & \lambda_1^* \sigma_w = \nabla \mu_w \end{array}$$

for some exterior covariant derivative ∇ on $\wedge^k E^*(W)$

c) $\sigma \in A^p(\wedge^k E^*(W))$ will be called λ invariant at $w \in W$ if $\delta_E(\lambda_2^* \sigma) = 0$ for any $w \in W$

d) If $\sigma \in A^p(\wedge^0 E^*(W))$ then $s \in A^0(E(W))$ will be called linear with respect to λ, σ, μ if $\langle \lambda_1^* \sigma\, s \rangle = d \langle \mu\, s \rangle$

1.5. Examples:

a) For any smooth vector field X on W we can define an action λ of W x R on W by setting $\lambda(r)_w = rX(w)$ for any $r \in R$. Let $\sigma \in A^2(W)$ then there always exists a smooth section of W x R^* such that:

$$1) \quad \lambda_2^* \sigma = \delta_E \mu = 0$$

$$2) \quad \lambda_1^* \sigma = \nabla \mu$$

for some exterior differentiation ∇ on W x R^*. The vector field X on W will be a Hamiltonian vector field (with respect to σ) if the above given exterior covariant derivative ∇ is just the standard flat exterior differentiation, i.e. the section (W,1) of W x R is linear with respect to λ, μ, σ

b) Let W = G be a Lie group. Let $X_1 \ldots X_n$ be a moving frame of linearly independent right invariant vector fields on W. Then we can define an action of G x g on G by setting: $\lambda(a, e_i) = -X_i(a)$, where $e_1 \ldots e_n$ is the basis for the Lie algebra of G. The - sign is taken because we are identifying g with the Lie algebra of the left invariant vector fields on G.

Now, let α be a left invariant one form on G. Then we have

$$\lambda_2(d\alpha) = -\delta_E(\lambda_1^* \alpha) \quad \text{and} \quad \lambda_1^*(d\alpha) = -d(\lambda_1^* \alpha)$$

In other words, $\lambda_1^* \alpha$ is the moment of $-d\alpha$ and λ_1.

1.6.

It is clear that if $\lambda_2^* \sigma = \delta_E \mu$ then $\delta_E (\lambda_2^* \sigma) = \delta_E^2 \mu = 0$. On the other hand, $\delta_E (\lambda_2^* \sigma) = 0$ does not imply the existence of the moment μ. Nevertheless, this obstruction can be lifted (for $\sigma \in A^2(W)$). The following theorem is a reformulation of a theorem due to Kostant and Souriau [10,14].

<u>Theorem:</u>

Let λ be an l.a.v.b. action of E(W) on W, $\sigma \in A^2(\Lambda^0 E(W))$ $\delta_E(\lambda_2^* \sigma)_w = 0$ for any $w \in W$, then there exists a central extension E(W) of $\tilde{E}(W) = E(W) \times R$ so that the lifted action $\tilde{\lambda}$ of $\tilde{E}(W)$ on W has a moment μ with respect to σ. In other words, we have the following commutative diagram

$$\begin{array}{ccccc} R(W) & \xrightarrow{i} & E(W) & \longrightarrow & E(W) \\ & & & \underset{\tilde{\lambda}}{\searrow} & \downarrow \lambda \\ & & & & T(W) \end{array}$$

where, in the upper row we have a short exact sequence of l.a.v.b. Also, any smooth section $\tilde{s}$ of $\tilde{E}(W)$ whose image under λ is a Hamiltonian vector field on W (with respect to σ) gives rise to a linear section s of E(W).

<u>Proof:</u> Let $\tilde{E}(W) = E(W) \oplus R(W)$ as vector bundles. Then we can use σ to define a Lie algebra bracket on $\tilde{E}(W)$ by setting $[(A,t),(B,s)]_w = ([A,B], (\lambda_2^* \sigma)(A \wedge B))_w$. From the fact that σ is λ invariant, i.e. $\delta_E (\lambda_2^* \sigma) = 0$, it follows that this bracket satisfies the Jacobi identity. Now, let us choose the section $(0.1) \in \tilde{E}^*(W)$ as μ. Then $\delta_E (\mu(W)) = \lambda_2^* \sigma_W$. In order to show that we can find a connection ∇ on $\tilde{E}^*(W)$ such that (1.b) is valid, we shall first choose a connection ∇ on $\tilde{E}(W)$ so that the corresponding exterior covariant derivative is given by $\nabla' s = (\nabla_1 s_1, ds_2)$ for an arbitrary linear connection ∇_1 on E(W), the standard flat exterior differentiation on R(W) and a smooth section $\tilde{s}$ of $\tilde{E}(W)$ written as (s_1, s_2). Then we can choose ∇ so that

$$(\nabla - \nabla')\, s = (0, \langle \lambda_1^* \sigma, s_1 \rangle \cdot s_2)$$

Let $\tilde{s} = (s_2, 1)$ then

$$(\nabla^* \mu, \tilde{s}) = ((\lambda_1^* \sigma\ s_1), 0) = \langle \hat{\lambda}_1^* \sigma, (s_1, 1) \rangle$$

So

$$\hat{\lambda}_1^* \sigma = \nabla^* \mu \quad \text{for} \quad \mu = (0, 1)$$

1.7.

An l.a.v.b. action λ of $E(W)$ on W will be called Hamiltonian (with respect to a two form σ) along $W_0 \subset W$ if there exists a smooth section μ of $E^*(W)$ such that at any $w \in W_0$ μ is a moment of λ, σ at $w \in W_0$.

1.8.

Given a Hamiltonian system (W, σ, X) and an l.a.v.b. action λ of $E(W)$ on W, we shall say that (W, σ, X) is λ invariant if $\delta(\lambda_2^* \sigma) = 0$ and $X \lrcorner \lambda(A) \lrcorner \sigma = 0$ on W. If λ is surjective we shall say that (W, σ, X) is an elementary system of λ.

Using the above given notation, we shall reformulate another standard result [9,11] concerning the moment map.

1.9. <u>Theorem:</u>

Let (W, σ, X) be a Hamiltonian system on W, and λ be an l.a.v.b. action of $E(W)$ on W. Furthermore, let -

a) σ be symplectic

b) there exist a submanifold W_0 of W, so that λ is Hamiltonian along W_0 and for any $w \in W_0$, $Y \in T_w(W)$ will belong to $T_w(W_0)$ if, and only if, there exists $B \in E_w(W)$ such that $Y \lrcorner \lambda(A) \lrcorner \sigma_w = \langle \mu_w [A,B] \rangle$ for any $A \in E_w(W)$. Here, μ is the corresponding moment section along W_0.

c) $\lambda(A) \lrcorner X \lrcorner \sigma = 0$ for any $A \in E_w(W)$ and $w \in W_0$.

d) there exist a submanifold $W_1 \subset W_0$ such that $\lambda(E_w) = T_w(W_1)$ and $X \lrcorner Y \lrcorner \sigma = 0$ for any $w \in W_1$ and $Y \in T_w(W_0)$.

Then W_0 is a coisotropic submanifold of W, and at any point $w \in W_0$, X is tangent to W_0. The restriction of σ and X to W_1 is an elementary Hamiltonian system for the action of the restriction of $E(W)$ to W_1 on W_1.

<u>Proof:</u> From b) and the definition of the moment map it follows that $\lambda(E)_w \subset T_w(W_0)$ for any $w \in W_0$ (hence d) makes sense). So if $\perp$ denotes the orthogonal complement with respect to σ then $\lambda^\perp(E(W)) \supset T_w^\perp(W_0)$. On the other hand, from b) it follows that $\lambda^\perp(E(W)) \subset T_w(W_0)$. So $T_w^\perp(W_0) \subset T_w(W_0)$ and W_0 is coisotropic submanifold of W. This means that the restriction of σ to W_0 is a presymplectic form. So c) implies that X at $w \in W_0$ is tangent to W_0. On the other hand, as we have seen above, $T_w^\perp(W) \subset \lambda(E)$, so any X at $w \in W_0$ which satisfies d) belongs to $\lambda(E_w)$. This in turn implies that X belongs to $T_w(W_1)$ for any $w \in W_1$ and X belongs to the kernel of the restriction of λ to W_1.

This theorem clearly generalizes the situation described in section 0. Indeed, we can reformulate the results of section 0 in terms of theorem 2 as follows.

1.10.
Let $W = T^*(G)$ be the cotangent bundle of a Lie group G, ω the canonical symplectic form on $T^*(G)$. Let θ be any nondegenerate one form on G with values in g, the Lie algebra of G and Φ the corresponding g^* valued function on $T^*(G)$ such that $\langle \Phi(p), v \lrcorner \theta \rangle = p(v)$ for any $p \in T_a^*(G)$, $v \in T_a(G)$ and any $a \in G$. Then we can define an action λ of g on $T_a^*(G)$ by setting

$$\lambda(p, A) \lrcorner \pi^* \theta = A \qquad \lambda(p, A) \lrcorner d\Phi = 0 \qquad p \in T_a^*(G) \quad A \in g$$

Here $\pi : T^*(G) \longrightarrow G$ is the canonical projection. Now, let us assume that $\langle Ad^{\#} b \cdot f, D\theta \rangle_a = 0$ for any $a,b \in G$, and some $f \in g^*$. Then the submanifold $G_{\mathcal{O}_f}$ of all $p \in T^*(G)$ so that $\Phi(a) \in \mathcal{O}_f$ the coadjoint orbit of G through f, can be identified with the submanifold W_0 of the previous theorem. Indeed, let us choose the map $\Phi : T(G) \longrightarrow g^*$ as the corresponding μ section. Then we have

$$(\lambda(A) \lrcorner \lambda(B) \lrcorner \omega)_p = \langle \Phi(p) \, [B, A] \rangle$$

$$(\lambda^* \circ \omega)_p = (\nabla \Phi)_p \, , \qquad \nabla = d + ad^{\#}(\pi^* \theta)$$

for any $p \in G_{\mathcal{O}_f}$. Furthermore

$$(Y \lrcorner \lambda(A) \lrcorner \omega) = \langle Y \lrcorner d\Phi \cdot A \rangle + \langle \Phi(p), [Y \lrcorner \pi^* \theta \, , A] \rangle$$

So $Y \in T_p G_{\mathcal{O}_f}$ iff $Y \lrcorner \lambda(A) \lrcorner \omega = \langle \Phi(p) \, [B, A] \rangle$, for any $A \in g$ and some $B \in g$. So $G_{\mathcal{O}_f}$ is a coisotropic submanifold of $T^*(G)$. Now, let H be any Hamiltonian function on $T^*(G)$ obtained by pulling back via Φ any smooth function on g^*. Then the corresponding Hamiltonian vector field X on $T^*(G)$ will be tangent to $G_{\mathcal{O}_f}$ at any point $p \in G_{\mathcal{O}_f}$. If H is a pull back of some d G invariant function on g^*, the level surface $\Phi(\cdot) = f'$, $f' \in \mathcal{O}_f$ corresponds to the submanifold W_1 of the theorem 2.

1.11.
It will be convenient to have a concept of a surjective Hamiltonian action without the condition (1.b) in the definition of the moment map. Given a Hamiltonian system (W, ω, X) and l.a.v.b. E(W), we shall call an l.a.v.b. action λ minimal with respect to (ω, X) at $w \in W$ if

a) $\lambda_1^* \omega = \nabla \mu$ for a smooth section μ of $E^*(W)$ and a connection ∇ on $E^*(W)$

b) $X \lrcorner \nabla\mu = 0$

c) λ is surjective.

In a certain sense, minimal systems are elementary systems with additional coupling through μ, i.e. $\delta = \delta' + \delta''$, $\delta(\lambda_2^* \delta') = 0$ and $\delta'' = \langle \mu \Sigma \rangle$ for some E valued two form Σ on W. Nevertheless, we shall be interested mainly in systems with $\Sigma = 0$, i.e. in elementary systems.

Section 2:

In this section we shall describe an abstract geometrical situation modelled on the Einstein theory and Yang-Mills field theories. The main purpose of this section is to connect the concept of minimal and elementary systems of the previous section with the notion of a classical matter moving under the influence of inertial and external forces. As in section 0, we would like to achieve a unified treatment of all the forces.

2.1.

Let M be an n-dimensional manifold, $E(M) \equiv \eta$ a fixed vector bundle over M. Then we shall denote by:

$\mathcal{E}$: a space of Lie algebra vector bundle structures on $E(M) = \eta$

$\mathcal{K}$: the space of all linear connections on $E(M) = \eta$

$\mathcal{B}$: some abstract space describing motions of the classical matter and fields in M

$\mathcal{D}$: $\mathcal{K} \times \mathcal{B} \times \mathcal{E}$

$A^o{}_c(\eta)$: is the vector space of all smooth sections of η which vanish outside of some compact subset of M.

$A_c^o(\eta) \times \mathcal{D}$: corresponding trivial bundle over $\mathcal{D}$.

$A_c^o(\eta)(\mathcal{D})$: is the Lie algebra vector bundle obtained by giving each $A_c^o(\eta) \times d$ a Lie algebra structure by setting: $[s_1, s_2](m) = [s_1(m), s_2(m)]_{\mathcal{E}(d)}$ where $\mathcal{E}(d)$ is the Lie algebra vector bundle structure on η entering into definition of d.

Let Λ be an l.a.v.b. action of $A_c^o(\eta)(\mathcal{D})$ on $\mathcal{D}$ and ν its moment (see 1.3!), then we shall call a point $d \in \mathcal{D}$ balance point of Λ, ν if (1) $\nu(d) = 0$

2.2.

Let $\Lambda(s,d) = \delta_s\nabla + \delta_s b + \delta_s \varepsilon$, where $\delta_s\nabla \in T_\nabla\mathcal{K}$, $\delta_s b \in T_b B$ and $\delta_s \varepsilon \in T_\varepsilon \mathcal{E}$. Then we can write

$$\nu(d) = \nu_1(d) + \nu_2(d) + \nu_3(d)$$

where $\nu_1(d)$, $\nu_2(d)$ and $\nu_3(d)$ are determined by linear functions on the images of $A_c^o(E(M))_d$ in $T_\nabla \mathcal{H}$, $T_b B$ and $T_\mathcal{E} \mathcal{E}$ respectively. Now, $T_\nabla \mathcal{H}$ at any ∇ can be identified with the space of $(E \otimes E^*)(M)$ valued one forms on M. So we can assume that for any $s \in A_c^o(\eta)_d$ we have

$$\text{(2)} \qquad \langle \nu_1(d) \cdot s \rangle = \int_M \langle Q_1(d) \wedge \delta_s \nabla \rangle$$

where $\langle Q_1(d) \wedge \delta_s \nabla \rangle$ is an n-form on M (vanishing outside of a compact subset of M) obtained by a pairing of the $(E^* \otimes E)(M)$ valued (n-1)-form $Q_1(d)$ with an $(E \otimes E^*)(M)$ valued one form $\delta_s \nabla$ on M. Let $\nu_4(d)$ be an auxiliary linear function on the image of $A_c^o(E(M))_d$ in $T_\nabla \mathcal{H}$ chosen so that

$$\nu_4(d)\, s = -\nu_2(d)\, s \quad \text{for any } d \in \mathcal{D}$$

Then we can rewrite equation (1) as

$$\text{(3)} \quad \begin{aligned} &\text{a)} \quad \int_M \langle (Q_4(d) - Q_1(d)) \wedge \delta_s \nabla \rangle = -\langle \nu_3(d) \cdot s \rangle \\ &\text{b)} \quad \int_M \langle Q_4(d) \wedge \delta_s \nabla \rangle = -\langle \nu_2(d) \cdot s \rangle \end{aligned}$$

for any $s \in A_c^o(\eta)_d$.

2.3.

We shall make some additional assumptions concerning the structure of Λ. Let

a) $(\delta_s \mathcal{E}) = 0 \quad$ for any $\quad d \in \mathcal{D}$ and $s \in \Gamma_c(E(M))$

b) $(\delta_s \nabla) = \chi_d(\nabla s) + \lambda(s) \lrcorner R(\nabla)$

Here ad : $E_d(M) \longrightarrow (E_d \otimes E_d^*)(M)$ is the strong vector bundle map defined by the adjoint representation on each fiber where ad is taken with respect to the Lie algebra vector bundle structure induced on E(M) by d, ∇ is the connection on E(M) defined by the point d, $R(\nabla)$ is its curvature form and $\lambda : E_1(M) \longrightarrow T(M)$ is an action of E(M) on M which can depend on d. We can use a) and b) together with partial integration to rewrite (3.a) and (3.b) as

$$(4)\quad \begin{aligned} &a)\quad \nabla^{*}(\chi^{*}(Q_1(d)) - \chi^{*}(Q_4(d)) + \langle Q_1(d) - Q_4(d) \wedge \lambda_1^{*}(R(\nabla))\rangle = 0 \\ &b)\quad \int_M (\nabla^{*}(\chi^{*}(Q_1(d)) - Q_4 \wedge \lambda^{*}(R(\nabla)))\wedge s = 0 \end{aligned}$$

Here, $\chi^{*} : (E^{*} \otimes E)(M) \longrightarrow E^{*}(M)$ is the dual of χ and $\langle Q_1(d) - Q_4(d) \wedge \lambda_1^{*} R(\nabla)\rangle$ is the pairing of an $(E^{*} \otimes E)$ valued (n-1) form with $((E \otimes E^{*}) \otimes E^{*})(M)$ valued one form which produces an $E^{*}(M)$ valued n form.

As we shall see in the next section, equation (4.a) is a weaker form of the Einstein equation

$$(5)\quad a)\quad \chi^{*}(Q_1(d)) = \chi^{*}(Q_4(d))$$

where (in the framework of the Einstein theory of relativity) $\chi^{*}(Q_1(d))$ is related to the Einstein tensor and $\chi^{*}(Q_4(d))$ to the stress energy momenta tensor.

We shall interpret $(Q_4(d)) = J(d)$ as giving the amount of charge $a_m \lrcorner J \in E_m^{*}$ contained in an infinitesimal n-1 volume $a_m \in \Lambda^{n-1} T_m(M)$, $m \in M$ [12]

We shall introduce the notion of passive matter into the theory by setting

$$(6)\quad \begin{aligned} &a)\quad G_1(d) = J_a(d), \quad G(d) \equiv \chi^{*}(Q_1(d)) \\ &b)\quad \int_M \langle J_p \wedge \delta_{\nabla} s\rangle = 0 \qquad \text{for any } s \in A_c^{0}(\eta)_d \\ &c)\quad J(d) = J_a(d) + J_p(d) \end{aligned}$$

whenever d is the balance point of Λ and ν.
The equation (6.b) constrains the possible motion of the passive matter.

2.4.
Let us consider a system describing a motion of a classical point particle. As a geometrical model of such an object we shall take a curve $\gamma : (\tau_1, \tau_2) \longrightarrow M$ given together with an assignment of an infinitesimal (n-1) volume $a_{\gamma(\tau)} \in \Lambda^{n-1}_{\gamma(\tau)} T(M)$ transversal to $\gamma_{*}(\frac{\partial}{\partial \tau})$ i.e.

$$(7)\qquad a_{\gamma(\tau)} \wedge \gamma_{*}(\frac{\partial}{\partial \tau}) \neq 0 \quad \text{for any } \tau \in (\tau_1, \tau_2)$$

Now, in order that $J_p(d)$ can be interpreted as describing (among other things) a classical point particle moving along γ, we shall have to have

$$(8)\quad \begin{aligned} &\text{a)}\quad \gamma_*(\tfrac{\partial}{\partial \tau}) \lrcorner\, J_p(d) = 0 \\ &\text{b)}\quad \int_\gamma \langle \mu\, \delta_s \nabla \rangle_d = 0 \end{aligned}$$

whenever d is a balance point of Λ and ν.
Here the condition $\gamma_*(\frac{\partial}{\partial \tau}) \lrcorner J_p(d) = 0$ means that we do not want any relative motion of the charge at $\gamma(\tau)$ with respect to $\gamma_*(\frac{\partial}{\partial \tau})$ (see ref. [15] for more details). $\mu(\ ) = a_{\gamma(\tau)} \lrcorner J_p(d)$ and condition (8.b) is the application of the general law to the case of point particles. (8.b) can be rewritten as

$$(9)\qquad \gamma_*(\frac{\partial}{\partial \tau}) \lrcorner \nabla' \mu_d = 0$$

for any balance point d of $\mathcal{D}$. Here ∇' depends on ∇ and $R(\nabla)$. Equation (9) is at least a second order equation because $a_{\gamma(\tau)}$ must depend at least on $\gamma_*(\frac{\partial}{\partial \tau})$ due to the transversality condition (7). We shall refer to equation (9) as the Newton equation of the theory. Now, we would like to convert equation (9) into a first order differential equation. This can be done by the usual procedure, i.e. we shall introduce an auxiliary fiber bundle (P, π, M) over M so that if we pull back equation (9) onto P, μ will become a smooth section of $E^*(P)$, the pull back of $E^*(M)$ onto P. On the other hand, if we have the equation $\tilde{\gamma}_*(\frac{\partial}{\partial \tau}) \lrcorner \tilde{\nabla} \mu = 0$ on P for a smooth section μ of $E^*(P)$, then we can look for a surjective action $\tilde{\lambda}$ of $E^*(P)$ on P such that

$$\tilde{\lambda}^* \sigma = \tilde{\nabla} \mu$$

for some two form σ on P. In this case we will obtain motion of a classical point particle as a minimal system defined in the previous section.

Section 3

In this section, we would like to give two examples of the general construction given above. The first example is the basic example of the subject, namely the Einstein theory of relativity. The second example is a Newtonian approximation to the Einstein theory. We are citing this example to illustrate that one can use the formalism of section 2 to mix internal and external symmetries in a rather nontrivial and physically relevant fashion. We shall keep all the notation of section 2.

3.1. Einstein theory of relativity, geometry

Let M be a 4 dimensional oriented manifold and $\mathcal{G}$ the space of all smooth Lorentzian metrics g of the signature (+,-,-,-) on M. Let T(M) be the tangent bundle of M and

$E(M) = \Lambda^2 T(M) \oplus T(M)$. Then for any $g \in \mathcal{G}$ we can give $E(M)$ a structure of a Lie algebra bundle $\mathcal{E}_g$ by setting:

$$[(x \wedge y, z), (x' \wedge y', z')]_{\mathcal{E}_g} = (S, z'') \qquad m,y,z,x',y',z',z'' \in T_m(M)$$

where $S = g(y,x')x \wedge y' + g(x,y')y \wedge x' + g(x,x')y' \wedge y + g(y,y')\, x \wedge x'$

and $z'' = x\, g(y,z') - y\, g(x,z') + yg'(x',z) - x'g(y,z)$

For any g we shall denote by K_g the space of all linear connections on E(M) which preserves g, i.e. $\nabla \in \mathcal{K}_g$ if and only if

$$\nabla [s_1, s_2]_{\mathcal{E}_g} = [\nabla s_1, s_2]_{\mathcal{E}_g} + [s_1, \nabla s_2]_{\mathcal{E}_g}$$

for smooth sections s_1, s_2 of E(M).
In the space $\mathcal{K}_g$ we have a preferred connection ∇_g obtained by choosing the Levi-Civita connection $\nabla_{g,n}$ on $T(M) \equiv n(M)$ and the corresponding induced connection $\nabla_{g,\mathfrak{h}}$ on $\Lambda^2 T(M) = \mathfrak{h}(M)$ and then setting

$$\nabla_g = (\nabla_{g,\mathfrak{h}} \oplus \nabla_{g,n}) + \mathrm{ad}\,\theta$$

where $(\nabla_{g,\mathfrak{h}} \oplus \nabla_{g,n})$ is the standard direct sum of $\nabla_{g,\mathfrak{h}}$ and $\nabla_{g,n}$ and ad θ is an $(E \otimes E^*)(M)$ valued one form on M obtained from the canonical E(M) valued one form θ (which identifies T(M) with a subbundle n(M) of E(M)) and the corresponding adjoint representation (associated to $\mathcal{E}_g$). It is easily seen that $\nabla_g \in \mathcal{K}_g$.

Let $\mathcal{D} = \mathcal{K} \times \mathcal{B} \times \mathcal{E}$, where $\mathcal{E}$ is some space of Lie algebra vector bundle structures on E(M), $\mathcal{K}$ is the space of all linear connections on E(M) and $\mathcal{B}$ is some auxiliary space of motions of the matter and fields in M. Then we shall denote by $\mathcal{D}_{\mathcal{G}}$ the subset of $\mathcal{D}$ consisting of all the points $d_g = (\nabla g, \mathcal{E} g, b)$. Clearly $\mathcal{D}_{\mathcal{G}}$ can be regarded as an imbedding of $\mathcal{G}$ into $\mathcal{D}$.

Let $\lambda : E(M) \longrightarrow T(M)$ be just the projection onto T(M) (i.e. $\lambda(\Lambda^2 T(M)) = 0$) then we can define:

$$(2) \qquad \Lambda(s, d) = \mathrm{ad}\,(\nabla s) + \lambda(s) \lrcorner R(\nabla)$$

Now, let $d_g = d$, then $R(\nabla_g)$ can be identified with an h(M) valued two form on M. The orientation on M together with g determine the star operator $\circledast : \Lambda T(M) \to \Lambda T(M)$. In particular, we can apply $\circledast$ to any element in $h(M) = \Lambda^2 T(M)$; $\circledast h(M) \subset h(M)$. So we can define $Q_1(d_g)$ as an $(E^* \otimes E)(M)$ valued 3 form on M by setting:

$$(3) \quad \int_M \langle Q_1(d_g) \wedge \delta_s \nabla \rangle = \int_M \bar{\theta} \wedge ([\circledast R(\nabla_g) \overset{\wedge}{,} \nabla s])$$

for any $s \in A^o_c(E(M))$.
Here $\bar{\theta}$ is an $E^*(M)$ valued one form on M, obtained by taking the adjoint of θ with respect to g, i.e. $\bar{\theta}$ is a $T^*(M) \equiv n^*(M)$ valued one form. $[\circledast R(\nabla_g) \overset{\wedge}{,} \nabla s]$ is a 3 form on M with values in E(M) obtained by pairing $R(\nabla_g)$ with ∇s in the indicated way, i.e. exterior product of forms and commutation relations of their values. This 3 form is paired with $\bar{\theta}$ to form a R-valued 4-form on M.

As $R(\nabla_g)$ has only h(M) values

$$(4) \quad \int_M \langle Q_1(d_g) \wedge \lambda(s) \lrcorner R(\nabla_g) \rangle = 0 \quad \text{for any } s \in A^o_c(E(M))$$

and $ad^*(Q_1(d)) = \langle ad^\#(\circledast R(\nabla_g) \wedge \bar{\theta})\rangle = G(d_g)$ is just the double dual of the Einstein tensor (see [12]). Here $ad^*(Q_1(d_g))$ is the corresponding $E^*(M)$ valued 3-form determined by $Q_1(d_g)$ and the dual of ad. Clearly, $G(d_g)$ has only $n^*(M)$ components. So let us assume that $ad^*(Q_4(d)) = J(d_g)$ has only $n^*(M)$ components as well. Then we can rewrite the equation $\nu(d) = 0$ as

$$(5) \quad \nabla_g^*(G(d_g) - J(d_g)) = 0$$

or

$$(5') \quad \begin{aligned} &a) \quad \nabla_g^*(G(d_g) - J(d_g))_h = ad^\# \theta \cdot (G(d_g) - J(d_g)) = 0 \\ &b) \quad \nabla_{g,n}(G(d_g) - J(d_g)) = 0 \end{aligned}$$

Now, from the fact that $\nabla_{g,n}(G(d_g)) = 0$ for any d_g, it follows that the above equation (5'.b) can be rewritten as

$$(5'') \quad b) \quad \nabla_{g,n} J(d_g) = 0$$

which is the usual conservation law of the Einstein theory.

3.2. Einstein theory of relativity, mechanics

Let $J(d_g)$ be a $n^*(M) \equiv T^*(M)$ valued 3-form on M so that $\nabla_g^* J = 0$ for some metric g on M. Let us assign to any curve $\gamma : (\tau_1, \tau_2) \longrightarrow M$ so that

$$g(\gamma_*(\tfrac{\partial}{\partial\tau}), \gamma_*(\tfrac{\partial}{\partial\tau})) > 0 \quad \text{for any } \tau \in (\tau_1, \tau_2),$$

$$a_{\gamma(\tau)} = \circledast \gamma_*(\tfrac{\partial}{\partial\tau}) \in \Lambda^3 T_{\gamma(\tau)} M$$

This $a_{\gamma(\tau)}$ is clearly transversal to $\gamma_*(\frac{\partial}{\partial\tau})$.
Furthermore, let us assume that for some such curve

a) $\gamma_*(\frac{\partial}{\partial\tau}) \lrcorner J = 0$

b) $\gamma_*(\frac{\partial}{\partial\tau}) \lrcorner \nabla_g (a_{\gamma(\tau)} \lrcorner J) = 0 \qquad a_{\gamma(\tau)} \lrcorner J \neq 0$

for any $\tau \in (\tau_1, \tau_2)$

Then $\gamma(\tau)$ is a geodesic (up to a reparametrization) and

$$\gamma_*(\frac{\partial}{\partial\tau}) = f(\tau)\, \overline{(a_{\gamma(\tau)} \lrcorner J)}$$

i.e. J determines $\gamma(\tau)$ (up to the physically irrelevant reparametrization). Indeed:

$$(\gamma_*(\frac{\partial}{\partial\tau}) \lrcorner \nabla_g (a_{\gamma(\tau)} \lrcorner J))_h = \mathrm{ad}^{\#}\gamma_*(\frac{\partial}{\partial\tau}) \quad (a_{\gamma(\tau)} \lrcorner J) = 0 .$$

This means that $a_{\gamma(\tau)} \lrcorner J = f(\tau)\ \overline{\gamma_*(\tau)}$ for some nonvanishing function f of τ.
On the other hand,

$$(\gamma_*(\frac{\partial}{\partial\tau}) \lrcorner \nabla_g (a_{\gamma(\tau)} \lrcorner J))_n = \gamma_*(\frac{\partial}{\partial\tau}) \lrcorner (\nabla^*_{g,n} (f(\tau)\ \overline{\gamma_*(\frac{\partial}{\partial\tau})})) = 0$$

So, up to reparametrization (i.e. setting $f(\tau) = 1$), the curve is a geodesic.

3.3. Extensions of the Einstein theory

Let M be a space-time as in 3.1. $E(M) = \Lambda^2 T(M) \oplus T(M)$, $\mathcal{G}$ is the space of all smooth Lorentzian metrics on M, F(M) some auxiliary vector bundle over M and S some auxiliary space chosen so that for any point $(g, s) \in \mathcal{G} \times S$ we are given

a) a Lie algebra vector bundle structure $f_{(g,s)}$ on F(M)

b) a strong vector bundle map $\varrho_{(g,s)}: E(M) \longrightarrow F(M)$ which injects E(M) into F(M).

We shall distinguish between two situations:

1) If $(\varrho(E(M)), \mathcal{E}_g)$ is a Lie algebra vector subbundle of $(F(M), f_{g,s})$, then we shall refer to the above situation as a trivial extension of the Einstein theory.

2) If $(\varrho(E(M)), \mathcal{E}_g)$ is not a Lie algebra vector subbundle of $F(M)\ f_{g,s})$, then we shall call the above given situation a nontrivial extension of the Einstein theory.

In this paper, we shall concentrate on one particular nontrivial extension of the

Einstein theory, namely, its Newtonian approximation. We are interested in this theory because it is an example of a theory in which internal symmetries are spontaneously generated by the correspondence principle. One can construct in a similar way a nontrivial conformal extension ([13,17,18]) or nontrivial graded Lie algebra extensions.

3.4. Newtonian relativity

Let M be a noncompact oriented space-time, and $\Gamma(M)$ the space of all smooth nowhere vanishing vector fields on M. Let $\Gamma(M)\,\hat{x}\,\mathcal{G}$ denote the space of pairs (g, X) so that $g(X\,X)_m = 1$ for any $m \in M$. Let $F(M) = E(M) \oplus R(M)$ where $R(M) = M \times R$ is the trivial R-bundle over M. Then for any $(g, X) \in \Gamma(M)\,\hat{x}\,\mathcal{G}$, we can give F(M) a Lie algebra vector bundle structure with typical fiber isomorphic to the Lie algebra of the central extension of the Galilei group. Indeed:

Let $j_{(g,x)} : F(M) \longrightarrow A(M) = (T \otimes T^*)(M) \oplus (R \otimes T^*)(M) \oplus T(M) \oplus R(M)$ be a strong vector bundle map given for any $m \in M$ by

$$j\,(X \wedge Y)_m = (1_5 \otimes \overline{Y} + Y \otimes \overline{X})_m$$

$$j\,(Y \wedge Z)_m = \frac{1}{2}(Y \otimes Z - Z \otimes \overline{Y})_m$$

$$j\,(X_m) = (X_m + 1_5) \quad \text{and} \quad j\,(1_5) = 1_5$$

where 1_5 is some fixed vector in R, $Y,Z \in X_m^{\perp}$ the orthogonal complement of X_m in T_mM with respect to g. As $\mathcal{A}(M)$ has a natural Lie algebra bundle structure, it induces a Lie algebra bundle structure $f_{(g\,,X)}$ on $j_{(g,X)}(F(M))$ and hence on F(M). On the other hand, we shall denote by $\varsigma_{(g,X)}$ the map $E(M) \longrightarrow F(M)$ given by

$$\varsigma\,(\alpha) = \alpha \quad \text{for any} \quad \alpha \in \Lambda^2 T(M)$$

$$\varsigma\,(X_m) = X_m + 1_5$$

$$\varsigma\,(Y) = Y \quad \text{for any} \quad Y \in X_n^{\perp}$$

Let $\mathcal{D}^N = \mathcal{E}^N \times \mathcal{K}^N \times \mathcal{B}$ where $\mathcal{E}^N$ is the space of all the Lie algebra vector bundle structures f(g,x) on F(M), $\mathcal{K}^N$ is the space of all linear connections on F(M) and $\mathcal{B}$ is again some auxiliary space of motions of the matter and fields in M. We would like to imbed $\Gamma(M)\,\hat{x}\,\mathcal{G}$ in $\mathcal{D}^N$ in a similar manner as we have done for $\mathcal{G}$ in 3.1. This can be done in the following way; for any $g \in \mathcal{G}$ let us denote by $L_g(M)$ the bundle of all g-orthonormal frames on M. Then the choice of a unit vector field X on M defines a reduction $L_X(M) \hookrightarrow L_g(M)$ of $L_g(M)$ to a principle $O(3,R)$ bundle $L_X(M)$ of all the orthonormal frames on M with X_m as the first frame vector.

Now $(E(M)) = L_X(M) \times E'$ where $E(M)$, $\mathcal{E}_g$ is vector bundle $E(M)$, with the g-induced Poincaré Lie algebra bundle structure and E' is Lie algebra of the Poincaré group identified with $\wedge^2 R^{1,3} \oplus R^{1,3}$ for the standard Minkowski space $R^{1,3}$. $L_X(M) \times E'$ denotes the associated bundle formed using the restriction of the adjoint action to the subgroup $O(3)$, defined by the space-time splitting of $R^{1,3}$. The map $\varrho_{g,x}$ can be lifted into a map $\hat{\varrho}_{g,x}$

$$\hat{\varrho}_{g,x} \; : \; L_X(M) \times E' \longrightarrow L_X(M) \times F'$$

where $F' = \wedge^2 R^{1,3} \oplus R^{1,3} \oplus R$ with its structure as a Lie algebra of the central extension of the Galilei group, induced by the choice of a space-time decomposition of $R^{1,3}$ given by the vector $(1,0,0,0)$.

Let ∇_g be the unique connection on E(M) associated to g in 3.1., then the pull back $\overline{\nabla}_g$ of ∇_g onto $L_X(M) \times E'$ has the exterior covariant derivative given by:

$$(d + \text{ad}\,\theta) = \overline{\overline{\nabla}}_g$$

for some E' valued one form θ on $L_X(M)$. On the other hand, let us denote by $\nabla_{g,X}$ the unique connection on F(M) whose pull back onto $L_X(M) \times F'$ has the exterior covariant derivative given by

$$(d + \text{ad}\,(\hat{\varrho}_{(g,x)}(\theta))) = \overline{\overline{\nabla}}_{g,x}$$

It is clear that $\nabla_{g,x}$ preserves $f_{(g,x)}$.

Now we would like to use the Einsteinian moments to construct the Newtonian moments. So let us first define a strong vector bundle map $\varphi_{g,x} : E^*(M) \longrightarrow F^*(M)$, so that

$$\text{(1)} \quad \begin{aligned} &\text{a)}\ \varphi_{g,x} \circ (\varrho_{g,x})^* = \text{identity on } E(M) \\ &\text{b)}\ \varphi_{g,x}(\alpha) \in \underline{X}^{\perp} \end{aligned}$$

where $\underline{X}$ is the smooth section of F(M) corresponding to the vector field X on M and $\underline{X}^{\perp}$ is its anihilator in $F^*(M)$. From the form of $\varrho_{g,x}$ it is clear that a) and b) determine $\varphi_{g,x}$ uniquely. So to any $E^*(M)$ valued p-form β on M we can assign an $F^*(M)$ valued p-form $\beta^N_{g,x}$ by setting $\varrho^*_{g,x}(\beta) = \beta^N_{g,x}$.

Let $F(M) = h(M) + n(M) = \Lambda^2 T(M) \oplus (T(M) \oplus R(M))$ and $F^*(M) = h^*(M) + n^*(M)$. Then $\langle \alpha^N_{g,x}, 1_5 \rangle = \langle \alpha, \underline{X} \rangle$, where $\underline{X}$ is now a section of E(M) and where 1_5 is the section of R(M) entering into the definition of $\varrho_{g,x}$. The maps $\varrho_{g,x}$ and $\varphi_{g,x}$ were chosen so that

$$(2) \qquad \langle \alpha \wedge R(\nabla_g) \rangle = \langle \alpha^N_{g,x} \wedge R(\nabla_{g,x}) \rangle = 0$$

for any n^* valued p-form α on M. In particular, this will be true for G and J of the paragraph 3.1. The left hand side is a consequence of the fact that $R(\nabla_g)_n = 0$. Now, the map $\varrho_{g,x}$ was chosen so that the only possible nonvanishing component of $R(\nabla_{g,x})$ is along the $\underline{X}$ direction. So the above given choice of $\varphi_{g,x}$ ensures (2). Equations (1) and (2) are part of the Newtonian correspondence principle. Conditions 1a) and 1b) can be formulated in the following physical terms. If a system is at rest with respect to the infinitesimal space-time splitting defined by X, then set its Newtonian mass (R(M) component) equal to its Einsteinian energy and its Newtonian energy set to zero, i.e. its Newtonian energy becomes the kinetic energy with respect to X. Equation (2) promises that if one does as one is told, the equations of motions of the system are particularly simple.

The last statement can be made more definite within the framework of our particle mechanics; every minimal system is elementary. This statement can be viewed as a reformulation of the Einstein correspondence principle.

REFERENCES

[1] Arnold, V.I., Mathematical Methods of Classical Mechanics, Springer Verlag.

[2] Bargman, V., Ann. Math. 59, 1 (1954).

[3] Cartan, E., Ann. Ecole Norm. Sup. 40 (1923).
Ann. Ecole Norm. Sup. 41 (1924).

[4] Cartan, E., La Méthode de Répère Mobile, la Théory de Groupes Continus et les Espaces Géneralisées, Herman, Paris, France.

[5] Duval, Ch. and Künzle, H., Rep. Math. Phys., Vol. 3 1978.

[6] Greub, W., Halpern, L., and Vanstone, R., Connections, curvature and cohomology, Vol. II, New York, Academic Press 1973.

[7] Guillemin, V. and Sternberg, S., Hadronic Journal, Vol. 1, No. 1, Nonatim Mass. 1978.

[8] Guillemin, V. and Sternberg, S., Ann. of Phys. 1980.

[9] Kazhdan, D., Kostant, B., and Sternberg, S., C.P.A.M. (1978).

[10] Kostant, B., Quantization and unitary representations, Lectures on Modern Analysis and Application III, Springer Verlag (1970).

[11] Marsden, J. and Weinstein, A., Rep. Math. Phys. 5 (1974).

[12] Misner, C.H.W., Thorne, K.S. and Wheeler, J.A., Gravitation, Freeman & Co. (1973).

[13] Segal, I., Mathematical Cosmology and Extragalactic Astronomy, New York, Academic Press, 1975.

[14] Souriau, J.M., Ann. Inst. H. Poinc. Sec A, Vol XX, No. 4, 1974.

[15] Souriau, J.M., Structure des Systèmes Dynamiques, Paris, Dunod, 1970.

[16] Sternberg, S., Differential Geometrical Methods in Mathematical Physics, Bonn, 1977, Lecture Notes in Mathematics, Vol. 676.

[17] Sternberg, S. and Ungar, T., Hadronic Journal, Vol. I, No. I, 1978, Nonatum, Mass.

[18] Sternberg, S. and Wolf, J.A., N. Cimento, Vol. 28A, No. 2, 1975.

WHAT KIND OF A DYNAMICAL SYSTEM IS THE RADIATING ELECTRON?

A.O. Barut

Department of Physics, The University of Colorado,
Boulder, Colorado 80309, U.S.A.

ABSTRACT

We show that both in classical and quantum theory of the relativistic electron there are three sets of independent dynamical variables: position, velocity and momentum. The independence of velocity and momentum is interpreted by internal degrees of freedom. The geometry of the internal phase-space is discussed. The close analogy between the classical and quantum equations and their algebraic and symplectic structures is shown.

1. Introduction

An accelerating electron radiates electromagnetic waves, it exchanges both reversible and irreversible energy with the electromagnetic field. Furthermore, charged particles do not simply interact via time-independent two-body potentials. Thus the equations of simple Hamiltonian dynamical systems have to be modified in the case of relativistic charged particles and electromagnetic interactions. In this work I discuss the new mathematical structures that arise in the dynamics of the electron both in classical and quantum theory. There is a remarkable parallelism between the Dirac equation of the quantum electron and the Lorentz-Dirac equation of the classical radiating electron. Both contain internal degrees of freedom and an internal dynamics. The origin of spin is explained as an orbital angular momentum of the internal motion. In the quantum case the internal dynamics is an example of Weyl's finite quantum mechanics, an oscillator with a compact phase-space $Sp(4) \sim O(5)$. The dynamical group $O(4,2)$ of the Dirac electron can be derived from this by "bosting".

2. The Classical Relativistic Radiating Electron

Let $z_\mu(s)$ be the world line of the electron in the Minkowski space, where s is the proper time. The classical motion of the electron in an external electromagnetic field F^{ext} (i.e. the field due to other charged particles) is governed by the Lorentz-Dirac equation [1]:

$$m \ddot{z}_\mu = e F^{ext}_{\mu\nu}(z) \dot{z}^\nu + k[\dddot{z}_\mu + (\ddot{z})^2 \dot{z}_\mu] \tag{1}$$

where m is the rest mass of the electron, e its charge and $k = \frac{2}{3} e^2$ (in units $c = \hbar = 1$); $\ddot{z}^2 = \ddot{z}_\mu \ddot{z}^\mu$.

The first term on the right hand side is the Lorentz force leading to the usual Hamiltonian system. The second term involves a part containing third order derivatives and another highly non-linear part. This term comes from the radiative effects [i.e. from the self-force of the electron: part of the self-force is in the form of an inertial force $\delta m \ddot{z}_\mu$. After this force has been put on the left hand side of eq. (1) and the mass has been renormalized, what is left is the second term in (1)].

First order Dynamical Equations

Since eq. (1) is of third order, we set

$$\dot{z}_\mu = u_\mu , \tag{2}$$

and introduce the new canonical momenta p_μ by

$$p_\mu \equiv m u_\mu + e A_\mu(z) - k \dot{u}_\mu , \tag{3}$$

where $A_\mu(x)$ is the electromagnetic potential from which the field $F_{\mu\nu}(x)$ in (1) is derived:

$$F = dA , \quad \text{i.e.} \quad F_{\mu\nu} = A_{\nu,\mu} - A_{\mu,\nu} \tag{4}$$

Then the equation of motion (1) can be written as

$$\dot{p}_\mu = e A_{\nu,\mu} u^\nu + k \dot{u}^2 u_\mu . \tag{5}$$

<u>Proof</u>: From (3):

$$\dot{p}_\mu = m \ddot{z}_\mu + e A_{\mu,\nu} \dot{z}^\nu - k \dddot{z}_\mu .$$

Replace the first term on the right hand side by eq. (1)

$$\dot{p}_\mu = e\,(A_{\nu,\mu} - A_{\mu,\nu})\,\dot{z}^\nu + k\,\dddot{z}_\mu + k\,\ddot{z}^2\dot{z}_\mu + e\,A_{\mu,\nu}\,\dot{z}^\nu - k\,\dddot{z}_\mu$$

Then cancellation of terms gives (5).

Equations (2), (3) and (4) are now our basic dynamical equations which we rewrite in the form

$$\dot{z}_\mu = u_\mu\,,$$

$$\dot{u}_\mu = \frac{1}{k}\,(m\,u_\mu + e\,A_\mu - p_\mu)\,, \tag{6}$$

$$\dot{p}_\mu = e\,A_{\nu,\mu}\,u^\nu + \frac{1}{k}\,(m\,u_\mu + e\,A_\mu - p_\mu)^2\,u_\mu\,.$$

It is more convenient to use new variables $\dot{u}_\mu \equiv v_\mu$ in which case we obtain

$$\dot{z}_\mu = u_\mu\,,$$

$$\dot{u}_\mu = v_\mu\,, \tag{7}$$

$$\dot{v}_\mu = -\frac{e}{k}\,F_{\mu\nu}\,u^\nu + \frac{m}{k}\,v_\mu - v^2\,u_\mu\,.$$

In these 12 equations we have two constraint equations which can be used to eliminate three variables, although sometimes it may be more convenient to use the equations in the above simpler form. The constraints are

$$u^2 = u_\mu\,u^\mu = 1\,, \tag{8}$$

which follows from the definition of u , and

$$u\,v = u_\mu\,v^\mu = 0\,, \tag{9}$$

which follows from the differentiation of (8). Using (8) and (9) we can express u_o and v_o in terms of the space-components $\vec{u}$ and $\vec{v}$:

$$u_o = (1 + \vec{u}^{\,2})^{\frac{1}{2}}\,, \tag{10}$$

$$v_o = \vec{u}\cdot\vec{v}\,/\,(1 + \vec{u}^{\,2})^{\frac{1}{2}}\,. \tag{11}$$

Then $\dot{v}_o$ can be expressed in terms of $\vec{u}$ and $\vec{v}$ alone:

$$\dot{v}_o = -\frac{e}{k}\vec{E}\cdot\vec{v} + \frac{m}{k}\frac{\vec{u}\cdot\vec{v}}{(1+\vec{u}^2)^{1/2}} - \left(\frac{(\vec{u}\cdot\vec{v})^2}{1+\vec{u}^2} - \vec{v}^{\,2}\right)\cdot(1+\vec{u}^{\,2})^{\frac{1}{2}} \tag{12}$$

Here we have put $F_{oi} = E_i$, the electric field.

Consequently we have the following nine linearly independent equations:

$$\begin{aligned}
\dot{z}_i &= u_i ,\\
\dot{u}_i &= v_i ,\\
\dot{v}_i &= \frac{e}{k}\left(F_{io}\,(1+\vec{u}^{\,2})^{\frac{1}{2}} + F_{ik}\,u^k\right) + \frac{m}{k}v_i - \left(\frac{(\vec{u}\cdot\vec{v})^2}{(1+\vec{u}^2)} - \vec{v}^{\,2}\right)u_i ,\\
i &= 1,2,3.
\end{aligned} \tag{13}$$

Unlike the usual dynamics, we have now three sets of dynamical variables $\vec{z}$, $\vec{u}$ and $\vec{v}$, and must treat the velocity $\vec{u}$ and the momentum $\vec{p}$ (or $\vec{v}$) to be independent of each other [2]. This startling fact, which also occurs in Dirac's theory of the quantum electron, will be interpreted later.

An Integral of the Motion

We will now show that the following function

$$\mathcal{H} = u^\mu (p_\mu - e A_\mu) - m\, u^\mu u_\mu \tag{14}$$

has the properties:

(i) $\mathcal{H} = 0$ by virtue of the equations of motion,

(ii) $\dot{\mathcal{H}} = 0$,

(iii) $\mathcal{H}$ generates a "Hamiltonian" p_o ,

(iv) $\mathcal{H}$ provides Hamilton's equations of motion for $\dot{z}$, $\dot{u}$ and for part of $\dot{p}$ (including $\dddot{z}$-term, but not including the nonlinear $(\ddot{z}^{\,2})\,\dot{z}$-term - see eq. (1)).

The first statement follows immediately from the second equation of (6) and the relation (9). The second property is proved as follows:

$$\begin{aligned}
\dot{\mathcal{H}} &= \dot{u}^\mu (p_\mu - e A_\mu) + u^\mu (\dot{p}_\mu - e A_{\mu,\nu} u^\nu)\\
&= v^\mu (m u_\mu - k v_\mu) + u^\mu (e A_{\nu,\mu} u^\nu - e A_{\mu,\nu} u^\nu + k v^2 u_\mu)\\
&= m\, u\cdot v - k v^2 + u^\mu (e F_{\mu\nu} u^\nu + k v^2 u_\mu) = 0 ,
\end{aligned}$$

where we have used, in the second line, eqs. (6), and in the third line the constraints (8) and (9).

Using $\mathcal{H} = 0$, we have from (14)

$$u^o (p_o - e A_o) - \vec{u}\cdot(\vec{p} - e\vec{A}) = m.$$

Writing $u^o = \frac{dz^o}{ds}$, we solve for p_o :

$$p_o = e A_o + \vec{u}\frac{ds}{dz^o}\cdot(\vec{p} - e\vec{A}) + \frac{ds}{dz^o}\, m. \tag{15}$$

It is remarkable that p_o has exactly the form of the quantum Dirac Hamiltonian p_o for the electron and that the velocity $\vec{u}\frac{ds}{dz} = \frac{d\vec{z}}{dt}$ corresponds to the Dirac matrix $\vec{\alpha}$ and $m\frac{ds}{dt}$ corresponds to the Dirac matrix β. We shall see in the next section that indeed $\vec{\alpha}$ is the velocity operator of the Dirac electron and is independent of the momentum $\vec{p}$.

Finally by property (iv) we mean the following equations:

$$\begin{aligned}
\dot{z}_\mu &= \frac{\partial\mathcal{H}}{\partial p^\mu} = u_\mu \\
\dot{u}_\mu &= -\frac{1}{k}\frac{\partial\mathcal{H}}{\partial u^\mu} = \frac{1}{k}(m\, u_\mu + e A_\mu - p_\mu) \\
\dot{p}_\mu &= -\frac{\partial\mathcal{H}}{\partial z^\mu} = e A_{\nu,\mu}\, u^\nu
\end{aligned} \tag{16}$$

These equations coincide with our dynamical equations (6) or (7) except the nonlinear term $v^2 u_\mu$ (which is the irreversible dissipative term). Note that z and p are conjugate to each other, and u is conjugate to a multiple of itself (k u) ! Eqs. (16) suggest to define a Poisson bracket by

$$\{f, g\} = \frac{\partial f}{\partial p_\nu}\frac{\partial g}{\partial z^\nu} - \frac{\partial f}{\partial z_\nu}\frac{\partial g}{\partial p^\nu} - \frac{1}{k}\left(\frac{\partial f}{\partial u_\nu}\frac{\partial g}{\partial u^\nu}\right)_{AS} \tag{17}$$

so that eqs. (16) can be expressed as

$$\dot{z}_\mu = \{\mathcal{H}, z_\mu\}, \quad \dot{u}_\mu = \{\mathcal{H}, u_\mu\}, \quad \dot{p}_\mu = \{\mathcal{H}, p_\mu\} . \tag{18}$$

However, in order to have a proper antisymmetric symplectic form in (17) the third term must be properly defined. This can be achieved by introducing an antisymmetric Poisson bracket for $\{u_\mu, u_\nu\}$, or treat u's as Grassmann variables. We see this also from a consideration of angular momenta:

Internal and External Angular Momentum

The orbital angular momentum $L_{\mu\nu} = (z_\mu p_\nu - z_\nu p_\mu)$ is not a constant of the motion for a free particle, but a total angular momentum defined by

$$J_{\mu\nu} = L_{\mu\nu} + S_{\mu\nu} \tag{19}$$

is conserved if the nonlinear $(\ddot{z}^2)\,\dot{z}_\mu$ -term is neglected:

$$\frac{d}{ds} L_{\mu\nu} = u_\mu p_\nu - u_\nu p_\mu \neq 0$$

If we add to this a term

$$\frac{d}{ds}(S_{\mu\nu}) = -\frac{d}{ds}\{u_\mu, u_\nu\} \overset{\text{def}}{=} -\dot{u}_\mu k\, u_\nu + u_\mu k\, \dot{u}_\nu$$

$$= u_\nu p_\mu - u_\mu p_\nu \tag{20}$$

Then the 'total angular momentum' is conserved.

These considerations suggest that the origin of spin goes back to the $\dddot{z}$-radiation term in eq. (1), hence need not be introduced *ad hoc* into the classical theory.

3. The Quantum Electron as a Dynamical System

We start from the Hamiltonian of the Dirac electron ($c = \hbar = 1$)

$$H = \vec{\alpha}\cdot(\vec{p} - e\vec{A}) + \beta m + V, \tag{21}$$

where $\vec{\alpha}$ and β commute with $\vec{x}$ and $\vec{p}$ and satisfy the anti-commutation relations

$$\{\alpha_i, \alpha_j\} = 2\delta_{ij} I, \quad \{\alpha_i, \beta\} = 0, \; \beta^2 = I; \; i,j = 1,2,3. \tag{22}$$

Using Heisenberg's equations of motion, we obtain analogous to (6) or (7):

$$\dot{\vec{x}} = i[H, x] = \vec{\alpha} = \frac{\partial H}{\partial \vec{p}}$$

$$\dot{\vec{\alpha}} = i[H, \vec{\alpha}] = i[-2(\vec{p} - e\vec{A}) + 2H\vec{\alpha} - 2V\vec{\alpha}] \tag{23}$$

$$\dot{\vec{p}} = i[H, \vec{p}] = ve\,\vec{\alpha}\cdot\nabla A - \nabla V = -\frac{\partial H}{\partial \vec{x}}$$

showing at once that $\vec{\alpha}$ is the velocity of the charge. (We could have written the more covariant form, using the Dirac wave operator $W = \gamma^{\mu}(p_{\mu} - e A_{\mu}) - m$, of the above equations, but this is not essential). For a free particle these equations simplify to

$$\dot{\vec{x}} = \vec{\alpha}$$

$$\dot{\vec{\alpha}} = 2i [H \vec{\alpha} - \vec{p}] \tag{24}$$

$$\dot{\vec{p}} = 0 , \quad (\dot{H} = 0) .$$

This system of dynamical equations is closed. We have a further dynamical variable ß satisfying

$$\dot{\beta} = i \, 2H \, \beta \tag{25}$$

which can be integrated separately.

For the system (24) we have to specify three initial values, $\vec{x}$ (o), $\vec{p}$ (o) and $\vec{\alpha}$(o) - and not just two as in the ordinary dynamics. The same is true for the classical system (6). Note that we have now the equation $\dddot{x} = i \, 2H \, \ddot{x}$ for a free particle, and not just $\ddot{x} = 0$. This and the form of the equations (6) and (23) shows that the Dirac equation is the quantized form of the Lorentz-Dirac equation (1) without the nonlinear term but with the third order term k $\dddot{z}_{\mu}$. It is this term which is responsible for the occurrence of a new independent variable besides the canonical pair ($\vec{x}$, $\vec{p}$), namely the velocity $\vec{\alpha}$. It can be interpreted as the self-force representing the continuous emission and absorption of radiation by the electron. In contrast the term k $(\ddot{z}^2) \, \dot{z}_{\mu}$ in (1) corresponds to the irreversible loss or gain of energy to or from infinity.

Internal Dynamical Structure of the Dirac Electron

The additional dynamical variables $\vec{\alpha}$, ß can be given a precise meaning as "internal dynamical coordinates" as follows [3]. If we solve eqs. (24) we can write the Heisenberg operator $\vec{x}$ (t) as

$$\vec{x}\,(t) = \vec{x}_A\,(t) + \vec{\xi}\,(t) , \tag{26}$$

where

$$\vec{x}_A\,(t) = \vec{a} + H^{-1} \vec{p}\, t,$$

$$\vec{\xi}\,(t) = \frac{1}{2} i [\vec{\alpha}(o) - H^{-1} \vec{p}] H^{-1} e^{-2iHt} ,$$

with $\vec{a}$ = a constant vector. We interpret $\vec{x}_A$ (t) as the position of the center of mass of the electron moving with a velocity $H^{-1} \vec{p}$ as it should. We interpret $\vec{\xi}$ (t), which oscillates violently around the center of mass, as the position of the charge relative to the center of mass. This oscillatory motion Schrödinger called the "Zitterbewegung" [4].

The old problem whether the 'correct' relativistic position operator should be $\vec{x}$ or $\vec{x}_A$ (or something related to $\vec{x}_A$) is solved by this distinction between the center of mass and the position of the charge; the two do not coincide!

In the center of mass frame, $\vec{p} = 0$, we can identify a finite quantum system of the type introduced by Weyl [5]. It is an oscillator with a compact phase space. In order to exhibit this, we define in this frame $\vec{p} = 0$, hence $H = m\,\beta$, the operators

$$\vec{P}\,(t) = m\,\vec{\alpha}(t)\;, \quad \vec{Q}\,(t) = \frac{i}{2m}\,\vec{\alpha}(t)\,\beta \tag{27}$$

satisfying the commutation relations

$$[Q_i, P_j] = -i\,\delta_{ij}\,\beta\;. \tag{28}$$

From eqs. (27) and (28) it is readily seen that $\vec{P}$, $\vec{Q}$ obey the equations of an oscillator with a frequency $\omega = 2m$.

Furthermore, introducing the spin $\vec{S}$

$$\vec{S} = -\frac{i}{4}\,\vec{\alpha}\wedge\hat{\alpha} \tag{29}$$

we find that the operators β, $\vec{Q}$, $\vec{P}$ and $\vec{S}$ generate the Lie algebra of Sp (4) ~ SO(5). This algebra may be compared with the usual algebra of the flat phase-space of the external dynamical variables I, $\vec{x}$, $\vec{p}$ and $\vec{L}$, comprising the Euclidean group E (3) and the Heisenberg algebra. The second algebra is a contraction of the first.

We have shown elsewhere [3] how to obtain starting from the finite quantum system representing the internal dynamics, and using the methods of relativistic dynamical groups, the equation for a moving electron, i.e. the Dirac equation. This procedure also shows how the dynamical algebra O (4,2) of the Dirac equation is related to the internal phase space algebra O (5).

In external fields, both the internal motion and the external motion undergo changes. The Heisenberg equations (23) have been explicitly solved in the case of a constant uniform magnetic field [6]. A precise meaning to the spin as the orbital angular momentum of the charge around its center of motion can also be given [7].

Conclusions

The main result of this paper is that the quantum Dirac equation is not a quantization of a classical relativistic Hamiltonian system but really corresponds to the quantization of the linearized Lorentz-Dirac equation of the radiating electron in which a third order term proportional to $\dddot{z}_\mu$ is present. This leads to a dynamical independence of the velocity u_μ and the momentum p_μ. A one-to-one algebraic correspondence is established between the classical and quantum theories. External and internal (spin) dynamics have been defined in both cases. The result seems to open a new view on the classical and quantum theories of spinning particles.

REFERENCES

[1] For a simple derivation of this equation by analytic continuation and references to earlier papers see Barut, A.O., Phys. Rev. D10, 3335 (1974).

[2] Wessel, W., Fortschritte der Physik 12, 409-440 (1964).

[3] Barut, A.O. and Bracken, A.J., The Zitterbewegung and the Internal Geometry of the Electron, Phys. Rev. D (1981); Proc. Group Theory Conference, Lecture Notes in Physics, Vol. 135, p. 206-211 (Springer, 1980).

[4] Schrödinger, E., Sitzungsb. Preuss. Akad. Wiss. Phys.-Math. Kl. 24, 418 (1930).

[5] Weyl, H., *The Theory of Groups and Quantum Mechanics*, (Dover, NY 1950), pp. 272-280.

[6] Barut, A.O. and Bracken, A.J., Exact Solutions of Heisenberg Equations and Zitterbewegung of the Electron in a Constant Uniform Magnetic Field, Phys. Rev. D (to be published).

[7] Barut, A.O. and Bracken, A.J., The Magnetic Moment Operator of the Relativistic Electron (to be published).

[8] In fact the radiative term in the Lorentz-Dirac equation (1) implies, in an external magnetic field, for example, a magnetic moment for the classical electron: Barut, A.O., Physics Letters 73B, 310 (1978).

III. DIFFERENTIAL OPERATORS ON MANIFOLDS

ASYMPTOTICS OF ELEMENTARY SPHERICAL FUNCTIONS

J.J. Duistermaat

Mathematisch Instituut der Rijksuniversiteit Utrecht,
Budapestlaan 6, 3508 TA Utrecht, NL.

In this talk I will report on some joint work with
J.A.C. Kolk and V.S. Varadarajan

1. Symmetric spaces of negative curvature

Let $S = K\backslash G$, with G a noncompact, connected, semisimple Lie group, acting to the right on S and K a maximal compact subgroup of G. A function on S is then a function on G which is left-K-invariant.

Any G-invariant continuous linear operator $\Phi : C^{\infty}(S) \longrightarrow D'(S)$ is of the form

$$C^{\infty}(K\backslash G) \ni f \longrightarrow \varphi * f \in C^{\infty}(K\backslash G)$$

where $\varphi \in \mathcal{E}'(K\backslash G/K)$ is uniquely determined and called a *spherical distribution* (with compact support). Furthermore, $(\mathcal{E}'(K\backslash G/K), *)$ is an algebra where $*$ is convolution $\approx$ composition of the corresponding operators.

Let $\mathfrak{g} = T_e G$, $\mathfrak{k} = T_e K$ be the Lie algebra of G, K,

and $\mathfrak{s}$ = orthogonal complement of $\mathfrak{k}$ in $\mathfrak{g}$ with respect to the *Killing form*

$$\kappa : (X, Y) \longrightarrow \mathrm{Tr}\,(\mathrm{ad}\, X \circ \mathrm{ad}\, Y)$$

so that $\mathfrak{g} = \mathfrak{k} \oplus \mathfrak{s}$, $\kappa > 0$ on $\mathfrak{s}$, $\kappa < 0$ on $\mathfrak{k}$.

Then $S \approx \exp s$ and geodesics in S correspond to the curves

$$t \longrightarrow x \cdot \exp tX \cdot x' ,$$

here $x \in G$, $X \in \mathfrak{s}$, x' = adjoint of x.

Pick now an $\mathfrak{a}$ which is maximal abelian $\subset \mathfrak{s}$ ($\Longleftrightarrow \exp \mathfrak{a} = A$ is a maximal flat subspace of S) and decompose

$$\mathfrak{g} = \sum_{\alpha \in \Delta}^{\oplus} \mathfrak{g}_\alpha \quad , \quad \Delta \subset \mathfrak{a}^*$$

where $(\mathrm{ad}\, X)(Y) = [X, Y] = \alpha(X) \cdot Y$ for $Y \in \mathfrak{g}_\alpha$, $X \in \mathfrak{a}$.

$$\mathfrak{g}_0 = \mathfrak{m} \oplus \mathfrak{a} \qquad \text{if } \mathfrak{m} = \mathfrak{g}_0 \cap \mathfrak{k}$$

$\mathfrak{a} \supset \operatorname{Ker} \alpha$, $\alpha \in \Delta$, $\alpha \neq 0$ are the root hyperplanes. The connected components of the complement are called *Weyl chambers*. If $k \in \tilde{M}$ = normalizer of $\mathfrak{a}$ in K, we get that $\mathrm{Ad}\, k : \mathfrak{a} \longrightarrow \mathfrak{a}$ permutes Weyl chambers, generated by orthogonal reflections in root hyperplanes. The *Weyl group* is $W = \tilde{M}/M$, where

$$M = \{ k \in K ; \ \mathrm{Ad}\, k\,(X) = X \quad \text{for all} \quad X \in \mathfrak{a} \}$$

and W acts simply transitively on $\{$Weyl chambers$\}$ (so W is finite).

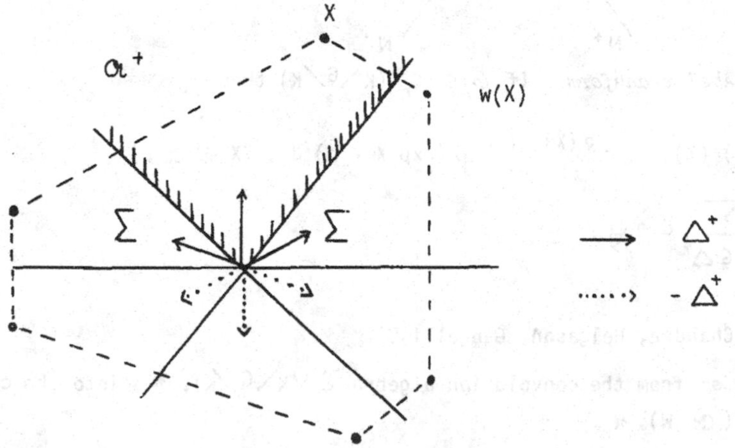

Example:

$G = SL(3, \mathbb{R})$
rank G = dim $\mathfrak{a}$ = 2

Now *choose* a Weyl chamber, called the *positive Weyl chamber* $\mathfrak{a}^+$ and correspondingly

$$\Delta^+ = \{\alpha \in \Delta ; \alpha(X) > 0 \quad \text{for some (all)} \quad X \in \mathfrak{a}^+\}$$

as the set of *positive roots*, so that $\Delta = \Delta^+ \cup \{0\} \cup (-\Delta^+)$. Denote by

$$\Sigma = \{\alpha \in \Delta^+; \text{ not } \alpha = \beta + \gamma \text{ for } \beta, \gamma \in \Delta^+\}$$

the set of *simple roots*. Σ is a basis of $\mathfrak{a}^*$ and $\Delta^+ \subset \mathbb{N}\cdot\Sigma$, consequently the $\operatorname{Ker}\alpha$, $\alpha \in \Sigma$ are the walls of $\mathfrak{a}^+$.

Now because $[\mathfrak{g}_\alpha, \mathfrak{g}_\beta] \subset \mathfrak{g}_{\alpha+\beta}$,

$$\mathfrak{n} = \sum_{\alpha \in \Delta^+} \mathfrak{g}_\alpha \qquad \text{is a nilpotent Lie algebra}$$

and $N = \exp \mathfrak{n}$ is a nilpotent Lie group. Furthermore $(k, X, n) \longrightarrow k \cdot \exp X \cdot n$ is a diffeomorphism: $K \times \mathfrak{a} \times N \longrightarrow G$ (*Iwasawa decomposition*) and $x \in K \cdot \exp H(x) \cdot N$, $H(x) \in \mathfrak{a}$ defines a fibration $H : G \longrightarrow \mathfrak{a}$ called the *Iwasawa projection*.

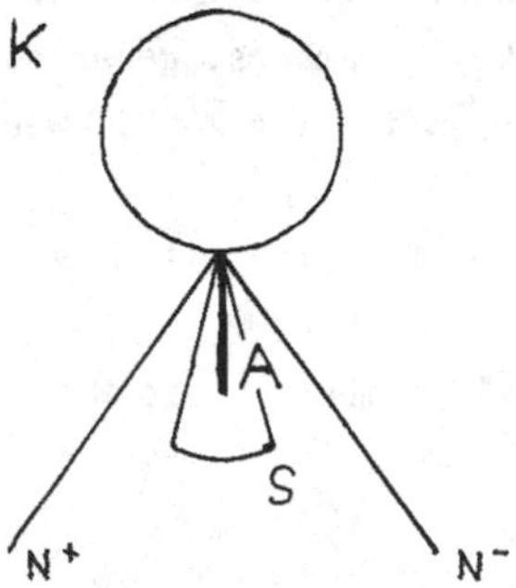

We also need the *Abel transform*. If $\varphi \in C_c^\infty(K\backslash G/K)$ then

$$(\mathcal{A}\varphi)(X) = e^{\rho(X)} \int_N \varphi(\exp X \cdot n)\, dn, \quad X \in \mathfrak{a},$$

here $\rho = \frac{1}{2} \sum_{\alpha \in \Delta^+} \dim \mathfrak{g}_\alpha \cdot \alpha$

Theorem (Harish Chandra, Helgason, Gangolli).

$\mathcal{A}$ is an *isomorphism* from the convolution algebra $\mathcal{E}'(K\backslash G/K), *$ into the convolution algebra $\mathcal{E}'(\mathfrak{a}\, W), *$.

Special case If $D \in \operatorname{Diff}(S)^G \longleftrightarrow \varphi \in \mathcal{E}'(K\backslash G/K)$ with $\operatorname{supp}\varphi \subset K \longleftrightarrow \operatorname{supp}\mathcal{A}\varphi \subset \{0\} \longleftrightarrow D_{\mathfrak{a}} \in \operatorname{Diff}(a)^{\mathfrak{a}, W}$

So if $u \in D'(S)$, $u \neq 0$ and $D u = \Lambda(D) \cdot u$ for all $D \in \mathrm{Diff}(S)^G$ then Λ is a homomorphism: $\mathrm{Diff}(S)^G \longrightarrow \mathbb{C}$. After the passage to $\mathfrak{a}$ it is given by

$$\Lambda(D) = D_{\mathfrak{a}}(X \longrightarrow e^{\lambda(X)})_{X=0}$$

for some $\lambda \in \mathfrak{a}^*_{\mathbb{C}}$, unique modulo W. The space of common eigendistributions for given λ contains a spherical one (unique up to a factor) given by

$$\varphi_\lambda(X) = \int_K e^{(\lambda-\rho)(H(xk))} \, dk, \quad x \in G$$

which is an *elementary spherical function* ;

$$\varphi_\lambda = \mathcal{A}^*(e^\lambda), \quad \mathcal{A}^*: C^\infty(\mathfrak{a}/W) \longrightarrow C^\infty(K\backslash G/K).$$

Fourier synthesis in $\mathfrak{a}$ yields via $\mathcal{A}$ that every $\varphi \in \mathcal{E}'(K\backslash G/K)$ can be synthesized from the $\varphi_{i\mu}$, $\mu \in \mathfrak{a}^*$

2. Asymptotics of φ_λ (exp X), as $X \longrightarrow \infty$ in $\mathfrak{a}^+$ (Harish Chandra)

The decomposition of Cartan:

$$x \in K \cdot \exp R(x) \cdot K, \quad R(x) \in \mathfrak{a}^+$$

defines a fibration R from an open dense subset U of G onto $\mathfrak{a}^+$. If $D \in \mathrm{Diff}(S)^G$ then

$$D_R : f \longrightarrow D(R^* f) \Big|_{\mathfrak{a}^+}$$

is a differential operator on $\mathfrak{a}^+$ called the *radial part of* D.

We also have the following *integration formula*

$$\int_G f(x)\,dx = \int_{\mathfrak{a}^+} \Delta(X) \int_{K\times K} f(k_1 \cdot \exp X \cdot k_2)\, dk_1\, dk_2\, dX,$$

where

$$\Delta(X) = \text{const.} \cdot \prod_{\alpha \in \Delta^+} (\sinh \alpha(X))^{m(\alpha)}, \quad m(\alpha) = \dim \mathfrak{g}_\alpha$$

Take now

$$\hat{D} = \Delta^{1/2} \cdot D_R \cdot \Delta^{-1/2},$$

then $D \longrightarrow \hat{D}$ is a homomorphism from Diff $(S)^G$ to the algebra of differential operators on $\mathfrak{a}^+$ with coefficients which are analytic functions of

$$z_\alpha = e^{-\alpha(X)} , \qquad \alpha \in \Sigma$$

with poles only at $z_\alpha = 1$ (corresponding to the walls of $\mathfrak{a}^+$).

The $\psi_\lambda = \Delta^{1/2} \varphi_\lambda \mid \mathfrak{a}^+$ are common eigenfunctions of the $\hat{D}$ with eigenvalues $\Lambda(D)$. Because the constant coefficient parts of the D (obtained by taking the coefficients at $z_\alpha = 0$) have the functions $X \longrightarrow e^{\lambda(w(X))}$, $w \in W$ as common eigenfunctions, it is likely that asymptotically

$$(*) \qquad \psi_\lambda(X) \sim c(\lambda) \sum_{w \in W} e^{\lambda(w(X))}$$

(using also that $\psi_{w*\lambda} = \psi_\lambda$, $w \in W$).

$(*)$ is related to the fact that $\mathcal{A}^* \circ \mathcal{F}^{-1}$ is a unitary embedding:

$$L^2 (\mathfrak{a}^*/W , \beta) \longrightarrow L^2 (K \backslash G / K) , \quad \text{with}$$

$\beta = \text{const} \cdot \left| c(i\mu) \right|^{-2} d\mu$ = *Plancherel measure for spherical functions.*
This statement is the main step in the proof that $\mathcal{A}$ is an isomorphism.

Proof of $(*)$

As also Kashiwara e.a.[5] observed, in the z_α-coordinates the ψ_λ satisfy a "*holonomic system with regular singularities at* $\prod_{\alpha \in \Sigma} z_\alpha = 0$".

This means that there exist differential operators $Q_1, \ldots, Q_{\#(W)}$ with $Q_1 = I$, such that if we write

$$u = \begin{pmatrix} Q_1 \psi_\lambda \\ \vdots \\ Q_{\#(W)} \psi_\lambda \end{pmatrix}$$

then u satisfies a system of *ordinary* linear differential equations along any curve in z-space, with regular singularities at $z_\alpha = 0$, viz.

$$t \frac{du}{dt} = L(t)\, u , \qquad t \longrightarrow L(t) \text{ regular}$$

The classical (19th century) theory of regular singularities then leads to an asymptotic expansion of u as $z_\alpha \longrightarrow 0$, which actually is a power series with radius of convergence equal to 1, because the solutions of the ordinary differential equations can be extended along any curve in the complement of $z_\alpha = 1$ (or $z_\alpha z_\beta = 1$ if $\alpha + \beta \in \Delta$).

For rank $S = \dim \mathfrak{a} = 1$ we have ordinary differential equations from the start - so the above result is especially spectacular for rank $S > 1$.

3. Asymptotics of $\varphi_{i\mu}(\exp X)$ as $\mu \longrightarrow \infty$.

We have to remember that λ is an eigenvalue parameter.
Here the approach is to view

$$\int_K e^{i\mu(H(xk))}\, g(k)\, dk$$

as an *oscillatory integral* with *phase function*

$$f_{\mu,X} : k \longrightarrow \mu(H(xk)), \qquad x = \exp X, \quad X \in \overline{\mathfrak{a}^+}.$$

(Actually $\varphi_{i\mu}$ is given as an integral over K/M rather than K. K/M is the *flag variety* of G : M is the stabilizer of a flag for a natural action of K on a certain flag space).

Then the asymptotics for $\mu = \omega \cdot \nu$, $\omega \in \mathbb{R}$, $\omega \longrightarrow \infty$ is given by the behaviour of g near the set $\Sigma_{\nu,X}$ of *stationary* points of $f_{\nu,X}$.

Theorem
$$\Sigma_{\nu,X} = K^X \cdot \tilde{M} \cdot K^\nu$$

where K^X, resp. K^ν is the centralizer of X, resp. ν in K. $\Sigma_{\nu,X}$ is a smooth submanifold of K and the Hessian of $f_{\nu,X}$ at $\Sigma_{\nu,X}$ has rank equal to the codimension d of $\Sigma_{\nu,X}$ in K. (*Clean stationary point set* in the sense of Bott.)

Of course $f_{\nu,X}$ is constant on the connected components of $\Sigma_{\nu,X}/M$, and because W acts transitively on the set of these, the *method of stationary phase* immediately yields:

$$\varphi_{i\mu}(X) \sim \omega^{-\frac{d}{2}} \sum_{w \in W} e^{i\mu(w^{-1}(X))} \sum_{j=0} c_{w,j}^{\nu,X}\, \omega^{-j}$$

as $\mu = \omega \cdot \nu$, $\omega \longrightarrow \infty$ and an explicit Hessian yields an explicit $c_{w,o}^{\nu,X}$.
If *both* ν *and* X *are regular* then

$$c^{\nu,X}_{w,o} = \prod_{\alpha\in\Delta^+} \left| \frac{\langle\alpha,\nu\rangle}{2\pi} \sinh \alpha(X) \right|^{-\frac{1}{2}m(\alpha)} \cdot e^{\frac{\pi i}{4}\delta_w},$$

$$\delta_w = \sum_{\alpha\in\Delta^+} m(\alpha)\,\mathrm{sgn}\,(\alpha,\nu)\cdot\alpha\,(w^{-1}(X))$$

Although the *proof* is quite different, the *result* is very similar to the Harish Chandra asymptotics, and in fact in an analogous fashion suggests the statement that

$\mathcal{F}\circ\mathcal{A}$ is a unitary embedding:

$$L^2\,(K\backslash G/K) \longrightarrow L^2\,(\mathfrak{a}^*/W\,,\,\beta\,).$$

In particular $\mathcal{A}^*\circ\mathcal{F}^{-1} = (\mathcal{F}\circ\mathcal{A})^*$ has dense range, completing the missing part in the proof that $\mathcal{A}$ is an isomorphism. (For us however the purpose was to improve our asymptotic analysis of spectra of G-invariant operators on S, pushed down to compact quotients S/Γ.)

Note also that $f_{\nu,X}$ is just "testing" the Iwasawa projection

$$k \longrightarrow H\,(xk) \quad : \quad K/M \longrightarrow \mathfrak{a}$$

by the linear form ν on $\mathfrak{a}$. It is a *fact* that the set of stationary points *does not move* with ν : K^ν only depends on the set of roots orthogonal to ν which varies within the finite set Δ, so the only thing which can happen is that the dimension of K^ν goes up when ν enters the intersection of more root hyperplanes.

For example, take $X\in\mathfrak{a}^+$. Then, if ν is regular, $\Sigma_{\nu,X}/M = W$ all the time so the maximum value of $f_{\nu,X}$ is equal to $\nu(w^{-1}(X))$ for some $w\in W$, *all the time.* It immediately follows that the image of K/M is contained in the convex hull of the $w^{-1}(X)$, $w\in W$. Expanding this argument a little further, one finds back the

Theorem (Kostant)

The image of K/M under the map $k \longrightarrow H(xk)$ is equal to the convex hull of the Weyl group orbit of X.

A very remarkable circumstance is that the image under a smooth mapping of the smooth manifold K/M without boundary is equal to a polyhedron!

REFERENCES

[1] Duistermaat, J.J., Kolk, J.A.C., and Varadarajan, V.S., Spectra of compact locally symmetric manifolds of negative curvature, Inv. Math. 52 (1979), 27-93.

[2] Gangolli, R., On the Plancherel formula and the Paley-Wiener theorem for spherical functions on semisimple Lie groups, Ann. of Math. 93 (1971), 150-165.

[3] Harish-Chandra , Spherical functions on a semisimple Lie group I, II, Am. J. Math. 80 (1958), 241-310, 553-613.

[4] Heckman, G.J., Projections of orbits and asymptotic behaviour of multiplicities for compact Lie groups, Thesis, Leiden,

[5] Kashiwara, M. e.a., Eigenfunctions of invariant differential operators on a symmetric space, Ann. of Math. 107 (1978), 1-39.

[6] Kostant, B., On convexity, the Weyl group and the Iwasawa decomposition, Ann. Sci. Ec. Norm. Sup. 6 (1973), 413-455.

HERMITIAN STRUCTURES ON SOLUTION VARIETIES OF NONLINEAR RELATIVISTIC WAVE EQUATIONS

Stephen M. Paneitz

Mathematics Department
University of California, Berkeley
Berkeley, California 94720

I. INTRODUCTION

In this lecture we describe some recent progress in the program of constructive quantum field theory, based on the differential geometry of classical solution varieties of partial differential equations, and initiated by Segal in, for example, [12] and [13]. For concreteness we discuss in detail the application to scalar relativistic wave equations with (local) self-interaction and positive mass. The results depend on the dimension n+1 of the space-time essentially only through the particular rate of L_∞-decay in time of solutions (generically $O(|t|^{-n/2})$), and are only improved with increased n. In the general context of evolution equations in Hilbert space, our methods are in part a generalization of the stability theory of M. G. Krein and collaborators [1].

One physical idea put forward in [12] which is illustrated here, is that the vacuum for the quantization of the tangential equation (linearized about a given background field) should be the presumably unique regular state of the Weyl algebra invariant under the scattering automorphism. Through the work of Weinless [17] the "presumably" was removed for linear temporally homogeneous equations by means of positive-energy considerations. We show here how it can be removed for a subvariety of (nonlinear) background fields (the *stable solution variety*, which includes at least all sufficiently small nonvanishing fields, cf. [7]) satisfying those field equations whose interaction term satisfies a certain inequality, roughly, that the "second variation of the interaction energy" be nonnegative.

Specifically, what is established here are Lorentz-invariant hermitian structures, necessarily nonlocal, in the stable solution varieties of relativistic wave equations of a general class, uniquely determined by the requirement that the differential scattering transformation be unitary. Until recently only the symplectic part of this structure was known, in the cases of nonlinear or time-dependent linear equations (including, say, the Klein-Gordon equation on a curved space-time background). In the terminology of "geometric quantization," what is introduced is the unique "totally complex positive polarization" (aside from integrability of the distribution, which may be dispensable for field-theoretical purposes) of the solution variety singled out by the dynamics.

This lecture represents part of the author's thesis and subsequent joint work with Irving Segal, to whom the author is greatly indebted for guidance and inspiration in mathematics and physics.

II. THE DIFFERENTIAL SCATTERING TRANSFORMATION

The nonlinear wave equations with which we are concerned have the form

$$\square\phi + m^2\phi + F(\phi) = 0 \tag{1}$$

in n+1 space-time dimensions, where $m^2 > 0$ and F is a given real function of a real variable, at least C^2 and satisfying $F(0) = F'(0) = 0$. By "solution of (1)" we mean a strict solution of the first-order equation in a Banach space

$$\frac{d}{dt}\begin{pmatrix}\phi\\ \dot\phi\end{pmatrix} = \begin{pmatrix}0 & I\\ \Delta-m^2 & 0\end{pmatrix}\begin{pmatrix}\phi\\ \dot\phi\end{pmatrix} + \begin{pmatrix}0\\ -F(\phi)\end{pmatrix} \tag{2}$$

obtained from the choice of a Lorentz frame. One has at least a hope for the existence of solutions global in time if $F = H'$ for some $H \geq 0$, in which case the energy

$$\int_{t=t_o}\{m^2|\phi|^2 + |\nabla\phi|^2 + |\dot\phi|^2 + 2H(\phi)\}\, d_n\vec{x}$$

is conserved and nonnegative. (Note: $\nabla \equiv$ grad is always only with respect to space variables.) Similarly, the method we develop to establish an hermitian structure in part of the solution variety of (1) is limited to cases in which $F' = H'' \geq 0$ also, for example, $F(\phi) = g\phi^p$, $g > 0$, p odd and >1.

We define the positive operator $B = \sqrt{m^2-\Delta}$ (having range $L_2(\mathbb{R}^n)$), $C = B^{\frac{1}{2}}$, the (free) energy norm

$$\|\phi\|_\varepsilon^2 = \|B\phi\|_2^2 + \|\dot\phi\|_2^2 \tag{3}$$

and the Lorentz-invariant norm

(4) $$\|\phi\|_L^2 = \langle \phi,\phi \rangle_L = \|C\phi\|_2^2 + \|C^{-1}\dot{\phi}\|_2^2 \quad ,$$

$\|\cdot\|_2$ denoting the L_2-norm over $\mathbb{R}^n$. (3) and (4) define real Hilbert spaces $\mathcal{H}_\varepsilon$ and $\mathcal{H}_L$ of finite-energy and -Lorentz norm solutions $\phi(t,\cdot)$ respectively (cf. [2] for details); note $\mathcal{H}_L$ contains $\mathcal{H}_\varepsilon$ properly as C^{-1} is bounded by $m^{-\frac{1}{2}}$. We take the point of view that elements $\phi(t,\cdot)$ of $\mathcal{H}_L$ are continuous curves in $H_{\frac{1}{2}}(\mathbb{R}^n)$ (that is, defined for all times) so that isomorphisms (determined by Cauchy data $(\phi(t,\cdot),\dot{\phi}(t,\cdot))$)

$$\mathcal{H}_\varepsilon \cong H_1(\mathbb{R}^n) \oplus L_2(\mathbb{R}^n) \ , \qquad \mathcal{H}_L \cong H_{\frac{1}{2}}(\mathbb{R}^n) \oplus H_{-\frac{1}{2}}(\mathbb{R}^n)$$

exist at each fixed time.

The symplectic structure

(5) $$a(\phi,\psi) = \int_{t=t_0} (\dot{\phi}\psi - \phi\dot{\psi})\, d\vec{x}$$

is defined for $\phi,\psi \in \mathcal{H}_L$, and $\langle\cdot,\cdot\rangle_L + ia(\cdot,\cdot)$ is a complex Hilbert structure on $\mathcal{H}_L$ with positive complex structure $\begin{pmatrix} 0 & -B^{-1} \\ B & 0 \end{pmatrix}$ at each time.

Control over solutions ϕ to nonlinear equations like (1) is based usually on estimation of the asymptotics (usually in the L_∞ and $\|\cdot\|_\varepsilon$-norms) of such ϕ at early and late times to solutions of the free equation $\ddot{\psi} + B^2\psi = 0$ (cf. [14] and references there). No one set of function spaces is appropriate for all purposes; however, a representative result is

<u>Theorem 4.4B</u> ([14II], p.491). Suppose $n=3$, $m>0$, and F is a C^2 function of the real variable ℓ such that

$$|F^{(j)}(\ell)| \leq g|\ell|^{p-j} \qquad (j = 0,1,2)$$

for some $p \geq 3$. Suppose also $F = H'$ where H is bounded from below.

Let ϕ_- be a given finite-energy solution of the free equation

(6) $$\Box\phi_- + m^2\phi_- = 0$$

such that $\mathrm{grad}\phi_-$ is also of finite energy, and suppose that the L_∞-norms (over space) of $\phi_-(t,\cdot)$ and $\mathrm{grad}\phi_-(t,\cdot)$ are bounded by const.$(1+|t|)^{-3/2}$. Then if either g, or ϕ_-, in a certain norm are sufficiently small, there exist unique solutions ϕ and ϕ_+ to (1) and (6), respectively, such that

(7) $$\|\phi - \phi_\pm\|_\varepsilon \to 0 \quad \text{as} \quad t \to \pm\infty \quad ,$$

and

(8) $$\phi, \phi_+ = O(|t|^{-3/2}) \quad \text{in} \quad L_\infty(\mathbb{R}^3) \quad .$$

It may also be discerned from these methods that in these and other similar situations

(9) $$|t|^{3/2} \|\phi - \phi_\pm\|_\infty \to 0 \quad \text{as} \quad t \to \pm\infty$$

a fact we will need later.

We define the wave and scattering operators $W_\pm, S$ by $W_\pm \phi_\pm = \phi$ and $S = W_+^{-1} \circ W_-$, so that $S\phi_- = \phi_+$. In the special case of $F(\phi) = g\phi^3$, nonperturbative results are known [4]; in particular, S is a (nonlinear) C^∞ homeomorphism of a linear space of free solutions $\mathfrak{F}$ complete in the norm

$$\|\phi\|^2_{\mathfrak{F}} = \|\phi\|^2_\varepsilon + \sup_{(t,x)} |\phi(t,x)|^2 + \int_{-\infty}^{\infty} \sup_x |\phi(t,x)|^2 \, dt \quad .$$

We will see that control over the latter two terms in this norm are precisely what is needed in order that our abstract stability result (Theorem 2) apply to this equation.

We define the variety

$$\mathfrak{M} = \{\phi : \Box\phi + m^2\phi + F(\phi) = 0\} \quad ,$$

and for $\phi \in \mathfrak{M}$

$$T_\phi(\mathfrak{M}) = \{\eta : \Box\eta + m^2\eta + F'(\phi)\eta = 0\} \quad ,$$

the tangent space at ϕ. $T_\phi(\mathfrak{M})$ also has a Lorentz-invariant symplectic structure $a(\cdot,\cdot)$ given by (5). Especially if ϕ is uniformly bounded in space-time, it is natural to require of $\eta \in T_\phi(\mathfrak{M})$ that $t \to (\eta(t,\cdot), \dot\eta(t,\cdot))$ be a continuous curve in $H_{1/2}(\mathbb{R}^n) \oplus H_{-1/2}(\mathbb{R}^n)$.

$\mathcal{H}_L$ is a linear space, so if $\phi_\pm \in \mathcal{H}_L$, the tangent spaces $T_{\phi_\pm}(\mathcal{H}_L) \cong \mathcal{H}_L$ canonically. Now $S = W_+^{-1} \circ W_-$, so

$$(dW_\pm)_{\phi_\pm} : T_{\phi_\pm}(\mathcal{H}_L) \to T_\phi(\mathfrak{M})$$

and

$$(dS)_{\phi_-} : T_{\phi_-}(\mathcal{H}_L) \to T_{\phi_+}(\mathcal{H}_L)$$

if $S\phi_- = \phi_+$ as above. We aim next to exhibit the differential scattering transformation $(dS)_{\phi_-}$ as the solving operator of (10) below;

by estimating the coefficients A(t) of this equation (which depend on ϕ) we will be able to show that $(dS)_{\phi_-}$ is "uniquely essentially unitarizable" (cf. Section III). Having show that $(dS)_{\phi_-}$ is unitary for a unique symplectic Hilbert structure on $\mathcal{H}_L$, we can then transfer this structure to $T_\phi(\mathfrak{M})$ by either of $(dW)_{\phi_\pm}$.

We exhibit the action of $(dS)_{\phi_-}$ as follows. Let us assume henceforth that we are in the situation of Theorem 4.4.B, and use the isomorphism $\mathcal{H}_L \cong H_{\frac{1}{2}} \oplus H_{-\frac{1}{2}}$ given by Cauchy data at time t=0; thus $\eta \in \mathcal{H}_L$ corresponds to

$$\begin{pmatrix} \eta(0) \\ \dot\eta(0) \end{pmatrix}, \quad \text{and} \quad \begin{pmatrix} \eta(t) \\ \dot\eta(t) \end{pmatrix} = e^{tQ} \begin{pmatrix} \eta(0) \\ \dot\eta(0) \end{pmatrix} \quad \text{where} \quad Q = \begin{pmatrix} 0 & I \\ \Delta - m^2 & 0 \end{pmatrix} .$$

Given any $\eta \in \mathcal{H}_L$ we can factor off the "free motion" (interaction representation) by defining $\omega(t) = e^{-tQ}\begin{pmatrix} \eta(t) \\ \dot\eta(t) \end{pmatrix}$; then ω satisfies

$$\frac{d}{dt}\,\omega = A(t)\omega \quad , \tag{10}$$

where

$$A(t) = e^{-tQ}\begin{pmatrix} 0 & 0 \\ -F'(\phi) & 0 \end{pmatrix} e^{tQ} \quad ,$$

a bounded operator in $\mathcal{H}_L$. Now $t \to A(t)$ is strongly continuous and norm-bounded since ϕ is uniformly bounded. Since $\|\phi(t,\cdot)\|_\infty = O(|t|^{-3/2})$, $\int_{-\infty}^{\infty} \|A(t)\|\, dt$ is finite, and one can prescribe Cauchy data for (10) at $t = \pm\infty$. Finally, one can check that $(dW_\pm)(\eta_\pm) = \eta$ and $(dS)_{\phi_-}(\eta_-) = \eta_+$, where

$$\begin{pmatrix} \eta_\pm(t) \\ \dot\eta_\pm(t) \end{pmatrix} = e^{tQ}\,\omega(\pm\infty) \quad .$$

Put another way (purely group-theoretically), if S(t) (for $-\infty \leqslant t \leqslant +\infty$) denotes the symplectic operator taking $\omega(-\infty)$ to $\omega(t)$, then $t \to S(t)$ is the norm-continuous curve which satisfies

$$\frac{d}{dt}\,S(t) = A(t)S(t) \quad , \qquad S(-\infty) = I \quad ,$$

and $(dS)_{\phi_-}$ is essentially just the endpoint of this curve.

Remark. For moderately regular ϕ, $(dS)_{\phi_-} = I + \text{cpt.}$; if $p \geqslant 5$ and with two space-time dimensions, $(dS)_{\phi_-} - I$ is Hilbert-Schmidt, but in physical space it is only 4th-power summable [15].

III. STABILITY THEORY FOR CAUSAL SYMPLECTIC EQUATIONS

Equation (10) (with $A(t)$ infinitesimally symplectic) includes as a special case Hill's equation in Hilbert space

$$y'' + p(t)y = 0 \quad , \tag{11}$$

where y takes values in a real Hilbert space H, and $t \to p(t)$ is a locally integrable map into the bounded symmetric operators on H. (The connection, of course, is via $\omega = \begin{pmatrix} y \\ \dot{y} \end{pmatrix}$ and $A(t) = \begin{pmatrix} 0 & I \\ -p(t) & 0 \end{pmatrix}$.)

The relation between stability theory for (11) and unitarizability can be illustrated as follows. Krein [1] has shown that if all $p(t)$ are nonnegative and periodic with period T $(p(t+T) = p(t) \geq 0)$, and

$$\varepsilon I < T \int_0^T p(t)dt < 4I$$

for some $\varepsilon > 0$, then all solutions of (11) on the real line are bounded; one says that the equation is *stable*.

Now there exist bounded operators $U(t)$ on $H \oplus H$ such that

$$\begin{pmatrix} y(t) \\ \dot{y}(t) \end{pmatrix} = U(t) \begin{pmatrix} y(0) \\ \dot{y}(0) \end{pmatrix} \quad \text{for all solutions } y(t).$$

The periodicity of p implies $U(t+nT) = U(t)U(T)^n$ (Floquet theory) for all integral n, and by the uniform boundedness principle, stability is equivalent to the uniform boundedness of all powers of $U(T)$. But by a general theorem of Nagy [5], the latter is equivalent to $U(T)$ being conjugate to a unitary operator.

It was necessary to generalize results of this type to deal with equations (10) not of Hill type. To describe this we need additional terminology and notation.

Let $\mathcal{H}$ be a real topological linear vector space on which is defined a continuous symplectic form $a(\cdot,\cdot)$, having the property that $\mathcal{H}$ is linearly and topologically isomorphic to a complex Hilbert space, in such a way that $a(\cdot,\cdot)$ is carried into the imaginary part of the inner product. $\mathcal{H}$ is called a *Hilbertizable symplectic space*. We let $Sp(\mathcal{H})$ ($sp(\mathcal{H})$) be the group (resp. Lie algebra) of bounded invertible (resp. bounded) operators on $\mathcal{H}$ preserving (resp. skew with respect to) the form $a(\cdot,\cdot)$. We define the set of (bounded positive) *complex structures*

$$\Gamma = \{J \in Sp(\mathcal{H}) : J^2 = -I, \text{ and } a(J\cdot,\cdot) \text{ is positive definite and defines the topology of } \mathcal{H}\} .$$

Let us disregard the P.D.E. origin of (10) and assume only that $A(t) \in Sp(\mathcal{H})$ for some Hilbertizable symplectic space $\mathcal{H}$. In generalizing Krein's theorem one finds that the nonnegativity condition on p(t) goes over to requiring that the A(t) lie in

$$\overline{C_o} = \{X \in sp(\mathcal{H}) : a(Xv,v) \geq 0 \quad \forall v \in \mathcal{H}\}$$

which (up to sign) is the unique $Sp(\mathcal{H})$-invariant convex cone in $sp(\mathcal{H})$ closed in the ultraweak or strong operator topologies. If $\mathcal{H}$ is separable and these closure conditions are relaxed to "norm-closed," one obtains only those further invariant convex cones trivially obtained from $\overline{C_o}$ by modification by the ideal of compact operators [10]. In finite dimensions this uniqueness is a recent independent observation of Vinberg [16]. Finally, one can show that there are exactly $4(2^n-1)$ invariant convex cones in $sp(n,\mathbb{R})$ with closure $\overline{C_o}$ [8].

$\overline{C_o}$ has an interior

$$C_o = \{X \in sp(\mathcal{H}) : a(Xv,v) \geq k\|v\|^2 \quad \forall v \in \mathcal{H}, \text{ for some } k > 0\}$$

which contains Γ, and in fact given $X \in C_o$ there exists $M > 0$ such that for all $r \geq M$, rX is the midpoint of two elements of Γ [8, section 8]. Each $X \in C_o$ commutes with a unique $J \in \Gamma$ [9], i.e., is "uniquely skew-hermitizable," hence generates a unitarizable one-parameter subgroup. However, it seems necessary for the applications to introduce also a weaker notion of unitarizability.

Definitions. Let $S \in Sp(\mathcal{H})$.

1) S is *unitarizable* provided $SJ = JS$ for a $J \in \Gamma$, and *uniquely unitarizable* if the J is unique.

2) S is *essentially unitarizable* provided there is a dense linear subspace $\mathfrak{D} \subseteq \mathcal{H}$ and pre-complex Hilbert structure $\langle \cdot,\cdot \rangle' + ia(\cdot,\cdot)$ on $\mathfrak{D}$ (with Hilbert completion $\mathcal{H}'$) such that

   i) $S(\mathfrak{D}) \subseteq \mathfrak{D}$

   ii) $\langle Sx,Sy \rangle' = \langle x,y \rangle' \quad \forall x,y \in \mathfrak{D}$

and iii) the injection $\mathfrak{D} \hookrightarrow \mathcal{H}'$ is a closed operator.

S is said to be *uniquely essentially unitarizable* if the $\mathfrak{D}$ and $\langle \cdot,\cdot \rangle'$ satisfying i)-iii) above are unique.

Theorem 1 [11]. 1) If $A \in \overline{C_o}$ then the $(I \pm A)^{-1}$ exist, and

$$S = (I + A)(I - A)^{-1} \in Sp(\mathcal{H}) \quad . \tag{12}$$

2) If the A in (12) is in C_o then S is uniquely unitarizable;

if the A is in $\{X \in \overline{C_0} : a(Xv,v) > 0 \quad \forall v \neq 0\}$, then S is uniquely essentially unitarizable.

The promised generalization of the Krein theorem is

Theorem 2 [8]. Let $t \in \mathbb{R} \to A(t) \in \overline{C_0}$ be a strongly continuous and norm-bounded map such that

$$\int_{-\infty}^{\infty} \|A(t)\| \, dt < 2 \quad .$$

Then the operator form of (10)

$$\frac{d}{dt} S(t) = A(t)S(t) \quad , \quad S(-\infty) = I \tag{13}$$

has a unique solution $S(t)$ $(-\infty \leq t \leq +\infty)$ which is norm-continuous, and all $(S(t)+I)^{-1}$ exist.

a) If $\int_{-\infty}^{\infty} A(t)dt \in C_0$, then $S = S(+\infty)$ is uniquely unitarizable.

b) If only $\bigcap_{t \in \mathbb{R}} \{\ker A(t)\} = \{0\}$, then $Sv = v$ implies $v=0$, and S is uniquely essentially unitarizable.

IV. APPLICATION TO NONLINEAR WAVE EQUATIONS

We return to the situation of Section II, and find conditions on the ϕ satisfying (1) so that Theorem 2 applies to (10). Now $\begin{pmatrix} 0 & -A \\ B & 0 \end{pmatrix} \in \overline{C_0}$ if and only if A,B are nonnegative, so $F' \geq 0$ guarantees $A(t) \in -\overline{C_0}$. Also, since e^{tQ} is unitary,

$$\|A(t)\|_{\mathcal{K}_L} dt \leq \int \left\| \begin{pmatrix} 0 & 0 \\ C^{-1}F'(\phi)C^{-1} & 0 \end{pmatrix} \right\|_{L_2 \oplus L_2} dt$$

$$= \int \|C^{-1}F'(\phi)C^{-1}\|_{L_2} dt \leq \frac{1}{m} \int \|F'(\phi(t,\cdot))\|_{\infty} dt \, .$$

We can now state the

Corollary. Take nonvanishing solutions $\phi, \phi_{\pm}$ as in Theorem 4.4B earlier (which implies in particular that ϕ is uniformly bounded and $\int_{-\infty}^{\infty} \|F'(\phi)\|_{\infty} dt < \infty$), and assume that F' is nonnegative and vanishes only at 0. If $\int_{-\infty}^{\infty} \|F'(\phi(t,\cdot))\|_{\infty} dt < 2m$ and if for some time t,

$$(m^2-\Delta)^{5/4} \phi_-(t,\cdot) \, , (m^2-\Delta)^{3/4} \dot{\phi}_-(t,\cdot) \in L_1(\mathbb{R}^3) \quad , \tag{14}$$

then $(dS)_{\phi_-} : \mathcal{K}_L \to \mathcal{K}_L$ is uniquely essentially unitarizable.

Proof. To show Theorem 2 b) applies, it remains only to show that all $-A(t) = e^{-tQ}\begin{pmatrix} 0 & 0 \\ F'(\phi) & 0 \end{pmatrix} e^{tQ}$ vanish on no nonzero vector $(u,v) \in H_{\frac{1}{2}} \oplus H_{-\frac{1}{2}}$. Now $e^{tQ}\binom{u}{v} = \binom{\eta}{\dot{\eta}}$ for η of finite Lorentz norm satisfying the Klein-Gordon equation, and $F'(\phi)\eta = 0$ is equivalent to $\phi\eta = 0$ since F' vanishes only at 0. The conclusion $\eta \equiv 0$ follows from (9), the asymptotics of $\phi_-(t,0)$ in some Lorentz frame [6, Cor. 2] obtainable from (14), hyperbolicity, and the vanishing of any $\eta \in \mathcal{K}_L$ which vanishes in a forward cone [2].

Theorem 2 may also be applied to the case of the Klein-Gordon equation on a curved space-time: if the metric perturbation is sufficiently small, nonvanishing, and "one-sided" ($A(t) \in \overline{C}_0$ translates into an inequality on the curvature; roughly, it must be appropriately bounded from above), the unique (free-field) quantization with vacuum of the equation, such that the scattering operator is unitary, is determined. Details and some explicit computations are given in [7].

V. FURTHER PROBLEMS

It is hoped that this almost Kähler structure on the stable solution variety will make possible the construction of some further elements of an interacting quantum field theory in physical space. The constructed metric is suggestive of a geodesic random walk on the variety which, as the step size goes to 0, can converge suitably (as in [3], for example) to a weak stochastic process defining an inner product on complex functionals on the variety. It is mathematically conceivable that such a random walk could be conditioned to remain in the stable solution variety, and perhaps such a restriction could be justified physically in certain situations.

Still, one would like to be able to quantize the entire solution variety, and on this question we should merely comment on what seems to be the behavior of the constructed riemannian structure as the boundary of the stable solution variety is approached.

As regards the apparent singularity at $\phi = 0$, it is clear that if 0 is approached along a fixed ray, the hermitian structure approaches a finite limit (which depends on the ray). And secondly, it appears from the computations in the section "Quantization in a Curved Space-Time" in [7] that, if the approach to 0 is made along a path increasingly along rays which represent perturbations less and less singular (perhaps better described in physical language: if the interaction is "turned on" gently over a long period of time), the hermitian structure approaches the usual free-field complex inner product.

However, as a nonzero point in the boundary of the stable solution variety is approached, the complex structure and symmetric form definitely diverge, but, at least in finite-dimensional situations [8, section 28], the polarization itself varies smoothly, and approaches at least partly real polarizations at the boundary.

REFERENCES

1. Ju.L.Daleckii and M.G.Krein, Stability of solutions of differential equations in Banach space, Trans. of Math. Monographs, Vol. **43**, 1971.
2. R.Goodman, "One-sided invariant subspaces and domains of uniqueness for hyperbolic equations," *Proc. Am. Math. Soc.* **15**, pp.653-660, 1964.
3. E.Jørgensen, "The central limit problem for geodesic random walks," *Zeit. für Wahrscheinlichkeitstheorie*, Vol. **32**, pp.1-64, 1975.
4. C.S.Morawetz and W.A.Strauss, I "Decay and scattering of solutions of a nonlinear relativistic wave equation," *Comm. Pure Appl. Math.* Vol. **25**, pp.1-31, 1972; II "On a nonlinear scattering operator," *Comm. Pure Appl. Math.* Vol. **26**, pp.47-54, 1973.
5. Reference 1, p.34.
6. S.Nelson, "L^2 asymptotes for the Klein-Gordon equation," *Proc. Amer. Math. Soc.* Vol. **27**, no. 1, pp.110-116, 1971.
7. S.Paneitz and I.Segal, "Quantization of wave equations and hermitian structures in partial differential varieties," *Proc. Nat. Acad. Sci. USA*, in press.
8. S.Paneitz, "Causal structures in Lie groups and applications to stability of differential equations," Doctoral dissertation, M.I.T. Dept. of Mathematics, June 1980.
9. S.Paneitz, "Unitarization of symplectics and stability for causal differential equations in Hilbert space," *J. Func. Anal.*, in press.
10. S.Paneitz, "Uniqueness of causal structures in sp($\mathcal{H}$)," unpublished, 1980.
11. S.Paneitz, "Essential unitarization of symplectic transformations," unpublished, 1980.
12. I.Segal, "Foundations of the theory of dynamical systems of infinitely many degrees of freedom, I," *Mat.-fys. Medd. Danske Vid. Selsk.* **31**, no. 12, 39 pp., 1959.

13. I.Segal, *J. Math. Phys.* **1**, pp.468.438.

14. I.Segal, "Dispersion for nonlinear relativistic equations" I, *Proc. Conf. Math. Th. Elem. Part.*, M.I.T. Press, pp.79-108, 1966; II, *Ann. Scient. de l'Ecole Norm. Sup.* (**4**)**1**, pp.459-497, 1968.

15. I.Segal, "Functional integration and interacting quantum fields," *Proc. Conf. on Func. Int.*, ed. by A.M.Arthur, 1973.

16. E.Vinberg, "Invariant convex cones and orderings in Lie groups," *Func. Anal. and Its Appl.* Vol. **14**, No. 1, pp.1-13, 1980.

17. M.Weinless, "Existence and uniqueness of the vacuum for linear quantized fields," *J. Func. Anal.* **4**, pp.350-379.

VECTOR BUNDLE CONNECTIONS AND LIFTINGS OF PARTIAL DIFFERENTIAL OPERATORS*

Stig I. Andersson

Institut für Theoretische Physik der TU
D-3392 Clausthal-Zellerfeld, FRG.

0. Introduction

Vector bundle connections are first order partial differential operators (PDO) of a rather special kind. This paper deals with the question of going in the opposite direction, i.e. attaching some kind of connections to a given PDO operating on the sections of a vector bundle.

Such a mutual relationship between connections and PDOs on vector bundles would be of considerable interest. Indeed, one would thereby expect to tie together intrinsic geometric characteristics of the PDO and the differential topology/geometry of the given vector bundle. Via curvature and torsion forms as well as the approach to characteristic classes which goes along with these concepts (Bott, Chern, Weil), one would be in a position to explore the interplay between operator properties and vector bundle geometry.

The results on propagation of singularities for scalar PDOs of fairly general type, which has accumulated in recent years, do offer an opportunity for constructing such a mutual relationship. Indeed, the transport equations along bicharacteristic bands essentially describe a connection. More generally though, for PDOs on vector bundles (i.e. systems of PDOs) this possibility is not available. This fact, combined with a feeling that sheaf theory is the natural frame for this question, led us to pursue an approach due to Akira Asada [As, 1978]. We extend and make this paper more precise in various regards. The results easily lend themselves for applications to liftings of PDOs to spin bundles and for the construction of complex (Kähler) structures

* Sponsored by the Swedish Natural Science Research Council (NFR), contracts F-3898-101 and R-RA 3898-104.

associated with the PDO under consideration.

The work reported on in this paper was begun in 1979/80 while visiting Dept. of Mathematics, Cal Tech, Pasadena and it was completed at the Institut für Theoretische Physik der TU Clausthal. We express our gratitude to both these institutions.

1. (Asada) Connections for PDOs on Vector Bundles. Basic Properties

In what follows, we basically use the notational conventions from [We, 1980]. So given $E \xrightarrow{\pi_E} X$, m-dimensional (complex) vector bundle over the n-dimensional, paracompact manifold X, we denote by $\mathcal{E}(X, E) = \mathcal{E}(E)$ the C^∞-sections of this vector bundle.

To describe the PDOs (or more generally pseudodifferential operators (Psdo)) acting on $\mathcal{E}(E)$, we need the following *local description of* E; let $(U_\alpha, \varphi_\alpha)_{\alpha \in I}$ be a local trivialization for E

$$\begin{array}{ccc} \pi_E^{-1}(U_\alpha) \equiv E\restriction U_\alpha & \xrightarrow{\varphi_\alpha} & \tilde{U}_\alpha \times \mathbb{C}^m \\ \downarrow \pi_E & & \downarrow \\ U_\alpha & \xrightarrow{\bar{\varphi}_\alpha} & \tilde{U}_\alpha \subset \mathbb{R}^n \end{array}$$

i.e. $\varphi_\alpha(z) = (\bar{\varphi}_\alpha(\pi_E(z)), \Phi_\alpha(z))$ (vector bundle homeomorphism) where $\Phi_\alpha := \text{proj} \circ \varphi_\alpha$ and $\Phi_\alpha^x := \text{proj} \circ \Phi_\alpha \restriction \pi_E^{-1}(x) : \pi_E^{-1}(x) \to \{x\} \times \mathbb{C}^m \to \mathbb{C}^m$ is a toplinear isomorphism. Define $\varphi_\alpha^* : \mathcal{E}(U_\alpha, E\restriction U_\alpha) \longrightarrow [\mathcal{E}(\tilde{U}_\alpha)]^m$ by

$$\mathbb{R}^n \supset \tilde{U}_\alpha \xrightarrow{(\bar{\varphi}_\alpha)^{-1}} U_\alpha \xrightarrow{f_\alpha} E\restriction U_\alpha \xrightarrow{\Phi_\alpha} \mathbb{C}^m$$

where $f_\alpha \in \mathcal{E}(U_\alpha, E\restriction U_\alpha)$ so that $f_\alpha^* := \varphi_\alpha^*(f_\alpha) = \Phi_\alpha \circ f_\alpha \circ (\bar{\varphi}_\alpha)^{-1}$. Let now $E \xrightarrow{\pi_E} X$, $F \xrightarrow{\pi_F} X$ be two vector bundles, F of dimension k and $(U_\alpha, \varphi_\alpha)_{\alpha \in I}$ a common local trivialization.

A PDO $\mathcal{L} : \mathcal{E}(E) \longrightarrow \mathcal{E}(F)$ has the following local description: let $\mathcal{L}_\alpha := \mathcal{L} \restriction \mathcal{E}(U_\alpha, E\restriction U_\alpha)$ so that (utilizing locality!)

$$\mathcal{L}_\alpha : \mathcal{E}(U_\alpha, E\restriction U_\alpha) \longrightarrow \mathcal{E}(U_\alpha, F\restriction U_\alpha)$$

The local corresponding expression $\tilde{\mathcal{L}}_\alpha$ is given by:

$$\tilde{\mathcal{L}}_\alpha : [\mathcal{E}(\tilde{U}_\alpha)] \xrightarrow{(\varphi_\alpha^*)^{-1}} \mathcal{E}(U_\alpha, E \upharpoonright U_\alpha) \xrightarrow{\mathcal{L}_\alpha} \mathcal{E}(U_\alpha, F \upharpoonright U_\alpha) \xrightarrow{\varphi_\alpha^*} [\mathcal{E}(\tilde{U}_\alpha)]^k$$

i.e. $\tilde{\mathcal{L}}_\alpha(f_\alpha^*) = (\mathcal{L}_\alpha f_\alpha)^*$ or, still more explicit:

$$\mathcal{L}_\alpha(\Phi_\alpha \circ f_\alpha \circ (\bar{\varphi}_\alpha)^{-1}) = \Phi_\alpha \circ \mathcal{L}_\alpha f_\alpha \circ (\bar{\varphi}_\alpha)^{-1}.$$

Let $(U_\alpha, \varphi_\alpha)$, (U_β, φ_β) be two locally trivializing charts with $U_{\alpha\beta} := U_\alpha \cap U_\beta \neq \emptyset$ and let $e_{\alpha\beta}$, $f_{\alpha\beta}$ be the transition functions of E and F respectively. Thus for local sections

$$e_\alpha^* \in [\mathcal{E}(\tilde{U}_\alpha)]^m, \quad e_\alpha^* = e_{\alpha\beta}\, e_\beta^*.$$

For $e_\alpha^* \in [\mathcal{E}(\tilde{U}_\alpha)]^m$, $e_\beta^* \in [\mathcal{E}(\tilde{U}_\beta)]^m$ we have $\tilde{\mathcal{L}}_\alpha(e_\alpha^*) \in [\mathcal{E}(\tilde{U}_\alpha)]^k$ and $\tilde{\mathcal{L}}_\beta(e_\beta^*) \in [\mathcal{E}(U_\beta)]^k$ i.e. on $U_{\alpha\beta} \neq \emptyset$ $\tilde{\mathcal{L}}_\alpha(e_\alpha^*) = f_{\alpha\beta}\, \tilde{\mathcal{L}}_\beta(e_\beta^*)$ i.e. on local E-sections, we have the *compatibility condition*

$$\tilde{\mathcal{L}}_\alpha\, e_{\alpha\beta} = f_{\alpha\beta}\, \tilde{\mathcal{L}}_\beta, \quad \text{on } U_{\alpha\beta} \neq \emptyset.$$

We denote by $\mathrm{Diff}_s(E, F)$ the (linear) PDOs $\mathcal{L} : \mathcal{E}(E) \longrightarrow \mathcal{E}(F)$ of order s, i.e. $\mathcal{L} \in \mathrm{Diff}_s(E, F)$ means that for any choice of local coordinates and local trivialization we have

$$\tilde{\mathcal{L}}_\alpha(e_\alpha^*)_i = \sum_{\substack{j=1 \\ |\gamma| \le s}}^{m} a_\gamma^{(ij)}\, D^\gamma (e_\alpha^*)_j, \quad i = 1, \ldots, k.$$

$$e_\alpha^* = ((e_\alpha^*)_1, \ldots, (e_\alpha^*)_m) \in [\mathcal{E}(\tilde{U}_\alpha)]^m, \quad \gamma = (\gamma_1, \ldots, \gamma_n) \in \mathbb{Z}_+^n$$

(multi-index) with the usual multi-index convention for D^γ. By $\sigma_s(\mathcal{L}) : \pi^*(E) \longrightarrow \pi^*(F)$ we denote the s-symbol of $\mathcal{L}$ ($\pi^*(E)$ = induced bundle of E over $T^*(X)$).

<u>Example</u> $E = \Lambda^p T^*(X)$, $F = \Lambda^{p+1} T^*(X)$ and let $\mathcal{E}^p(X) = C^\infty(X, \Lambda^p T^*(X))$ (C^∞ p-forms on X). Then exterior differentiation $d : \mathcal{E}^p(X) \longrightarrow \mathcal{E}^{p+1}(X)$ is an element in $\mathrm{Diff}_1(E, F)$ with 1-symbol $\sigma_1(d)(x,\xi) : \Lambda^p T_x^*(X) \longrightarrow \Lambda^{p+1} T_x^*(X)$ given by $\sigma_1(d)(x,\xi) f = \xi \wedge f$, $(x,\xi) \in T_x^*(X)$.

Now a (vector bundle) connection ∇_E on $E \xrightarrow{\pi_E} X$ is a linear map

$$\nabla_E : \mathcal{E}(X, E) \longrightarrow \mathcal{E}^1(X, E)$$

$$(\mathcal{E}^p(X,E) = \mathcal{E}^p(E) = C^\infty(X, \wedge^p T^*(X) \otimes E) \text{ , E-valued } C^\infty \text{ p-forms})$$

such that

$$\nabla_E(f\,\phi) = df \otimes \phi + f\,\nabla_E(\phi) \;,\quad f \in C^\infty(X) \;,\quad \phi \in \mathcal{E}(E)$$

and with symbol mapping $E_x \longrightarrow T_x^*(X) \otimes E_x$, $E_x = \pi^{-1}(x)$ by

$$\sigma_1(\nabla_E)\,(x, \xi)\, f \;=\; \xi \otimes f \,.$$

So in particular $\nabla_E \in \mathrm{Diff}_1(E, T^*(X) \otimes E)$. Similarly, for the naturally induced map

$$\nabla_E^{(p)} \;:\; \mathcal{E}^p(E) \longrightarrow \mathcal{E}^{p+1}(E) \,,$$

the symbol maps

$$\wedge^p T_x^*(X) \otimes E_x \longrightarrow \wedge^{p+1} T_x^*(X) \otimes E_x$$

by

$$\sigma_1(\nabla_E^{(p)})\,(x, \zeta)\,(\eta \otimes f) \;=\; (\zeta \wedge \eta) \otimes f \,.$$

So in this case the symbol satisfies

$$\sigma_1(\nabla_E^{(p)}) \;=\; \sigma_1(d) \otimes \mathrm{Id}_E \tag{1}$$

where Id_E = identity map in $\pi^*(E)$.

Now, the definition of $\nabla_E^{(p)}$ can be rephrased by saying that it is a lift of the differential operator $d : \mathcal{E}^p(X) \longrightarrow \mathcal{E}^{p+1}(X)$ to an operator $\mathcal{E}^p(E) \longrightarrow \mathcal{E}^{p+1}(E)$ such that the symbols fulfil (1). More generally, we could start from an arbitrary $P : \mathcal{E}^p(X) \longrightarrow \mathcal{E}^{p+1}(X)$ instead of d or, still more generally, instead of the special bundles $\wedge^p T^*(X)$ we might start with an arbitrary $P \in \mathrm{Diff}_s(E_1, E_2)$, for vector bundles E_1, E_2 over X and consider a lift $\tilde{P} \in \mathrm{Diff}_s(E_1 \otimes F, E_2 \otimes F)$ such that

$$\sigma_s(\tilde{P}) \;=\; \sigma_s(P) \otimes \mathrm{Id}_F \,. \tag{2}$$

It is tempting to call the lower order part of $\tilde{P}$ a connection of P with respect to F or, shorter, an F-*connection of* P.

This is the basic idea in the Asada approach; let $P \in \mathrm{Diff}_s(E_1, E_2)$ i.e.

$P = \{P_\alpha\}$, P_α local PDO $E_1 \restriction U_\alpha \longrightarrow E_2 \restriction U_\alpha$ and

$$P_\alpha \, e^1_{\alpha\beta} = e^2_{\alpha\beta} \, P_\beta \tag{3}$$

($e^i_{\alpha\beta}$ = transition functions in E_i, $i = 1,2$). The extension $\tilde{P} = \{\tilde{P}_\alpha\}$ is defined locally, by suitably defining $\tilde{P}_\alpha : \mathcal{E}(U_\alpha, (E_1 \otimes F)\restriction U_\alpha) \longrightarrow \mathcal{E}(U_\alpha, (E_2 \otimes F)\restriction U_\alpha)$ so as to have (2). This however, in general, means that we have violated the compatibility condition (3), i.e. in general $\{\tilde{P}_\alpha\}$ don't patch together, so $\tilde{P} \notin \mathrm{Diff}_s(E_1 \otimes F, E_2 \otimes F)$. The connections are then introduced as a global measure of the obstruction to the compatibility condition (3) for the $\{\tilde{P}_\alpha\}$.

This situation, having local objects (the perturbed local differential operators $\tilde{P}_\alpha$) and asking when this local information patch together to yield global data, is exactly what sheaf theory is designed to handle. More precisely, given $P = \{P_k\} \in \mathrm{Diff}_s(E_1, E_2)$ where P_k acts on $\mathcal{E}(U_k, E_1 \restriction U_k)$ and is given by

$$P_k := \sum_{|\gamma| \le s} a_{\gamma,k} \, D^\gamma(k) , \quad \gamma = (\gamma_1, \dots, \gamma_n) \in \mathbb{Z}^n_+$$

$D_j(k) = \frac{1}{i} \frac{\partial}{\partial x_j(k)}$, $x(k) = (x_1(k), \dots, x_n(k))$ local coordinates in U_k and $D^\gamma(k) = D_1^{\gamma_1}(k) \dots D_n^{\gamma_n}(k)$ with the usual multi-index conventions. We define the local extensions by

$$\tilde{P}_k := \sum_{|\gamma| \le s} (a_{\gamma,k} \otimes I_F) \, D^\gamma(k) \tag{4}$$

(I_F = identity map on the fibres of F).

The symbol relation (2) holds, since for $g \in C^\infty(U_k)$, $v = dg = \sum_{l=1}^n \xi_l \, dx^l$, $\alpha \in \mathcal{E}(U_k, E_1)$, $\beta \in \mathcal{E}(U_k, F)$ such that $\alpha(x) = e \in \pi^{-1}_{E_1}(x)$ and $\beta(x) = f \in \pi^{-1}_F(x)$ we get:

$$\sigma_s(\tilde{P}_k)(x,v)(e \otimes f) = \sum_{|\gamma| \le s} (a_{\gamma,k}(y) \otimes I_F) \, D^\gamma(k) \left(\frac{1}{s!} (g(y)-g(x))^s \, (\alpha \otimes \beta)(y) \right) \Big|_{y=x} =$$

$$= \sum_{|\gamma| = s} (a_{\gamma,k}(x) \otimes I_F) \, \xi^\gamma \, (e \otimes f) = (\sigma_s(P_k) \otimes Id_F)(e \otimes f) .$$

Now, in case F is trivial, $F \cong X \times \mathbb{F}$ ($\mathbb{F}$ could here and furtheron be infinite dimensional)

we will have $\tilde{P}_i (e^1_{ij} \otimes f_{ij}) = (e^2_{ij} \otimes f_{ij}) \tilde{P}_j$, ($e^1_{ij} \otimes f_{ij}$ = transition function for $E_1 \otimes F$ etc.) so that in this case by a partition of unity the $\{\tilde{P}_i\}$ patch together and yield a unique extension $\tilde{P} \in \mathrm{Diff}_s(E_1 \otimes F, E_2 \otimes F)$. More generally we will have

$$(e^2_{ij} \otimes f_{ij})\, \tilde{P}_j - \tilde{P}_i\, (e^1_{ij} \otimes f_{ij}) = -\sum_{|\gamma| = s-1} \left[\sum_{l=1}^{n} (e^2_{ij}\, a_{\gamma+1_l, j} \otimes \frac{1}{i} \frac{\partial f_{ij}}{\partial x_l(j)}) \right] D^{\gamma}(j)$$

modulo terms of lower order

where $\gamma + 1_l = (\gamma_1, \ldots, \gamma_{l-1}, \gamma_l + 1, \gamma_{l+1}, \ldots, \gamma_n)$.

So, in general, if we by a partition of unity try and patch these local objects together, we will end up with a non-unique $\tilde{P}$, *the highest order part of which will be unique though, i.e. these non-equal extensions all have the same index!* These obstructions to uniqueness which enters via lower order terms are thus the objects to be measured in terms of generalized curvature (connections).

Def. 1: A local differential operator $\{\theta_i\}$, $\theta_i : \mathcal{E}(U_i, (E_1 \otimes F) \restriction U_i) \longrightarrow \mathcal{E}(U_i, (E_2 \otimes F) \restriction U_i)$ is an *F-connection of* P if

a) $\tilde{P}_\theta := \tilde{P} + \theta \equiv \{\tilde{P}_i + \theta_i\} \in \mathrm{Diff}_s(E_1 \otimes F, E_2 \otimes F)$ i.e.

$$(e^2_{ij} \otimes f_{ij})\,(\tilde{P}_j + \theta_j) = (\tilde{P}_i + \theta_i)\,(e^1_{ij} \otimes f_{ij}) \tag{5}$$

b) degree $\theta_i \leq s - 1$

Remark: $\{\theta_i\}$ measure the obstructions to $\{\tilde{P}_i\} \in \mathrm{Diff}_s(E_1 \otimes F, E_2 \otimes F)$ and trivially the local obstructions patch together so that to any $P \in \mathrm{Diff}_s(E_1, E_2)$ and any vector bundle F there exists an F-connection of P.

Before studying the properties of these connections closer, let us remark that there is a natural concept of *equivalence of F-connections of* P.

Let $h = \{h_i\}$ be a vector bundle automorphism in F. The transformed bundle F^1 under h being isomorphic to F, transition functions are cohomologous $f^1_{ij} = h_i f_{ij} h_j^{-1}$ are the transition functions for F^1. Similarly we have $(\tilde{P}_j)^* = (I_{E_2} \otimes h_j)\, \tilde{P}_j\, (I_{E_1} \otimes h_j^{-1})$ for the lifted operator mapping $\mathcal{E}(E_1 \otimes F^1) \longrightarrow \mathcal{E}(E_2 \otimes F^1)$. Writing

$G^{1,2}_{ij} = e^{1,2}_{ij} \otimes f_{ij}$ and $H^{1,2}_{ij} = e^{1,2}_{ij} \otimes f^1_{ij} = (I_{E_{1,2}} \otimes h_i)\, G^{1,2}_{ij}\, (I_{E_{1,2}} \otimes h_j^{-1})$ and

changing F to F^1 by the vector bundle automorphism $h = \{h_i\}$, (5) takes the form

$$H^2_{ij}\, (\tilde{P}_j)^* - (\tilde{P}_i)^*\, H^1_{ij} = \theta^*_i\, H^1_{ij} - H^2_{ij}\, \theta^*_j \; . \tag{5'}$$

The right-hand side could be written $(I_{E_2} \otimes h_i) \{\Theta_i G^1_{ij} - G^2_{ij} \Theta_j\} (I_{E_1} \otimes h_j^{-1})$ and $(\tilde{P}_j)^*$ can be made explicit as follows:
let

$$P = \sum_{|\gamma| \leq s} a_\gamma(x) D^\gamma$$

be a (locally defined) PDO and h a local bundle automorphism. Separating the highest order term of the PDO we could write

$$Phf = \sum_{|\gamma| \leq s} a_\gamma h \; D^\gamma f + \sum_{|\gamma| \leq s} \sum_{|\alpha| > 0} a_\gamma (D^\alpha h) \, D^{\gamma - \alpha} f$$

(modulo combinatorial factors) for any section f, so we write $Ph = {}^hP + P^h$ where

$${}^hP := \sum_{|\gamma| \leq s} a_\gamma h \; D^\gamma \; ; \; P^h := \sum_{|\gamma| \leq s} \sum_{|\alpha| > 0} a_\gamma \, (D^\alpha h) \; D^{\gamma - \alpha} \text{ mod combinatorial factors}$$

Furthermore, if g is another local bundle automorphism, we have

$${}^hPgf = \sum_{|\gamma| \leq s} a_\gamma \, hg \, D^\gamma f + \sum_{|\gamma| \leq s} \sum_{|\alpha| > 0} a_\gamma \, h \, (D^\alpha g) \, D^{\gamma-\alpha} f$$

i.e. ${}^hPg = {}^{hg}P + {}^hP^g$. Consequently $P = Pg^{-1}g = ({}^{g^{-1}}P + P^{g^{-1}}) \, g = {}^{g^{-1}}Pg + P^{g^{-1}} g =$
$= P + {}^{g^{-1}}P^g + P^{g^{-1}} g$ so that finally

$$P^{g^{-1}} = - \, {}^{g^{-1}}P^g \; g^{-1} \tag{6}$$

Applying this to $(\tilde{P}_j)^*$ we obtain $(\tilde{P}_j)^* = (I_{E_2} \otimes h_j) \, \tilde{P}_j (I_{E_1} \otimes h_j^{-1}) =$

$$= (I_{E_2} \otimes h_j) \left\{ {}^{(I_{E_1} \otimes h_j^{-1})} \tilde{P}_j + \tilde{P}_j^{(I_{E_1} \otimes h_j^{-1})} \right\} =$$

$$= (I_{E_2} \otimes h_j) \left\{ {}^{(I_{E_1} \otimes h_j^{-1})} \tilde{P}_j - {}^{(I_{E_1} \otimes h_j^{-1})} \tilde{P}_j^{(I_{E_1} \otimes h_j)} (I_{E_1} \otimes h_j^{-1}) \right\}$$

using (6) .

From (4) we get furthermore

$${}^{(I_{E_1} \otimes h_j^{-1})} \tilde{P}_j = \sum_{|\gamma| \leq s} (a_{\gamma,j} \otimes I_F) (I_{E_1} \otimes h_j^{-1}) \, D^\gamma(j) = (I_{E_2} \otimes h_j^{-1}) \, \tilde{P}_j$$

so that finally $(\tilde{P}_j)^* = \tilde{P}_j - \tilde{P}_j^{(I_{E_1} \otimes h_j)} (I_{E_1} \otimes h_j^{-1})$.

Inserting the above into (5') and rearranging

$$H^2_{ij}\{\tilde{P}_j - \tilde{P}_j^{(I_{E_1}\otimes h_j)}(I_{E_1}\otimes h_j^{-1})\} - \{\tilde{P}_i - \tilde{P}_i^{(I_{E_1}\otimes h_i)}(I_{E_1}\otimes h_i^{-1})\}\, H^1_{ij} =$$

$$= (I_{E_2}\otimes h_i)\left\{\theta_i\, G^1_{ij} - G^2_{ij}\,\theta_j\right\}(I_{E_1}\otimes h_j^{-1}),$$

we obtain

$$H^2_{ij}\,\tilde{P}_j - \tilde{P}_i\, H^1_{ij} = H^2_{ij}\,\tilde{P}_j^{(I_{E_1}\otimes h_j)}(I_{E_1}\otimes h_j^{-1}) - (I_{E_2}\otimes h_i)\, G^2_{ij}\,\theta_j\,(I_{E_1}\otimes h_j^{-1})$$

$$- \tilde{P}_i^{(I_{E_1}\otimes h_i)}(I_{E_1}\otimes h_i^{-1})\, H^1_{ij} + (I_{E_2}\otimes h_i)\,\theta_i\, G^1_{ij}\,(I_{E_1}\otimes h_j^{-1})$$

so that the rôle of connection is now played by the quantity θ^1_j given by

$$- H^2_{ij}\,\theta^1_j = H^2_{ij}\,\tilde{P}_j^{(I_{E_1}\otimes h_j)}(I_{E_1}\otimes h_j^{-1}) - (I_{E_2}\otimes h_i)\, G^2_{ij}\,\theta_j\,(I_{E_1}\otimes h_j^{-1})$$

i.e.

$$\theta^1_j = (I_{E_2}\otimes h_j)\,\theta_j\,(I_{E_1}\otimes h_j^{-1}) - \tilde{P}_j^{(I_{E_1}\otimes h_j)}(I_{E_1}\otimes h_j^{-1}) \qquad (7)$$

<u>Def. 2</u>: Two F-connections $\theta^1 = \{\theta^1_i\}$ and $\theta^2 = \{\theta^2_i\}$ are equivalent if for a common locally trivializing $\{U_i\}$ we can find a vector bundle automorphism $h = \{h_i\}$ such that (7) holds.

Formally, the orbit space looks the same as for an ordinary vector bundle connection. Also, the set of F-connections of P could be topologized in a fairly natural way which will not be used here though.

This concept of equivalence will be used when we now pass over to a description of connections in terms of obstruction classes. This description will include non-abelian cohomology as the main difference to the standard theory.

2. Connections and Obstruction Classes

As was made clear in the Remark following Def. 1, the existence of connections (i.e. operators of order $\leq$ s-1) puts no restrictions either on P or on the geometry. Requiring connections of order s - j, j $\geq$ 2, will however mean non-trivial restrictions which are best expressed in cohomology terms.

Rewriting (5) we have

$$\theta_i - G^2_{ij}\ \theta_j\ G^1_{ji} = - \tilde{P}_i + G^2_{ij}\ \tilde{P}_j\ G^2_{ji}\ ,$$

where the right side is easily computable. Denoting by $\lambda_{ij} = \theta_i - G^2_{ij}\ \theta_j\ G^1_{ji}$ we have $\lambda_{ji} = - G^2_{ji}\ \lambda_{ij}\ G^1_{ij}$ and by definition the algebraic 1-cocycle relation

$$\lambda_{ij} + G^2_{ij}\ \lambda_{jk}\ G^1_{ji} + G^2_{ik}\ \lambda_{ki}\ G^1_{ki} = 0 \tag{8}$$

To bring in cohomology properly, we compute λ_{ji} as follows:
let

$$\theta_i = \sum_{|\gamma|=s-1} \theta_{\gamma,i}\ (x)\ D^{\gamma}(i) + \text{lower order}$$

and assume $\theta = \{\theta_i\}$ and $\theta' = \{\theta'_i\}$ are equivalent. By (7) we get

$$\theta'_j = (I_{E_2} \otimes h_j)\ \theta_j\ (I_{E_1} \otimes h_j^{-1}) - \tilde{P}_j^{\,(I_{E_1} \otimes h_j)}\ (I_{E_1} \otimes h_j^{-1})$$

for a vector bundle automorphism $h = \{h_i\}$ in F. On the symbol level this means

$$\theta'_{\gamma,j} = \theta_{\gamma,j} - \begin{matrix}\text{principal part}\\ \text{of symbol of}\end{matrix} \Big\{ \sum_{|\gamma| \leq s}\ \sum_{|\alpha| > 0} (a_{\gamma,j} \otimes I_F)(D^{\alpha}(I_{E_1} \otimes h_j))D^{\gamma-\alpha}(I_{E_1} \otimes h_j^{-1}) \Big\}$$

$$= \theta_{\gamma,j} - \sum_{l=1}^{n} (a_{\gamma+1_l,j} \otimes I_F)\ (I_{E_1} \otimes \frac{1}{i} \frac{\partial h_j}{\partial x_j(i)})\ (I_{E_1} \otimes h_j^{-1}) \qquad |\gamma| = s - 1$$

and

$$\gamma + 1_l = (\gamma_1, \ldots, \gamma_{l-1}, \gamma_l + 1, \gamma_{l+1}, \ldots, \gamma_n)\ .$$

Thus we get ($|\gamma| = s - 1$)

$$\theta'_{\gamma,i} = \theta_{\gamma,i} - \sum_{j=1}^{n} (a_{\gamma+1_j,i} \otimes \frac{1}{i} \frac{\partial h_i}{\partial x_j(i)}\ h_i^{-1}) \tag{9}$$

Furthermore

$$\lambda_{ij} = \theta_i - G^2_{ij}\ \theta_j\ G^1_{ji} = - \tilde{P}_i + G^2_{ij}\ \tilde{P}_j\ G^1_{ji}$$

which we compute by writing

$$\tilde{P}_i = \sum_{|\gamma|=s-1} \Big\{ \sum_{l=1}^{n} (a_{\gamma+1_l,i} \otimes I_F)\ D^{\gamma}(i)\ \frac{1}{i}\ \frac{\partial}{\partial x_l(i)} \Big\} + \text{lower order}\ .$$

First,

$$P = \{P_i\} \text{ is a PDO so } 0 = (e^2_{ij} P_j - P_i e^1_{ij})\, h =$$

$$= \sum_{|\gamma|=s-1} \left\{ \sum_{l=1}^{n} \left[e^2_{ij}\, a_{\gamma+1_l,j}\, D^\gamma(j) \frac{1}{i} \frac{\partial h}{\partial x_l(j)} - a_{\gamma+1_l,i} \left(\frac{\partial x(j)}{\partial x(i)}\right)^{\gamma+1_l} D^\gamma(j) \frac{1}{i} \frac{\partial}{\partial x_l(j)} (e^1_{ij} h) \right] \right.$$

$$\left. + e^2_{ij}\, a_{\gamma,j}\, D^\gamma(j) h - a_{\gamma,i} \left(\frac{\partial x(j)}{\partial x(i)}\right)^{\gamma} D^\gamma(j)\, (e^1_{ij}\, h) \right\} + \text{lower order.}$$

This gives ($|\gamma| = s - 1$)

$$\sum_{l=1}^{n} e^2_{ij}\, a_{\gamma+1_l,j} = \sum_{l=1}^{n} a_{\gamma+1_l,i} \left(\frac{\partial x(j)}{\partial x(i)}\right)^{\gamma+1_l} e^1_{ij} \tag{10}$$

$$-\sum_{l=1}^{n} a_{\gamma+1_l,i} \left(\frac{\partial x(j)}{\partial x(i)}\right)^{\gamma+1_l} \frac{1}{i} \frac{\partial e^1_{ij}}{\partial x_l(j)} + e^2_{ij}\, a_{\gamma,j} - a_{\gamma,i} \left(\frac{\partial x(j)}{\partial x(i)}\right)^{\gamma} e^1_{ij} = 0 \tag{11}$$

Using (10), (11) we now compute for $h = (h_1 \otimes h_2) \in \mathcal{E}(E_1) \otimes \mathcal{E}(F)$

$$\lambda_{ij}(h_1 \otimes h_2) = - \sum_{|\gamma|=s-1} \left\{ \sum_{l=1}^{n} (a_{\gamma+1_l,i} \otimes I_F)\, D^\gamma(i) \frac{1}{i} \frac{\partial}{\partial x_l(i)} \right\} (h_1 \otimes h_2) +$$

$$+ G^2_{ij} \sum_{|\gamma|=s-1} \left\{ \sum_{l=1}^{n} (a_{\gamma+1_l,j} \otimes I_F)\, D^\gamma(j) \frac{1}{i} \frac{\partial}{\partial x_l(j)} \right\} G^1_{ji}\, (h_1 \otimes h_2) -$$

$$- (a_{\gamma,i} \otimes I_F)\, D^\gamma(i)\, (h_1 \otimes h_2) + \sum_{|\gamma|=s-1} G^2_{ij}\, (a_{\gamma,j} \otimes I_F)\, D^\gamma(j)\, G^1_{ji}\, (h_1 \otimes h_2)$$

$$+ \text{ lower order}$$

To highest orders

$$D^\gamma(j) \frac{1}{i} \frac{\partial}{\partial x_l(j)} G^1_{ji}\, (h_1 \otimes h_2) = \left(G^1_{ji}\, D^\gamma(j) \frac{1}{i} \frac{\partial}{\partial x_l(j)} + \frac{1}{i} \frac{\partial G^1_{ji}}{\partial x_l(j)} D^\gamma(j)\right)(h_1 \otimes h_2)$$

$$= \left(G^1_{ji}\, D^\gamma(j) \frac{1}{i} \frac{\partial}{\partial x_l(j)} + \frac{1}{i} \left(\frac{\partial e^1_{ji}}{\partial x_l(j)} \otimes f_{ji}\right) D^\gamma(j) + \frac{1}{i} \left(e^1_{ji} \otimes \frac{\partial f_{ji}}{\partial x_l(j)}\right) D^\gamma(j)\right) (h_1 \otimes h_2)$$

and

$$D^\gamma(j)\, G^1_{ji}\, (h_1 \otimes h_2) = G^1_{ji}\, D^\gamma(j)\, (h_1 \otimes h_2)\,.$$

Inserting these expressions and using (11) we obtain:

$$\lambda_{ij} = \sum_{|\gamma|=s-1} \sum_{l=1}^{n} G_{ij}^{2} (a_{\gamma+1_l,j} \otimes I_F) (e_{ji}^{1} \otimes \frac{1}{i} \frac{\partial f_{ji}}{\partial x_l(i)}) (\frac{\partial x(i)}{\partial x(j)})^{\gamma+1_l} D^{\gamma}(i)$$

$$+ \text{ lower order}$$

We have

$$\theta_i = \sum_{|\gamma|=s-1} \theta_{\gamma,i} D^{\gamma}(i) + \text{lower order},$$

thus

$$\lambda_{ij} = \theta_i - G_{ij}^{2} \theta_j G_{ji}^{1} = \sum_{|\gamma|=s-1} \{ \theta_{\gamma,i} - G_{ij}^{2} \theta_{\gamma,j} (\frac{\partial x(i)}{\partial x(j)})^{\gamma} G_{ji}^{1} \} D^{\gamma}(i)$$

$$+ \text{ lower order },$$

yielding

$$\theta_{\gamma,i} - G_{ij}^{2} \theta_{\gamma,j} (\frac{\partial x(i)}{\partial x(j)})^{\gamma} G_{ji}^{1} = \sum_{l=1}^{n} (a_{\gamma+1_l,i} \otimes f_{ij} \frac{1}{i} \frac{\partial f_{ji}}{\partial x_l(i)}) \qquad (12)$$

for $|\gamma| = s - 1$, where we have again used (10) on the right hand side. We thus have

$$\delta(\lambda_{ij}) = (I_{E_1} \otimes f_{ij}) \sum_{|\gamma|=s-1} \sum_{l=1}^{n} (a_{\gamma+1_l,i} \xi^{\gamma}(i) \otimes \frac{1}{i} \frac{\partial f_{ji}}{\partial x_l(i)}) := (I_{E_1} \otimes f_{ij}) A_i(f_{ji})$$

where

$$A_i(h) = \sum_{|\gamma|=s-1} \sum_{l=1}^{n} (a_{\gamma+1_l,i} \xi^{\gamma}(i) \otimes \frac{1}{i} \frac{\partial h}{\partial x_l(i)})$$

independent of the choice of connection. Here $(\xi_1(i),\ldots,\xi_n(i))$ is dual to $\left\{\frac{1}{i}\frac{\partial}{\partial x_l(i)}\right\}_{l=1}^{n}$.

<u>Lemma</u> $\delta(\lambda_{ij})$ is independent of the choice of connection and

$$T^*(U_{ij}) \ni (x,\xi) \longrightarrow \delta(\lambda_{ij})(x,\xi) \in \mathrm{Hom}_x (E_1 \otimes F, E_2 \otimes F)$$

is a 1-cochain with values in $\pi^*(\mathrm{Hom}(E_1 \otimes F, E_2 \otimes F))$ (= the induced bundle of $\mathrm{Hom}(E_1 \otimes F, E_2 \otimes F)$ over $T^*(X)$) for $U_{ij} \equiv U_i \cap U_j \neq \emptyset$.

In analogy to (8) we have the 1-cocycle relation

$$\delta(\lambda_{ij}) + G^2_{ij}\,\delta(\lambda_{jk})\,G^1_{ji} + G^2_{ik}\,\delta(\lambda_{ki})\,G^1_{ki} = 0 \qquad (8')$$

By (9) we have for equivalent connections θ_i and θ'_i,

$$\delta(\theta'_i) = \delta(\theta_i) - \sum_{|\gamma|=s-1}\sum_{l=1}^{n}\left(a_{\gamma+1_l,i}\,\xi^{\gamma}(i)\otimes\frac{1}{i}\frac{\partial h_i}{\partial x_l(i)}\,h_i^{-1}\right) =$$

$$= \delta(\theta_i) - A_i(h_i)\,(I_{E_1}\otimes h_i^{-1})\,. \qquad (13)$$

Expressing the equivalence of $\{\theta_i\}$ and $\{\theta'_i\}$ in forms of the 1-cocycles $\delta(\lambda_{ij})$ we obtain on $T^*(U_{ij})$;

$$\delta(\lambda'_{ij}) - \delta(\lambda_{ij}) = \{\delta(\theta'_i) - G^2_{ij}\,\delta(\theta'_j)\,G^1_{ji}\} - \{\delta(\theta_i) - G^2_{ij}\,\delta(\theta_j)\,G^1_{ji}\}$$

$$= \{\delta(\theta'_i) - \delta(\theta_i)\} - G^2_{ij}\,(\delta(\theta'_j) - \delta(\theta_j))\,G^1_{ji} = \text{(by 13)}$$

$$= -A_i(h_i)\,(I_{E_1}\otimes h_i^{-1}) + G^2_{ij}\,A_j(h_j)\,(I_{E_1}\otimes h_j^{-1})\,G^1_{ji}\,. \qquad (14)$$

In particular, $\delta(\theta)$ is given as an invariantly defined function on the cotangent bundle if and only if for suitable vector bundle automorphism $h = \{h_i\}$ we have

$$\delta(\lambda_{ij}) = -A_i(h_i)\,(I_{E_1}\otimes h_i^{-1}) + G^2_{ij}\,A_j(h_j)\,(I_{E_1}\otimes h_j^{-1})\,G^1_{ji} \qquad (15)$$

in $U_{ij} \equiv U_i \cap U_j$.

Furthermore if P has an F-connection θ such that $\delta(\theta)$ is defined, it necessarily follows that there is an F-connection θ_o of P with $\deg\theta_o \leq s-2$. This is because $\delta(\theta)$ defined, means that there is a $\eta \in \mathrm{Diff}_{s-1}(E_1\otimes F,\ E_2\otimes F)$ with $\delta(\theta) = \delta(\eta)$ and since F-connections always exist to any differential operator we have the existence of θ_o.

Now, in cohomology terms $T^*(U_i) \ni (x,\xi) \longrightarrow A_i \in \mathcal{E}(U_i, \mathrm{Hom}(E_1, E_2)\otimes T(X))$ is a 0-cocycle with values in $\mathrm{Hom}(E_1, E_2)\otimes T(X)$ and so $A_i(h_i) \in \mathcal{E}(U_i, \mathrm{Hom}(E_1, E_2)\otimes \mathrm{Hom}(F, F))$ and the same holds for $A_i(h_i)(I_{E_1}\otimes h_i^{-1})$. This is because $T(X) : \mathcal{E}(X, \mathrm{Hom}(F, F)) \longrightarrow \mathcal{E}(X, \mathrm{Hom}(F, F))$ by $Y(f) = df(Y)$ for $f \in \mathcal{E}(X, \mathrm{Hom}(F, F))$ and $Y \in T(X)$. Thus, in the range of the local operator $A_i(\cdot)(I_{E_1}\otimes h_i^{-1})$

$$\mathfrak{X}_i := \{A_i(h_i)(I_{E_1}\otimes h_i^{-1}) \,/\, h_i \in \mathcal{E}(U_i, \mathrm{Hom}(F, F))\}$$

let $\tilde{\mathfrak{X}}_i$ be the local sheaf of germs and $\tilde{\mathfrak{X}}$ the sheaf of germs. Then, since $\mathfrak{d}(\lambda_{ij}) = \mathfrak{d}(\theta_i) - G^2_{ij}\,\mathfrak{d}(\theta_j)\,G^1_{ji} = A_i\,(f_{ij})(I_{E_1} \otimes f_{ji})$ (a short calculation of the same type as before) we have in particular that $\mathfrak{d}(\lambda_{ij}) \in \tilde{\mathfrak{X}}$ and by (8') $\mathfrak{d}(\lambda_{ij})$ defines an element in $H^1(X, \tilde{\mathfrak{X}})$. We denote this cohomology class by $[\mathfrak{d}(\lambda_{ij})]$. By the Lemma above, $[\mathfrak{d}(\lambda_{ij})]$ is independent of choice of $\{\theta_i\}$ and by (14) this class is the same for equivalent connections.

The result above could then be restated in cohomology terms as follows: $\mathfrak{d}(\theta)$ defined $\Longleftrightarrow$ (15) i.e. $\mathfrak{d}(\lambda_{ij})$ is a 1-coboundary. So there is an F-connection θ of P with $\deg\theta \leq s-2$ if and only if $[\mathfrak{d}(\lambda_{ij})] = 0$. We illustrate this by the following example due to Asada and which generalizes directly to elliptic complexes:

$$s = 2,\ \dim E_1 = \dim E_2 = 1 \quad\text{and}\quad P = \sum_{i,j} A_{ij}(x)\,\frac{\partial^2}{\partial x_i \partial x_j} + \text{lower order}$$

with $A_{ij} = A_{ji}$

$$A_i(f_i) = \sum_{|\gamma|=1} \Big(\sum_{j=1}^{n} A_{i,\gamma+1_j}\, \xi^{\gamma}(i) \otimes \frac{\partial f_i}{\partial x_j(i)}\Big) = \sum_{l=1}^{n}\Big(\sum_{j=1}^{n} A_{k,j}(x)\,\frac{\partial f_i}{\partial x_j(i)}\Big)\, dx^l(i)$$

so that in $\mathfrak{X}_i$ we have the element

$$A_i(f_i)\,(I_{E_1} \otimes f_i^{-1}) = \sum_{l=1}^{n}\Big(\sum_{j=1}^{n} A_{k\,j}(x)\,\frac{\partial f_i}{\partial x_j(i)}\, f_i^{-1}\Big)\, dx^l(i)$$

Assume now F is a complex line bundle and $\{A_{ij}(x)\}$ regular for each x, then $B = \operatorname{Ker}\{A(h)\,(I_{E_1} \otimes h^{-1})\}$ is the sheaf of complex numbers over X with the following trivial resolution

$$0 \longrightarrow B \longrightarrow \mathcal{E}(X, \operatorname{Hom}(F,F)) \xrightarrow{A(h)\,(I_{E_1} \otimes h^{-1})} \tilde{\mathfrak{X}} \longrightarrow 0$$

which generates

$$0 \longrightarrow H^0(X,B) \longrightarrow H^0(X, \mathcal{E}(X, \operatorname{Hom}(F,F)) \longrightarrow H^0(X, \tilde{\mathfrak{X}}) \longrightarrow$$

$$\longrightarrow H^1(X,B) \longrightarrow H^1(X, \mathcal{E}(X, \operatorname{Hom}(F,F)) \longrightarrow H^1(X, \tilde{\mathfrak{X}})$$

Now using [De 1, 1960] and denoting by Φ the automorphisms of $\mathcal{E}(X, \operatorname{Hom}(F,F))$ leaving B invariant (or equivalently, those automorphisms of B that could be extended to automorphisms of $\mathcal{E}(X, \operatorname{Hom}(F,F))$), we have a map

$$\delta : H^1(X, \tilde{\mathfrak{X}}) \longrightarrow H^2(X, \Phi)$$

for a 2-cohomology set $H^2(X, \Phi)$.

Under this map, $\delta([\text{ಹ}(\lambda_{ij})])$ corresponds to the 1st Chern class.

Now, the above structure of which we have given but a short outline, can be extended considerably. This extension to spin complexes and to pseudodifferential operators is the subject of a continuation of this paper. It will involve a more substantial use of non-abelian cohomology theory and will also give rise to new questions in this theory.

REFERENCES

[As, 1978] Akira Asada; Connections of Differential Operators, J. Fac. Sci., Shinshu Univ. Vol 13, 87-102.

[We, 1980] R.O. Wells; Differential Analysis on Complex Manifolds, GTM 65, Springer Verlag, 1980.

[De, 1960] P. Dedecker; Sur la cohomologie non abelienne I, Can. J. Math. 12, 231-251.

PHASE SPACE OF THE COUPLED VECTORIAL KLEIN-GORDON-MAXWELL EQUATIONS

Pedro L. García

Universidad de Salamanca

"To Professor KONRAD BLEULER on the occasion of his 70th birthday"

Introduction

The geometric presentation of the theory of Yang-Mills fields, "free" and in "minimal coupling" with other fields, is at present a well-known topic for a wide audience of mathematicians and physicists. One aspect of special interest of this doctrine is the study of the relation between the orbits of the "gauge group" in the manifold of solutions of the field equations and the radical of the canonical pre-symplectic metric on the said manifold. The reason for this interest may be found in the desire to get the "phase space" of the field as the set of the said orbits. Seven recent references are: *P. García* [2], *V. Moncrief* [6], *P. García - A. Pérez-Rendón* [4], *I. Segal* [9], *T. Branson* [1], *V. Moncrief* [7], *P. García* [3].

In the free case, the problem can be established as follows. Let $p:P \longrightarrow X$ be a principal bundle with structural group G with Lie algebra L. The connections on P can be identified with the global sections of the affine bundle $\pi:E \longrightarrow X$ corresponding to the vector bundle Hom(T(X),Ad P), where Ad P is the bundle associated to P with respect to the adjoint representation of G on L. The group of vertical automorphisms on P (gauge group), when acting in the natural way on the connections on P, induces a representation $s \in \Gamma(X,\mathrm{Ad}\,P) \longmapsto D^s$ of its "Lie algebra" by vector fields on E. A Yang-Mills field is a variational problem over the fibered manifold E whose lagrangian admits the vector fields D^s as infinitesimal symmetries. In [2], the following general result is proved: *if* (V, Ω_2) *is the pre-symplectic manifold of solutions of an arbitrary Yang-Mills field, then the orbits of the gauge group in* V *are tangents to the radical of the metric* Ω_2. The question is now the reciprocal: is every vector in the radical of Ω_2 tangent to an orbit of the gauge group? The trivial example of the elliptic version of the electromagnetic field shows that this result is false in general. However in [1], [3], [7], [9] it is proved that, under certain conditions, this is the case for ordinary Yang-Mills fields (those defined by the square of the norm of the curvature) over a Lorentz manifold, with the agreeable result that *the "phase space" of such fields is then the set of orbits of the gauge group in the corresponding manifold of solutions*. The crucial point of this research is, on one hand, a study of the Cauchy problem for the corresponding field equations; on the other, an adequate application of the theory of elliptic operators on the space-like surface which carries the initial data.

Now, if E' is a vector bundle associated to P with respect to a linear representation of G on a vector space, it is possible to define, in a general and understandable way, the notion of "minimal coupling" between a Yang-Mills field over the affine bundle E and a field over the vector bundle E'; the result being a variational problem over the fibered product $E \times_X E'$ whose lagrangian is invariant by the natural action of the gauge group on $E \times_X E'$ (see, for example, *A. Pérez-Rendón* [8], and [4]). As in the free case, in [4] it is proved, with all generality, that *the orbits of the gauge group in the pre-symplectic manifold of solutions* (V, Ω_2) *of such variational problems are tangent to the radical of the metric* Ω_2. The question is now, as in the free case, the following: is every vector in the radical of Ω_2 tangent to an orbit of the gauge group? In [3] it is proved that this is the case for a complex scalar Klein-Gordon field over a Lorentz manifold in minimal coupling with the electromagnetic field generated by itself (this is, probably, a Corollary of more general results obtained in [6] for the standard Yang-Mills-Higgs lagrangian). Under the same hypothesis, this is also the case for the coupled Dirac-Yang-Mills equations (informal notes of the author, Salamanca 1980).

In this paper, we shall illustrate this question with another example: a complex vectorial Klein-Gordon field over a Lorentz manifold in minimal coupling with the electromagnetic field generated by itself. In particular, in the *massless case*, the radical of the metric is *bigger* than the tangent space to the orbits of the gauge group. So, this example, apart from its own interest, also points the way to the exploration of the general situation. Using the approach of [3], the results of this example may be generalized in a straighforward way to the coupled vectorial Klein-Gordon-Yang-Mills fields.

Concepts and notations of Variational Calculus and Symplectic Geometry in this paper are the same as the ones in [3] of which this paper is a natural continuation.

1. Klein-Gordon 1-forms with values in a hermitian line bundle.

Let L be a complex line bundle over an orientable pseudoriemannian manifold (X,g) with pseudoriemannian volume element η , endowed with a hermitian metric h and a hermitian connection A with respect to h. The metrics g and h define a natural norm $\| \ \|$ over the L-valued forms on X. This allows us to globalize the ordinary notion of a complex vectorial Klein-Gordon field by considering the variational problem defined over the fiber bundle $J^1(T^*(X) \otimes L)$ of the 1-jets of local sections ω of $T^*(X) \otimes L$, by the lagrangian density $\mathcal{L}\eta$, where $\mathcal{L}$ is the function given by:

$$\mathcal{L}(j^1_x\omega) = \frac{1}{4} \| d_A\omega\|^2_x - \frac{1}{2} m^2 \cdot \| \omega\|^2_x$$

where d_A is the exterior differential for L-valued forms on X with respect to the connection A.

The corresponding Euler equations then become:

$$\delta_A d_A\omega - m^2\omega = 0 \qquad (1)$$

where $\delta_A = *^{-1} d_A *$ is the codifferential with respect to d_A and the metric g.

Physically, one can speak of an "L-valued Klein-Gordon covector field in the external electromagnetic field defined by the connection A".

From (1) it follows for $m \neq 0$:

$$\delta_A\omega = \frac{i}{m^2} \langle \mathrm{Curv}\, A, d_A\omega \rangle$$

where $\langle \mathrm{Curv}\, A, d_A\omega\rangle$ is the scalar product of the ordinary 2-form, Curv A, and the L-valued 2-form, $d_A\omega$, with respect to the metric g and the module structure of $\Gamma(X,L)$. This formula generalizes the well-known *Lorentz constraint*.

The tangent space $T_\omega(V)$ at a point ω to the manifold of solutions V of the equations (1) can be identified with the vector space of L-valued 1-forms ω' on X, verifying the same equations (1), as one could expect from their linearity:

$$\delta_A d_A\omega' - m^2\omega' = 0 \qquad (2)$$

With respect to this parametrization of $T_\omega(V)$, the corresponding pre-symplectic metric $(\Omega_2)_\omega$ is given by:

$$(\Omega_2)_\omega (\omega', \overline{\omega}') = \mathrm{Re}(\langle\overline{\omega}', d_A\omega'\rangle - \langle\omega', d_A\overline{\omega}'\rangle)\cdot\eta \qquad (3)$$

where $\overline{\omega}'$ is considered contravariated with the metric g, $\langle\overline{\omega}', d_A\omega'\rangle$ denotes h-hermitian contraction with the first covariant index of $d_A\omega'$, Re( ) is considered contravariated again with the metric, and the dot means contraction with the first covariant index of η.

Notation. In the above formulas, forms appear which are obtained by operations of L-valued exterior differential calculus over X. Similar expressions will appear in what follows. We shall use the notation for this differential calculus employed by *J. Koszul* [5]; thus, for example, if A' and ω are respectively an ordinary 1-form and a L-valued 1-form, $A'\wedge\omega$ will denote the L-valued 2-form obtained by exterior product of A' and ω with respect to the bilinear product defined by the module structure of $\Gamma(X,L)$.

As is usual, when we speak of the contraction of an index of a form with that of the other, it should be understood that the first is considered contravariated with the metric g. Two of this type of contractions which will appear frequently in this paper are the following.

If α and β are, respectively, an ordinary form and a L-valued form, $\langle\alpha,\beta\rangle$ will denote a L-valued form obtained by contraction of some indices of α with corresponding ones of β with respect to the bilinear product defined by the module structure of $\Gamma(X,L)$. This is the case, for example, of the L-section $\langle \mathrm{Curv}\, A, d_A \rangle$ obtained above and similarly the L-sections $\langle A',\omega\rangle$, $\langle dA', d_A\omega\rangle$ and $\langle \mathrm{Curv}\, A, A'\wedge\omega\rangle$ and the L-valued 1-form $\langle A', d_A\omega\rangle$.

On the other hand, if α and β are now L-valued forms, $\langle\alpha,\beta\rangle$ will denote an ordinary form obtained by contractions of indices of α with corresponding ones of β combined with the hermitian scalar product defined by h. This is the case, for example, of the 1-form $\langle\overline{\omega}', d_A\omega'\rangle$ obtained above in this way by contraction of the index of $\overline{\omega}'$ with the first of $d_A\omega'$.

2. Minimal Coupling with the induced electromagnetic field.

2.1 The above hermitian connections, A, are the connections over the principal U(1)-bundle defined by the elements $e\in L$ such that $h(e,e) = 1$. Let E be the affine bundle whose global sections are these connections. The ordinary Yang-Mills field defined on E can be interpreted as the "electromagnetic field induced by the electric particle described by the above Klein-Gordon field".

Now, the minimal coupling between these two fields is defined by the variational problem on $J^1((T^*(X) \otimes L) \times_X E)$ whose lagrangian is the sum of the corresponding lagrangians, i.e.:

$$\mathcal{L}(j^1_x(\omega, A)) = \frac{1}{4} \| d_A \omega \|^2_x - \frac{1}{2} m^2 \cdot \| \omega \|^2_x + \frac{1}{4} \| \text{Curv } A \|^2_x$$

The corresponding Euler equations become the coupled vectorial Klein-Gordon-Maxwell equations in the following global version:

$$\left.\begin{aligned} \delta_A d_A \omega - m^2 \omega &= 0 \\ \delta \, \text{Curv } A &= \text{Re} \langle i\omega, d_A \omega \rangle \end{aligned}\right\} \quad (1')$$

The term $\text{Re} \langle i\omega, d_A\omega \rangle$ being the *current* 1-*form* associated to this minimal coupling.

The linearization of the equations (1') in one of its solutions (ω, A), i.e., the tangent space $T_{(\omega,A)}V$ at a point (ω, A) to the manifold of solutions V of the equations (1'), can be identified with the vector space of all couples (ω', A') (ω' = L-valued 1-form and A' = ordinary 1-form over X) such that:

$$\left.\begin{aligned} \delta_A d_A \omega' - m^2 \omega' &= i(\delta_A(A' \wedge \omega) + \langle A', d_A\omega \rangle) \\ \delta dA' &= \text{Re}(\langle i\omega', d_A\omega \rangle + \langle i\omega, d_A\omega' \rangle - \langle A', \omega \rangle \omega) + \|\omega\|^2 A' \end{aligned}\right\} \quad (2')$$

For $m \neq 0$, one has the Lorentz constraint:

$$\delta_A \omega = \frac{i}{m^2} \langle \text{Curv } A, d_A \omega \rangle$$

The linearization of this equation in the solution $(\omega, A) \in V$ is:

$$\delta_A \omega' = \frac{i}{m^2} (\langle \text{Curv } A, d_A\omega' \rangle + m^2 \langle A', \omega \rangle + \langle dA', d_A \omega \rangle - i \langle \text{Curv } A, A' \wedge \omega \rangle)$$

With respect to the above parametrization of $T_{(\omega,A)}V$, the pre-symplectic metric $(\Omega_2)_{(\omega,A)}$ is given by the formula:

$$(\Omega_2)_{(\omega,A)}((\omega'A'),(\bar{\omega}'\bar{A}')) = -\operatorname{Re}(\langle\bar{\omega}', d_A\omega'\rangle - \langle\omega', d_A\bar{\omega}'\rangle)\cdot\eta -$$

$$-\operatorname{Re}(\langle i\bar{\omega}', A'\wedge\omega\rangle - \langle i\omega', \bar{A}'\wedge\omega\rangle)\cdot\eta + \langle A', d\bar{A}'\cdot\eta\rangle - \langle\bar{A}, dA'\cdot\eta\rangle \qquad (3')$$

where the dot means contraction with the first covariant index of η. The last two terms define the pre-symplectic metric of the free electromagnetic field.

Proposition. The lagrangian $\mathcal{L}$ is invariant with respect to the natural action of the vertical automorphisms of P on the 1-jet fiber bundle $J^1((T^*(X)\otimes L)\times_X E)$.

Proof. Let ϕ be the differentiable complex function on X defined by the natural action on L of a vertical automorphism τ of P. One has $\phi\phi^* = 1$; and if ω and A are, respectively, a L-valued 1-form on X and a connection on P, one has: $\bar{\omega} = \tau\omega = \phi\omega$ and $\bar{A} = \tau A = A - i\,\frac{d\phi}{\phi}$. Thus, one has: $\operatorname{Curv}\bar{A} = \operatorname{Curv}A - d(i\,\frac{d\phi}{\phi}) = \operatorname{Curv}A$ and $d_{\bar{A}} = d_A - \frac{d\phi}{\phi}\wedge$. In particular:

$$d_{\bar{A}}\bar{\omega} = d_A(\phi\omega) - \frac{d\phi}{\phi}\wedge\phi\omega = d\phi\wedge\omega + \phi d_A\omega - d\phi\wedge\omega = \phi\, d_A\omega$$

So, by the condition $\phi\phi^* = 1$, one has:

$$\mathcal{L}(j^1_x(\bar{\omega},\bar{A})) = \frac{1}{4}\| d_{\bar{A}}\bar{\omega}\|^2_x - m^2\|\bar{\omega}\|^2_x + \frac{1}{4}\|\operatorname{Curv}\bar{A}\|^2_x =$$

$$= \frac{1}{4}\|\phi\, d_A\omega\|^2_x - \frac{1}{2}m^2\|\phi\omega\|^2_x + \frac{1}{4}\|\operatorname{Curv}A\|^2_x =$$

$$= \frac{1}{4}\| d_A\omega\|^2_x - \frac{1}{2}m^2\|\omega\|^2_x + \frac{1}{4}\|\operatorname{Curv}A\|^2_x = \mathcal{L}(j^1_x(\omega,A)) \qquad //$$

Thus, the natural action of the gauge group on $(T^*(X)\otimes L)\times_X E$ induces a representation of its "Lie algebra" (= differentiable functions on X with Lie product zero) by infinitesimal symmetries of the variational problem under consideration. By the general formalism of the Variational Calculus, such infinitesimal symmetries define vector fields on the manifold of solutions. This can be seen directly, by observing that the vertical vector field of $(T^*(X)\otimes L)\times_X E$, defined along a section (ω,A) by one of these infinitesimal symmetries corresponding to a function f on X, is $(if\omega, df)$, which obviously satisfies the equations (2').

As mentioned in the Introduction, all these vector fields belong to the radical of the pre-symplectic metric. We shall prove this directly in our example as an application of the explicit formulas we have just obtained.

Theorem 1. The orbits of the gauge group in the pre-symplectic manifold of solutions $(\mathcal{V}, \Omega_2)$ of the coupled vectorial Klein-Gordon-Maxwell equations are tangent to the radical of the metric Ω_2.

Proof. If $(if\omega, df)$ is a vector tangent to the orbit of the gauge group through the point (ω, A), and $(\omega', A') \in T_{(\omega,A)}\mathcal{V}$ is arbitrary, one has:

$$(\Omega_2)_{(\omega,A)}((if\omega,df),(\omega'A')) = -\mathrm{Re}(\langle\omega', id_A(f\omega)\rangle - \langle if\omega, d_A\omega'\rangle + \langle i\omega', df\wedge\omega\rangle +$$

$$+\langle f\omega, A'\wedge\omega\rangle)\cdot\eta + \langle df, dA'\cdot\eta\rangle = -\mathrm{Re}(\langle\omega', idf\wedge\omega\rangle + \langle\omega', ifd_A\omega\rangle - \langle if, d_A\omega'\rangle +$$

$$+\langle i\omega' df\wedge\omega\rangle + f\langle A',\omega\rangle\omega - f\,\|\omega\|^2 A')\cdot\eta + .d^{\nabla}(fdA'\cdot\eta) - f(.d^{\nabla}(dA'\cdot\eta))$$

where $.d^{\nabla}$ denotes contraction of the covariant index of the differentiation with the unique contravariant index of the tensor on which it acts.

Bearing in mind that $.d^{\nabla}(dA'\cdot\eta) = [.d^{\nabla}(dA')]\cdot\eta = (\delta dA')\cdot\eta$, the above expression becomes:

$$-f[\delta dA' - \mathrm{Re}(\langle i\omega', d_A\omega\rangle + \langle i\omega, d_A\omega'\rangle - \langle A',\omega\rangle\omega + \|\omega\|^2 A')]\cdot\eta + .d^{\nabla}(fdA'\cdot\eta)$$

From the second of the equations (2'), which must be satisfied by the tangent vector (ω', A'), the first term is zero. The result can be now obtained from the following calculus of the second term:

$$.d^{\nabla}(fdA'\cdot\eta) = [.d^{\nabla}(fdA')]\cdot\eta = [\delta(fdA')]\cdot\eta = *\delta(fdA') = **^{-1}d*(fdA') =$$

$$= d*(fdA') \quad //$$

2.2 We now see how one can achieve the determination of the radical of $(\Omega_2)_{(\omega,A)}$ for the case of a Lorentz manifold $(X, {}^{(4)}g)$. For simplicity's sake, we shall consider the following situation.

Let $S \subset X$ be a space-like compact connected surface. If one takes a normal gaussian coordinate system along S, then in a tubular neighbourhood of S one has $X = (-\varepsilon, \varepsilon) \times S$, and ${}^{(4)}g$ becomes:

$${}^{(4)}g = -dt^2 + {}^{(3)}g_t$$

where "t" is the natural coordinate in $(-\varepsilon, \varepsilon)$ and ${}^{(3)}g_t$ is the riemannian metric defined on S by the restriction of ${}^{(4)}g$ to the surface $\{t\} \times S$.

With respect to the local decomposition $X = (-\varepsilon, \varepsilon) \times S$, an ordinary 1-form A' and its exterior differential dA' may be expressed as follows: $A' = A'_o(t)dt + A'_t$, $dA' = dt \wedge E'_t + H'_t$, where $A'_o(t)$, A'_t, E'_t and H'_t are, for each t, respectively, a function, two 1-forms and a 2-form on S defined by the restriction to the surface $\{t\} \times S$ of $A'(\frac{\partial}{\partial t})$, A', $dA'(\frac{\partial}{\partial t},)$ and dA'. In physical terminology they are the "variations" of the scalar and vectorial potentials and the electric and magnetic fields. In a similar way, if one identifies the bundles $L_{\{t\} \times S}$ and L_S by the parallel translation with respect to the connection "A" along the curves $\{y = \text{const.}, y \in S\}$, an L-valued 1-form ω' and its exterior differential $d_A\omega'$ may be expressed as follows: $\omega' = \omega'_o(t)dt + \omega'_t$, $d_A\omega' = dt \wedge \mathcal{E}'_t + \mathcal{H}'_t$, where $\omega'_o(t)$, ω'_t, $\mathcal{E}'_t$ and $\mathcal{H}'_t$ are, for each t, respectively, a function, two 1-forms and a 2-form on S with values in L_S defined by the restriction to the surface $\{t\} \times S$ of $\omega'(\frac{\partial}{\partial t})$, ω', $d_A\omega'(\frac{\partial}{\partial t},)$ and $d_A\omega'$.

In terms of these new objects, equations (2') give rise to four first order evolution equations and the four constraint conditions:

$$\left.\begin{aligned} &\mathcal{H}'_t = d_{A_t}\omega'_t, \quad \delta_{A_t}\mathcal{E}'_t = -m^2\omega'_o(t) - i\,\delta_{A_t}(\omega_o(t)A'_t - A'_o(t)\omega_t) + i\langle A'_t, \mathcal{E}_t\rangle \\ &H'_t = dA'_t, \quad \delta_t E'_t = \mathrm{Re}(\langle i\,\omega'_t, \mathcal{E}_t\rangle + \langle i\,\omega_t, \mathcal{E}'_t\rangle + \langle A'_t, \omega_t\rangle\,\omega_o(t)) - \|\omega_t\|^2 A'_o(t) \end{aligned}\right\} \quad (4)$$

where, for each t, $\omega_o(t), \omega_t$, and $\mathcal{E}_t$ are obtained from the component ω of the solution (ω, A) in a similar way to $\omega'_o(t)$, ω'_t, and $\mathcal{E}'_t$ from ω'; d_{A_t} is the exterior differential for L_S-valued forms on S with respect to the connection defined on $L_S \simeq L_{\{t\} \times S}$ by restricting A to $L_{\{t\} \times S}$; and finally δ_t and δ_{A_t} are, respectively, the ordinary codifferential with respect to ${}^{(3)}g_t$ and the L_S-valued codifferential with respect to ${}^{(3)}g_t$ and d_{A_t}.

Remark. For $m \neq 0$, there is one more evolution equation, which is easily obtained from the linearization of the *Lorentz constraint*.

The solution of the Cauchy problem for the first order partial differential equations system obtained above, allows us to establish a canonical projection $\pi: T_{(\omega,A)}\mathcal{V} \longrightarrow \mathcal{F}$ from the tangent space $T_{(\omega,A)}\mathcal{V}$ onto the space $\mathcal{F}$ defined by the elements on S, $(\omega'_o, \omega', \mathcal{E}', A', A', E')$, which satisfy the second and the fourth equations in (4) for $t = 0$; i.e.:

$$\left.\begin{aligned}\delta_A E' &= -m^2\omega_0' - i\,\delta_A(\omega_0 A' - A'\omega) + i\,\langle A', E\rangle \\ \delta E' &= \mathrm{Re}(\langle i\omega', E\rangle + \langle i\omega, E'\rangle + \langle A', \omega\rangle\omega_0) - \|\omega\|^2 A'\end{aligned}\right\} \qquad (5)$$

The subspace $\{if\omega, df\}$ of $T_{(\omega,A)}V$ defined by the vectors tangent to the orbit of the gauge group through (ω,A) is projected by π onto the subspace of F defined by the vectors of type

$$(ig\omega_0, ig\omega, i(h\omega + gE - \omega_0 dg), h, dg, 0) \qquad (6)$$

h and g being arbitrary differentiable real functions on S.

On the other hand, the pre-symplectic metric $(\Omega_2)_{(\omega,A)}$, when interpreted as a real 2-form via the mapping $\int_S : H^3(X, \mathbb{R}) \longrightarrow \mathbb{R}$, is projected by π on the following 2-form $\hat{\Omega}_2$:

$$\hat{\Omega}_2((\omega_0', \omega', E', A_0', A', E'), (\bar{\omega}_0', \bar{\omega}', \bar{E}', \bar{A}_0', \bar{A}', \bar{E}')) = \int_S \mathrm{Re}(\langle\bar{\omega}', E'\rangle - \langle\omega', \bar{E}'\rangle)\,{}^{(3)}\eta\ +$$

$$+ \int_S \mathrm{Re}(\langle i\bar{\omega}', A_0'\omega - \omega_0 A'\rangle - \langle i\omega', \bar{A}_0'\omega - \omega_0\bar{A}'\rangle)\,{}^{(3)}\eta + \int_S (\langle A', \bar{E}'\rangle - \langle\bar{A}', E'\rangle)\,{}^{(3)}\eta \qquad (7)$$

where ${}^{(3)}\eta$ is the volume element on S canonically associated to ${}^{(3)}g$.

Under all the aforementioned conditions, we can prove the following:

<u>Theorem 2</u>. For $m \neq 0$, the radical of the pre-symplectic metric Ω_2 on the manifold of solutions V of the coupled vectorial Klein-Gordon-Maxwell equations on a Lorentz manifold *coincides*, in each point $(\omega,A) \in V$, with the subspace tangent at such a point to the orbit of the gauge group through (ω,A). Consequently, the *phase space* of the minimal coupling defined by these equations is the set of orbits of the corresponding gauge group.

<u>Proof</u>. It will be enough to prove that every vector in the radical of $\hat{\Omega}_2$ is of the type (6). Let $(\omega_0', \omega', E', A', A', E') \in \mathrm{rad}\,\hat{\Omega}_2$, i.e. for all $(\bar{\omega}_0', \bar{\omega}', \bar{E}', \bar{A}', \bar{A}', \bar{E}') \in F$ one has:

$$\int_S \mathrm{Re}(\langle\bar{\omega}', E'\rangle - \langle\omega', \bar{E}'\rangle + \langle i\bar{\omega}', A_0'\omega - \omega_0 A'\rangle - \langle i\omega', \bar{A}_0'\omega - \omega_0\bar{A}'\rangle)^{(3)}\eta \; +$$

$$+ \int_S (\langle A', \bar{E}'\rangle - \langle\bar{A}', E'\rangle)^{(3)}\eta = 0 \qquad (8)$$

Taking $\bar{\omega}_0' = 0$, $\bar{\omega}' = 0$, $\bar{E}' = 0$, $\bar{A}_0' = 0$ and $\bar{A}' = 0$, then, for every 1-form $\bar{E}'$ such that $\delta\bar{E}' = 0$, one has:

$$\int_S \langle A', \bar{E}'\rangle^{(3)}\eta = 0$$

which implies, by the Hodge decomposition theorem:

$$A' = dg \qquad (a)$$

where g is an arbitrary differentiable real function on S.

Using this in (8) one has:

$$\int_S \mathrm{Re}(\langle\bar{\omega}', E'\rangle - \langle\omega', \bar{E}'\rangle + \langle i\bar{\omega}', A_0'\omega - \omega_0 dg\rangle - \langle i\omega', \bar{A}_0'\omega - \omega_0\bar{A}'\rangle)^{(3)}\eta \; +$$

$$+ \int_S g\left[\mathrm{Re}(\langle i\bar{\omega}', E\rangle + \langle i\omega, \bar{E}'\rangle + \langle\bar{A}', \omega\rangle\omega_0) - \|\omega\|^2 \bar{A}_0'\right]^{(3)}\eta - \int_S \langle\bar{A}', E'\rangle^{(3)}\eta = 0$$

$$(9)$$

Taking $\bar{\omega}' = 0$, $\bar{E}' = 0$ and $\bar{A}_0' = 0$ and bearing in mind the constraint equations (5), one has:

$$\int_S \langle\bar{A}', \mathrm{Re}\left[(g\omega + i\omega')\otimes\omega_0\right] - E'\rangle^{(3)}\eta = 0$$

for all 1-forms $\bar{A}'$ such that

$$\int_S \langle\bar{A}', \mathrm{Re}(\omega\otimes\omega_0)\rangle^{(3)}\eta = 0$$

where $\omega\otimes\omega_0$ is the ordinary 1-form defined by the tensorial product of the L-valued 1-form, ω, and the L-section, ω_0, with respect to the hermitian metric on L.

This implies $E' = \mathrm{Re}\left[(g\,\omega + i\,\omega')\otimes\omega_0\right] + \lambda\,\mathrm{Re}(\omega\otimes\omega_0)$, where λ is a real constant.

Using this in (9) one has:

$$\int_S \mathrm{Re}(\langle \bar{\omega}', E'\rangle - \langle \omega', \bar{E}'\rangle + \langle i\bar{\omega}', A_0'\omega - \omega_0\, dg\rangle - \langle i\,\omega', \bar{A}_0'\omega\rangle)^{(3)}\eta \; + \tag{10}$$

$$+ \int_S g\left[\mathrm{Re}(\langle i\bar{\omega}', E\rangle + \langle i\omega, \bar{E}'\rangle) - \|\omega\|^2 \bar{A}_0'\right]^{(3)}\eta - \lambda\int_S \langle \bar{A}', \mathrm{Re}(\omega\otimes\omega_0)\rangle^{(3)}\eta = 0$$

Now taking $\bar{\omega}' = 0$, $\bar{A}_0' = 0$ and $\bar{A}' = 0$, one has:

$$\int_S \mathrm{Re}\,\langle ig\,\omega - \omega', \bar{E}'\rangle^{(3)}\eta = 0$$

for all 1-forms $\bar{E}'$ such that

$$\int_S \mathrm{Re}\,\langle i\omega, \bar{E}'\rangle^{(3)}\eta = 0$$

which implies:

$$\omega' = i(g-\mu)\omega\,, \text{ and then } E' = (\lambda+\mu)\mathrm{Re}(\omega\otimes\omega_0) \tag{b}$$

where μ is a real constant.

Returning to (10) one has:

$$\int_S \mathrm{Re}(\langle \bar{\omega}', E'\rangle + \mu\langle i\omega, \bar{E}'\rangle + \langle i\bar{\omega}', A_0'\omega - \omega_0\, dg\rangle - \mu\langle \omega, \bar{A}_0'\omega\rangle)^{(3)}\eta \; +$$

$$+ \int_S g\,\mathrm{Re}\langle i\bar{\omega}', E\rangle^{(3)}\eta - \lambda\int_S \langle \bar{A}', \mathrm{Re}(\omega\otimes\omega_0)\rangle^{(3)}\eta = 0 \tag{11}$$

Finally, taking $\bar{E}' = 0$, $\bar{A}_0' = 0$ and $\bar{A}' = 0$, one has:

$$\int_S \mathrm{Re}\,\langle \bar{\omega}', E' - iA_0'\omega + i\,\omega_0 dg - ig\,E\rangle^{(3)}\eta = 0$$

for all 1-forms $\bar{\omega}'$ such that:

$$\int_S \mathrm{Re}\langle i\,E, \bar{\omega}'\rangle^{(3)}\eta = 0$$

which implies:

$$E' = i\left[A_0'\omega + (g+\nu)E - \omega_0\, dg\right] \tag{c}$$

where ν is a real constant.

Now, as the constraint equations (5) must be satisfied by the vector $(\omega_0', \omega', E', A', A', E')$, one has:

$$\left.\begin{aligned} & i\,\delta_A[A_0'\omega + (g+\nu)E - \omega_0 dg] = -m^2\omega_0' - i\,\delta_A(\omega_0 dg - A_0'\omega) + i\langle dg, E\rangle \\ & (\lambda+\mu)\,\delta\,\mathrm{Re}(\omega\otimes\omega_0) = \mathrm{Re}(\langle(\mu-g)\omega, E\rangle + \langle i\omega, i[A_0'\omega + (g+\nu)E - \omega_0 dg] + \langle dg,\omega\rangle\omega_0) - \\ & - \|\omega\|^2 A_0' \end{aligned}\right\} \tag{12}$$

From the first equation it follows:

$$i(g+\nu)\,\delta_A E = -m^2\omega_0'$$

But, as the field equations (1') must be satisfied by the solution $(\omega, A) \in V$, one has the constraint condition $\delta_A E = -m^2\omega_0$, from where:

$$\omega_0' = i(g+\nu)\,\omega_0 \tag{d}$$

The second of the equations (12) implies:

$$(\lambda+\mu)\,\delta\,\mathrm{Re}(\omega\otimes\omega_0) = (\mu+\nu)\,\mathrm{Re}\,\langle\omega, E\rangle \tag{$*$}$$

On the other hand (returning to (11)), in order for a vector $(\omega_0', \omega', E', A_0', A', E')$ defined by the equalities (a), (b), (c) and (d) to belong to rad $\hat{\Omega}_2$,

$$(\lambda+\mu)\int_S \langle \bar{A}', \mathrm{Re}(\omega\otimes\omega_0)\rangle^{(3)}\eta + (\mu+\nu)\int_S \mathrm{Re}\,\langle i\bar{\omega}', E\rangle^{(3)}\eta = 0 \tag{$**$}$$

must be verified for all vectors $(\bar{\omega}_0', \bar{\omega}', \bar{E}', \bar{A}', \bar{A}', \bar{E}') \in F$.

The following cases may be presented:

1) <u>$\mathrm{Re}\,\langle\omega, E\rangle$, $\delta\,\mathrm{Re}(\omega\otimes\omega_0)$</u> R-linearly independent: Then, the equation ($*$) implies $\lambda+\mu = 0$ and $\mu+\nu = 0$, from where the vector defined by the equalities (a), (b), (c) and (d) is of the type (6).

2) <u>$\mathrm{Re}\,\langle\omega, E\rangle \equiv 0$, $\delta\,\mathrm{Re}(\omega\otimes\omega_0) \not\equiv 0$.</u> In this case ($*$) implies $\lambda+\mu = 0$, from where ($**$) implies in turn:

$$(\mu + \nu) \int_S \mathrm{Re}\langle i\bar{\omega}', E\rangle^{(3)}\eta = 0$$

for all $\bar{\omega}'$ such that $(\bar{\omega}_0', \bar{\omega}', \bar{E}', \bar{A}_0', \bar{A}', \bar{E}') \in F$. In particular:

2') If $E \not\equiv 0$, for all $\bar{\omega}'$ such that $(\bar{\omega}_0', \bar{\omega}', 0, 1, 0, \bar{E}') \in F$, i.e.:

$$\left.\begin{aligned} 0 &= -m^2\bar{\omega}_0' + i\delta_A\omega \\ \delta\bar{E}' &= \mathrm{Re}\langle i\bar{\omega}', E\rangle - \|\omega\|^2 \end{aligned}\right\}$$

a system which has solutions if and only if:

$$\int_S (\mathrm{Re}\langle i\bar{\omega}', E\rangle - \|\omega\|^2)\,^{(3)}\eta = 0$$

Then, for such vectors one has:

$$\int_S \mathrm{Re}\langle i\bar{\omega}', E\rangle^{(3)}\eta = \int_S \|\omega\|^{2}\,^{(3)}\eta > 0$$

So, $\mu + \nu = 0$, and the vector defined by (a), (b), (c) and (d) is of the type (6).

2") If $E \equiv 0$, the equation $\delta_A E = -m^2\omega_0$ implies $\omega_0 \equiv 0$, and then (a), (b), (c) and (d) give rise to a vector of type (6).

3) <u>$\mathrm{Re}\langle\omega, E\rangle \not\equiv 0$, $\delta\,\mathrm{Re}(\omega\otimes\omega_0) \equiv 0$.</u> In this case (*) implies $\mu + \nu = 0$, from where (**) implies in turn:

$$(\lambda + \mu) \int_S \langle\bar{A}', \mathrm{Re}(\omega\otimes\omega_0)\rangle^{(3)}\eta = 0$$

for all $\bar{\omega}'$ such that $(\bar{\omega}_0', \bar{\omega}', \bar{E}', \bar{A}_0', \bar{A}', \bar{E}') \in F$. In particular:

3') If $\mathrm{Re}(\omega\otimes\omega_0) \not\equiv 0$, for all $\bar{\omega}'$ such that $(\bar{\omega}', 0, 0, 1, \bar{A}', \bar{E}') \in F$, i.e.:

$$\left.\begin{aligned} 0 &= m^2\bar{\omega}_0' - i\delta_A(\omega_0\bar{A}' - \omega) + i\langle\bar{A}', E\rangle \\ \delta\bar{E}' &= \langle\bar{A}', \mathrm{Re}(\omega\otimes\omega_0)\rangle - \|\omega\|^2 \end{aligned}\right\}$$

a system which has solutions if and only if:

$$\int_S (\langle\bar{A}', \mathrm{Re}(\omega\otimes\omega_0)\rangle - \|\omega\|^2)\,^{(3)}\eta = 0 \qquad (13)$$

Then, for such vectors one has:

$$\int_S \langle \bar{A}', \mathrm{Re}(\omega \otimes \omega_0) \rangle \, ^{(3)}\eta = \int_S \| \omega \|^2 \, ^{(3)}\eta > 0$$

So, $\lambda + \mu = 0$, and the vector defined by (a), (b), (c) and (d) is of the type (6).

3") If $\mathrm{Re}(\omega \otimes \omega_0) \equiv 0$, (a), (b), (c) and (d) give rise to a vector of the type (6).

4) $\underline{\mathrm{Re}\langle \omega, E \rangle \not\equiv 0,\ \delta\,\mathrm{Re}(\omega \otimes \omega_0) \not\equiv 0}$ and R-linearly dependent. Let α be a non-zero real number such that $\delta\,\mathrm{Re}(\omega \otimes \omega_0) = \alpha\,\mathrm{Re}\langle \omega, E \rangle$. Then, if $\lambda + \mu \neq 0$, $(*)$ implies $\mu + \nu = \alpha(\lambda + \mu)$, from where $(**)$ implies in turn:

$$\int_S \langle \bar{A}', \mathrm{Re}(\omega \otimes \omega_0) \rangle \, ^{(3)}\eta + \alpha \int_S \mathrm{Re}\langle i\bar{\omega}', E \rangle \, ^{(3)}\eta = 0 \qquad (14)$$

for all $\bar{A}'$ and $\bar{\omega}'$, such that $(\bar{\omega}_0', \bar{\omega}', \bar{E}', \bar{A}_0', \bar{A}', \bar{E}') \in F$. In particular, for all $(\bar{\omega}_0', 0, 0, 1, \bar{A}', \bar{E}')$ verifying (13). But, for such vectors (14) is $\neq 0$. Then $\lambda + \mu = 0$, from where $(**)$ implies $\mu + \nu = 0$, and the equalities (a), (b), (c) and (d) give rise to a vector of the type (6).

5) $\underline{\mathrm{Re}\langle \omega, E \rangle \equiv 0,\ \delta\,\mathrm{Re}(\omega \otimes \omega_0) \equiv 0.}$ In this case $(*)$ is an identity. Let us see what $(**)$ implies.

5') If $\mathrm{Re}(\omega \otimes \omega_0) \not\equiv 0$, applying $(**)$ for all $(\bar{\omega}_0', 0, 0, 1, \bar{A}', \bar{E}')$ verifying (13), one has:

$$(\lambda + \mu) \int_S \| \omega \|^2 \, ^{(3)}\eta = 0$$

Then $\lambda + \mu = 0$, from where $(**)$ is converted in to:

$$(\mu + \nu) \int_S \mathrm{Re}\langle i\bar{\omega}', E \rangle \, ^{(3)}\eta = 0$$

and, by the argument employed in 2), the vector defined by the equalities (a), (b), (c) and (d) is also of the type (6).

5") Finally, the case $\mathrm{Re}(\omega \otimes \omega_0) \equiv 0$ gives rise to the situation 2) again //

3. The massless case.

3.1 The argument used in the above proof for obtaining the equalities (a) and (d) may be applied to this case with the same result. However, everything else cannot be applied because it depends on $m \neq 0$ essentially. So, the procedure to be followed in order to determine rad $\hat{\Omega}_2$ in this case will have to be substantially different.

For the moment, we shall limit ourselves to seeing what conditions a vector of the type $(\phi, d_A\phi, 0,0,0,E')$ must verify in order to belong to rad $\hat{\Omega}_2$.

Above all, as the constraint equations (5) (with $m = 0$) must be satisfied by the said vector, we should have:

$$\delta E' = \mathrm{Re}\langle i d_A\phi, E\rangle \tag{15}$$

On the other hand, the condition of belonging to rad $\hat{\Omega}_2$ implies, for all vectors $(\bar{\omega}', \bar{\omega}_0', \bar{E}', \bar{A}_0', \bar{A}', \bar{E}') \in F$:

$$0 = -\int_S \mathrm{Re}(\langle d_A\phi, \bar{E}'\rangle + \langle i d_A\phi, \bar{A}_0'\omega - \omega_0\bar{A}'\rangle)^{(3)}\eta - \int_S \langle \bar{A}', E'\rangle^{(3)}\eta =$$

$$= -\int_S \mathrm{Re}[\phi(\delta_A\bar{E}' + i\delta_A(\omega_0\bar{A}' - \bar{A}_0'\omega))]^{(3)}\eta - \int_S \langle \bar{A}', E'\rangle^{(3)}\eta =$$

$$= -\int_S \mathrm{Re}(\phi\langle i\bar{A}', E\rangle)^{(3)}\eta - \int_S \langle \bar{A}', E'\rangle^{(3)}\eta = \int_S \langle \bar{A}', \mathrm{Re}(iE\otimes\phi) - E'\rangle^{(3)}\eta$$

In particular, $E' = \mathrm{Re}(iE\otimes\phi)$ satisfies the condition. If this kind of solution is considered, equation (15) becomes an identity due the constraint condition $\delta_A E = 0$.

Bearing all this in mind, we can enunciate the following:

<u>Theorem 2'.</u> For $m = 0$, the radical of the pre-symplectic metric $\hat{\Omega}_2$ of the manifold of solutions V of the coupled vectorial Klein-Gordon-Maxwell equations on a Lorentz manifold is, in each point $(\omega, A) \in V$, *strictly bigger* than the subspace tangent at such a point to the orbit of the gauge group through (ω, A). More precisely:

$$\pi(\text{subspace tangent to the orbit}) + \{(\Phi, d_A\phi, 0; 0,0, \mathrm{Re}(iE\otimes\phi))\} \subseteq \mathrm{rad}\,\hat{\Omega}_2$$

(+)

where Φ and ϕ are arbitrary sections of L_S. Consequently, the *phase space* of the minimal coupling defined by these equations is a quotient of the set of orbits of the corresponding gauge group.

3.2 The reason for this result must be sought after in the reducibility of the pre-symplectic metric of the free massless vectorial Klein-Gordon field. As is easily seen, this does not happen in any of the other examples mentioned. In fact, this observation was what suggested the reasoning in §3.1 to us.

Using the same notation as §2.2, the solution of the Cauchy problem for the equations $\delta_A d_A \omega' = 0$ of §1, allows us to establish a canonical projection $\pi : T_\omega(V) \longrightarrow F_{free}$ from the tangent space $T_\omega(V)$ onto the space F_{free} defined by the elements on S, (ω_o', ω', E'), such that $\delta_A E' = 0$. The projection of the presymplectic metric (3) onto this space is:

$$(\hat{\Omega}_2)_{free}((\omega_o', \omega', E'),(\bar{\omega}_o', \bar{\omega}', \bar{E}')) = \int_S \mathrm{Re}(\langle\bar{\omega}', E'\rangle - \langle\omega', \bar{E}'\rangle)\ {}^{(3)}\eta \qquad (16)$$

If $(\omega_o', \omega', E') \in \mathrm{rad}(\hat{\Omega}_2)_{free}$, taking $\bar{E}' = 0$, then for all 1-forms $\bar{\omega}'$ one has:

$$\int_S \mathrm{Re}\langle\bar{\omega}', E'\rangle\ {}^{(3)}\eta$$

which implies $E' = 0$.

Using this in (16) one now has for all $\bar{E}'$ such that $\delta_A \bar{E}' = 0$:

$$\int_S \mathrm{Re}\langle\omega', \bar{E}'\rangle\ {}^{(3)}\eta = 0$$

from where, applying the Fredholm decomposition to the operator $d_A : \Lambda^0(S,L) \longrightarrow \Lambda^1(S,L)$ adjoint to $\delta_A : \Lambda^1(S,L) \longrightarrow \Lambda^0(S,L)$, it follows that $\omega' = d_A \phi$.

Bearing now in mind that $\omega_o' = \Phi$ has remained arbitrary, one has:

$$\mathrm{rad}(\hat{\Omega}_2)_{free} = \{(\Phi, d_A\phi, 0)\} \qquad (17)$$

where Φ and ϕ are arbitrary sections of L_S.

Returning again to the coupled equations, for each solution $(\omega, A) \in V$, it is possible to establish a mapping $j : \mathrm{rad}(\hat{\Omega}_2)_{free} \longrightarrow F$ by the rule:

$$j(\Phi, d_A\phi, 0) = (\Phi, d_A\phi, 0; 0, 0, \mathrm{Re}(i\, E \otimes \phi)) \qquad (18)$$

Thus the formula $(^+)$ may be rewritten as follows:

$$(\text{subspace tangent to the orbit}) + j\ \mathrm{rad}(\hat{\Omega}_2)_{free} \subseteq \mathrm{rad}\,\hat{\Omega}_2 \qquad (^{+'})$$

Even though this result illustrates well how the reducibility of the free field produces an "increase" of the rad $\hat{\Omega}_2$ in the massless case, it is not satisfactory enough, because above all it is not shown that the said "increase" is the maximum possible. The difficulty found in completely determining rad $\hat{\Omega}_2$ in this case lies in the appearance of a constraint condition which is defined by a differential operator to which, in principle, we cannot apply a Fredholm-type decomposition theorem. Independent of this, and perhaps previous to it, it would be desirable to have a geometric characterization of the above "increase" previous to the Cauchy problem. This would require a more detailed study of the presymplectic structure of the free field in order to later attempt to interpret geometrically the (non trivial) "increase" defined in ($^{+'}$), reflected in the term $\mathrm{Re}(iE \otimes \Phi)$ of the mapping j.

REFERENCES

[1] Branson, T., The Yang-Mills equations: quasi-invariance, special solutions, and Banach manifold geometry (Thesis), M.I.T., 1979.

[2] García, P., Reducibility of the Symplectic Structure of Classical fields with gauge-symmetry. Lect. Not. in Math., Springer-Verlag, 570, 1977.

[3] García, P., Tangent structure of Yang-Mills equations and Hodge Theory, Lect. Not. in Math., Springer-Verlag, 836, 1980.

[4] García, P. - Pérez-Rendón, A., Reducibility of the Symplectic Structure of Minimal Interactions, Lect. Not. in Math., Springer-Verlag, 676, 1978.

[5] Koszul, J., Lectures on Fibre Bundles and Differential Geometry, Tata Inst. of Fund. Research, Bombay, 1960.

[6] Moncrief, V., Gauge Symmetries of Yang-Mills Fields, Annals of Physics, Vol. 108, no. 2, 1977.

[7] Moncrief, V., Reduction of the Yang-Mills equations, Lect. Not. in Math., Springer-Verlag, 836, 1980.

[8] Pérez-Rendón, A., A Minimal Interaction Principle for Classical Fields, Symposia Mathematica, Vol. 14, 1974.

[9] Segal, I., General properties of the Yang-Mills equations in physical space (non linear wave equation/guage-invariance/Cauchy problem/Symplectic Structure). Proc. Natl. Acad. Sic. U.S.A., Vol. 75, no. 10, 1978.

IV. SPACE-TIME GEOMETRY AND GENERAL RELATIVITY

PARTICLE THEORY AND GLOBAL GEOMETRY

I.E. Segal

M.I.T., Dept. of Mathematics,
Cambridge, MA 02139, U.S.A.

Introduction

Suppose one wants to do particle theory in a general cosmos (= space-time), in a way that can lead to concrete predictions. With what format does one begin, and how does one proceed towards experimentally accessible conclusions?

Obviously the usual assumption that space is Minkowskian is established only locally, and even then may be an inadequate approximation. On the other hand, if one takes a very general point of view, e.g. that the cosmos has a non-static curvature, there is no apparent unique and effective definition of energy, or of temporal duration; the usual group-theoretic apparatus of quantum particle theory is lacking; empirically meaningful predictions are hardly visible even on the horizon. Until recently, in any event, little consideration was given to possible global effects on ultramicroscopic physics, on the intuitive grounds that the curvature of space, even if non-zero, was too small to produce observable results in laboratory particle experiments. But it is impossible to be sure of this until one has at least a well-defined procedure for determining empirical implications.

Fortunately, it seems possible to take as a starting point some very general structural elements, causality and symmetry, and in spite of this generality of format, arrive at fairly specific physical results with the prospect of such implications. A very natural locus for particle theory appears to be a manifold M endowed with causality in the sense of a convex cone field given in the tangent space at each point, representing physically the future directions at the point; and with sufficient symmetry that at least energy and duration are definable in conjunction with the simple group theory usually involved in their treatment. It is sufficient for the latter to have a 'covariant clock', this being a 'time' parameter τ, $M \longrightarrow R^1$, that meshes with an evolutionary group on M, the latter being a one-parameter subgroup T_t of the group G(M) of all casual (i.e. causal-cone-field-preserving) transformations on M, that displaces each point into its 'future' (definable in a natural way from the cone field), and satisfies the 'interwining relation' $\tau(T_t p) = \tau(p) + t$ for all points p of M. For a covariant clock there is a corresponding 'energy' and a well-defined notion of 'temporal duration', unaffected (except in scale) by reparametrization.

In this context it is natural to postulate that a massless particle is represented by a cross section of a homogeneous vector bundle over M that has positive energy for the G-action. A causal group has a 'positive semigroup' (if one makes the natural assumption of the absence of closed time-like curves), consisting of the transformations that displace into the future; this semigroup corresponds to an invariant convex cone C in the Lie algebra of G; and the usual notion of positive-energy for the Lorentz group is reproduced if one defines as a positive-energy representation one that to each direction in C makes correspond an operator with non-negative spectrum. The Maxwell and massless Dirac equations define such representations of the Poincaré group (the latter equation after second quantization, or related redefinition of the group action).

A *massive* particle will be represented similarly, but only the 'mass-conserving' subgroup will leave invariant the space of cross-sections corresponding to a particular particle species. Whether massless or massive, the cross-sections should be determined by their restrictions to space at a fixed time, and propagated causally in accordance with the given cone field, which requires that the sections satisfy a hyperbolic partial differential equation whose associated cone field is that given.

If one assumes a very modest amount of isotropy, then as follows from Tits' and Vinberg's work the 4-dimensional cosmos are highly restricted. The chronometric cosmos then provides the simplest, and one of only two, alternatives to Minkowski space; it thus seems of interest as a prototype for possible alternatives, and for other physical reasons.

The next section develops these points further, and relates them to cosmology and the 'chronometric principle'. This principle derives essentially from the circumstance that the causal group has both 'extensive' and 'intensive' generators, the former having expectation values that depend on the *global* character of the wave function (i.e. section), while the latter depend only on the *local* character of the wave function. Only the latter can be directly observed, but the possibility appears that the extensive generators may nevertheless be the dynamically effective ones. This distinction between the local and the global can be realized in any curved cosmos, in a way made familiar by Einstein's theory of general relativity. The chronometric principle in particular is a specialization of the idea proposed by Minkowski, that difficulties in fundamental physics may be explicable by passage from one geometry to another that is less singular, and has the original one as a 'limiting case'; he replaced the Galilean by the Poincaré group, and altered the casual structure of the cosmos, but not its geometry otherwise; the chronometric cosmos is causally locally Minkowskian, but has otherwise a different geometry, and a causal group of which the Poincaré group is a limiting case. Our new group has certain terminal properties, and emerged from Minkowski-like considerations in particle physics (Segal, 1951, et seq.); its empirically tangible application was however first made in cosmology. Here I enlarge on points already made in joint work with H.P. Jakobsen, B. Ørsted, and B. Speh, directed back towards particles. In particular, we indicate model-dependence of the apparent non-conservation of parity; linear aspects of parity in general spaces were treated by Lichnerowicz, 1964; here it is the relation to the Fermi and Yukawa interactions that is involved.

The chronometric principle

A change in the space-time geometry or symmetry by itself is essentially kinematical, and does not directly imply dynamical consequences subject to experimental test. It is only when we ask the question: how could one experimentally distinguish between a putative space-time model and a conventional one that we are led to what may serve as a dynamical principle.

The chronometric space-time, or 'cosmos' for short, is the unique one other than Minkowski space that enjoys comparable properties of causality, symmetry, and separability of time from space, as detailed elsewhere (cf. Segal, 1976 and 1980). At the same time it is also the simplest elementary conceptual variant of Minkowski space, being representable in the form $R^1 \times S^3$, with its natural causal (conformal) structure. But as noted, the attainment of definite physical predictions requires more than a pure geometry; the *chronometric principle* supplies this need. If we ask how we might decide whether the cosmos is $R^1 \times S^3$ or $R^1 \times R^3$, we are first forced to admit that our direct observations are inherently local; and then are led to realize, that our data reduction, and ultimately our measurements in laboratories, proceed on the assumption that space is locally R^3. Very succinctly, the chronometric principle states that in microscopic observation we see not necessarily the actual large-scale global variables or states, but rather their closest 'flat' approximations.

It is a natural adaptation of a general relativistic principle, but is more concerned with symmetry than with curvature, and since groups and their representations are more rigid than curved spaces, has more explicit implications. The central one regards the form of the energy, which is assumed locally observed as the generator of temporal displacement in Minkowski space, rather than as the analogous generator in chronometric space.

More specifically, both spaces are locally Minkowskian in their causal (conformal) structure, i.e. there is a local isomorphism that maps the future cone in the tangent space at a point in one space onto that at the image point. But within the group of local causality-preservation transformations in Minkowski space, there are two distinct one-parameter subgroups that have all of the general physical properties of temporal evolution, but are non-conjugate. These are the two respective groups defining temporal evolution in the respective spaces.

Concretely, Minkowski space is locally causally the same as U(2) with the invariant causal structure that which at the identity I defines the future as the positive-definite hermitian matrices, relative to the usual correspondence between the space H(2) of 2 x 2 hermitian matrices and the Lie algebra u(2) of U(2). It is globally the same as H(2). In H(2), the one-parameter group: $H \longrightarrow H + tI$ has all the natural properties of temporal evolution; in U(2), the one-parameter group $U \longrightarrow e^{it}U$ likewise has these properties. However, the two groups are fundamentally inequivalent, although locally H(2) and U(2) are causally equivalent, via the Cayley transform $H \rightarrow (1+iH/2)(1-iH/2)^{-1}$. (The exponential mapping is not a local causal equivalence.)

The chronometric principle leads in a clear-cut way to the separation of a class of generators of the local causal group (which is the same for both cosmos, and is global rather than local for the chronometric cosmos) into a scale-covariant part, which according to the principle is the local part directly observed in the laboratory, and an anti-scale-covariant part, which is delocalized as far as observation by homo sapiens is concerned. (Those generators that are unaffected by scale transformations - space rotations and pure Lorentz transformations - are however devoid of any such separation.) Because light theory depends only on the causal structure, and because observations on light are physically so basic, particularly in astronomy, and finally because no new unobserved physical parameters are involved, the principle is natural and appealing but a priori highly vulnerable to comparison with redshift observations. However, it has fit objective galaxy and quasar data very well, without the extensive ancillary hypotheses required for a fit to the expanding-universe hypothesis. The usual special relativistic energy operator forms the scale-covariant part of the chronometric energy operator (Segal, 1976), forming together with Maxwell's equations (or equivalently, for present purposes, the unitary representation of the causal group containing both generators), the basis for redshift predictions.

What this means more specifically will be illustrated by consideration of spinor fields and the Dirac equations on our alternative cosmos, to which we now turn.

Fields on the chronometric cosmos

Our alternative cosmos is describable covariantly as the homogeneous space $\tilde{M}$ of $\widetilde{SU}(2,2)$ obtained by forming its quotient with respect to its natural Poincaré subgroup $\tilde{P}^e$, where the tildes denote the universal covers, and the exponent e indicates extension by the one-dimensional group of scale transformations. This space $\tilde{M}$ is also describable as $R^1 \times S^3$, or as the universal cover $\tilde{U}(2)$ of the conformal compactification $\bar{M} \simeq U(2)$ of Minkowski space M. All these spaces have defined on them natural causal structures, with respect to which they are globally equivalent; and $\widetilde{SU}(2,2)$ is simply the universal, and 2-fold, cover of the group of all causality-preserving transformations on $\tilde{M}$.

Within the special relativistic framework, particles are classified by their spin, which define finite-dimensional representations of $\tilde{P}^e$ that are trivial on the translation subgroup. By induction, these representations give rise to bundles over $\tilde{M}$, the cross-sections of which form in the most natural way chronometric analogues of the relativistic wave functions. For example, the general spinor field over $\tilde{M}$ is definable in this way by the bundle induced from the usual four-dimensional spin representation of the Poincaré group.

Particles are of course characterized by mass as well as spin, and it is with the consideration of particles of non-vanishing mass that new physical theory is required. The Maxwell and other equations for particles of vanishing mass are conformally covariant, and lift up directly to $\tilde{M}$. Although considerable further analysis remains to deal with such equations in the curved setting appropriate to the chronometric theory, they present no immediate physical problem. However, massive particles raise a number of fundamental, interlocking, physical issues.

For brevity, suffice it to say here that mathematically, there are three subgroups of $\widetilde{SU}(2,2)$ which are a priori tenable as symmetry groups of massive particles, without doing violence to broad formal physical ideas. These are the groups $\tilde{P}^e$, $\widetilde{SO}(3,2)$, and $R^1 \times \widetilde{SO}(4)$, the latter being the maximal essentially compact subgroup. All of these groups admit conformally equivalent invariant metrics. Taking as base point the origin, or point of observation as interpreted physically, and positioning them in $G = \widetilde{SU}(2,2)$ so that they locally osculate to the extent that their different dimensions permit, their respective orbits are Minkowski space M; a de Sitter type space M'; and chronometric space $\tilde{M}$. The space M' has virtually all of the general features of the other two, except that it is not globally factorizable as time x space, and the region of influence of a compact set propagates peculiarly for long times, but these might quite reasonably be regarded as physically expendable. A more serious basis for the elimination of M' is that its volume relations bear no natural relation to cosmological observational counts that, on the assumption of spatial homogeneity, might be expected to yield results approximating proportionality to volume. Those for $\tilde{M}$ - essentially simply the volume relations on S^3 - fit quasar counts in this sense (Segal et al., 1980a), and are in satisfactory agreement with the less clear-cut case (lacking redshift observations) of radio source counts. In addition, there is an important general theoretical indication for the subgroup $\tilde{K}$, namely that it is sure to have an invariant state, unlike the open simple group SO(2,3). Such a state may be formulated in a very general way as a fixed point within a convex set in function space (the so-called state space of an algebra of observables) under the action of the mass-preserving subgroup of the fundamental group acting as automorphisms of the algebra, or formally equivalently, leaving invariant the fundamental equations defining particle interactions. Very briefly, the physical 'vacuum', crucial for microscopic theory, is the lowest-energy such invariant state.

Let us then fix the vacuum (and mass) conserving subgroup as $\tilde{K}$, and investigate the consequences.

Parallelization

A special feature of the chronometric geometry is the possibility of a simple identification of the space $R^1 \times SU(2)$ with a subgroup of the symmetry group, $\tilde{SU}(2,2)$; or, after covering transformations, of U(2) with a subgroup of the adjoint group of SU(2,2). It is the exploitation of this feature that leads to a simple explicit formula for the parallelization via group translation within the subgroup. In order to present the idea without distracting particularities, it is given in a general form applicable to the similar treatment of SU(n,n) and other cases.

LEMMA. *Let* N *be a subgroup of the connected Lie group* G. *Let* G *act as a transformation group on* N: $(g,x) \longrightarrow g(x)$; $(gg')(x) = g(g'(x))$, *in such a way that if* $x \in N$, *then* $x(y) = xy$, $y \in N$.

Let R *be a finite-dimensional representation of the subgroup* G_{x_0} *of* G *leaving fixed a point* x_0 *of* N; *let* $g \longrightarrow V(g)$ *denote the action of* G *in the induced R-bundle B over* N, $V(g)$ *being formulated as the operator on* C^∞ *sections:* $\phi(x) \longrightarrow S(g,x)\phi(g^{-1}(x))$, *where* $S(g,x)$ *is an operator from the fiber* $F_{g^{-1}(x)}$ *to* F_x (F_x *denoting the fiber over* x), *and also a smooth function on* $G \times N$ *such that* $S(g_0,x_0) = R(g_0)$, $g_0 \in N$. *Let* L *denote the 'parallelization' map on* B: $\emptyset \longrightarrow \tilde{\emptyset}$, *where* $\tilde{\emptyset}(x) = S(xx_0^{-1}, x)^{-1}\emptyset(x)$.

Then the parallelized action V, $V(g) = LV(g)L^{-1}$, *takes the form*

$$\tilde{V}(g): \tilde{\emptyset}(x) \longrightarrow R(x_0x^{-1}g \cdot g^{-1}(x)x_0^{-1}),$$

where $\cdot$ *indicates multiplication in* G.

Proof. By virtue of the representation property of V, the following holds for arbitrary $a,b \in G$ and $x \in N$:

(✱) $$S(ab,x) = S(a,x)S(b,a^{-1}x).$$

Now to compute $LV(g)L^{-1}$, take an arbitrary cross-section $\emptyset(.)$; then

L^{-1}: $\tilde{\emptyset}(x) \longrightarrow S(xx_0^{-1},x)\tilde{\emptyset}(x)$;

$V(g)L^{-1}$: $\tilde{\emptyset}(x) \longrightarrow S(g,x)S(g^{-1}(x)\cdot x_0^{-1}, g^{-1}(x))\tilde{\emptyset}(g^{-1}(x))$;

$LV(g)L^{-1}$: $\tilde{\emptyset}(x) \longrightarrow S(xx_0^{-1},x)^{-1}S(g,x)S(g^{-1}(x)\cdot x_0^{-1}, g^{-1}(x)\tilde{\emptyset}(g^{-1}(x))$.

The question is whether the multiplier here is the same as $R(g')$, where $g' = x_0x^{-1}g\cdot g^{-1}(x)\cdot x_0^{-1}$. Note to begin that $g' \in G_{x_0}$, since

$$\emptyset(g')x_0 = \emptyset(x_0)\emptyset(x^{-1})\emptyset(g)\ (\emptyset(g^{-1})x) = x_0$$

Now by defining property of the induced bundle, $R(g',x_0) = S(g',x_0)$ for $g' \in G_{x_0}$. Thus the question is equivalent to the equality

$$S(xx_0^{-1},x)^{-1}S(g,x)S(g^{-1}(x)\cdot x_0^{-1}, g^{-1}(x)) = S(x_0x^{-1}g\cdot g^{-1}(x)\cdot x_0^{-1}, x_0).$$

To show this, apply equation (✱) to

$$S(g,x)S(g^{-1}(x)\cdot x_0^{-1}, g^{-1}(x)),$$

obtaining $S(g\cdot g^{-1}(x)\cdot x_0^{-1},x)$. Now multiplying on the left by $S(xx_0^{-1},x)$, it suffices to show that

$$S(g\cdot g^{-1}(x)\cdot x_0^{-1},x) = S(xx_0^{-1},x)S(x_0x^{-1}g\cdot(g^{-1}(x))x_0^{-1},x_0).$$

This follows by another application of equation (✱) to the right side of this putative equality.

The application to the immediate context may be more succinctly indicated on the minimal covered space U(2), with G limited to the group SU(2,2). The subgroup N may then be described as that consisting of the elements $g \in G$ of the form $g = \begin{pmatrix} A & 0 \\ 0 & D \end{pmatrix}$, where D is a scalar matrix. The map $g \longrightarrow AD^{-1}$ is a homomorphism of this subgroup N into U(2); this homomorphism is a local isomorphism, and it will suffice here to work locally in G (but globally on SU(2)), to establish a local representation of G, which may then be appropriately lifted. The inverse map from U(2) into G takes the form $Z \longrightarrow (\det Z)^{-1/4} \begin{bmatrix} Z & 0 \\ 0 & I \end{bmatrix}$.

This needs to have the property that if $g \in N$, then g acts on $x \in N$ just as left multiplication. Now if $g = \begin{bmatrix} A' & 0 \\ 0 & D' \end{bmatrix}$, then this carries $x = U \in U(2)$, say, into $A'UD'^{-1}$. On the other hand, as an element of SU(2,2), $x = U$ corresponds to $(\det U)^{-1/4} \begin{bmatrix} U & 0 \\ 0 & I \end{bmatrix}$, and left multiplication by g gives $g\,U = (\det U)^{-1/4} \begin{bmatrix} A'U & 0 \\ 0 & D' \end{bmatrix}$. The latter matrix in SU(2,2) corresponds to $(A'U)\,(D')^{-1}$ in U(2), i.e. to $(A'D'^{-1})U$, which is left multiplication in U(2) by the element of U(2) corresponding to g.

Thus the Lemma is applicable. To apply it requires the computation of $x_0 x^{-1} g \cdot g^{-1}(x) x_0^{-1}$. Here it is naturally convenient to choose x_0 so that its isotropy group G_{x_0} corresponds to the usual physical Poincaré group via the equivalence between the causal group on M and SU(2,2) modulo its center on U(2). The physical Poincaré group does not leave any point of M fixed, and so x_0 corresponds to a point of the compactification $\overline{M}$ external to M. It corresponds in fact to the element -I of U(2). Computation now leads to the

COROLLARY. *The parallelized action of* G *on the R-bundle induced from the physical Poincaré group takes the form*

$$\tilde{\phi}(x) \longrightarrow R(g^*)\tilde{\phi}(g^{-1}(x)),$$

where

$$g^* = (\det ZW^{-1})^{1/4} \begin{bmatrix} Z^{-1}AW & -Z^{-1}B \\ -CW & D \end{bmatrix}$$

and

$$W = (A'Z + B')\,(C'Z + D')^{-1};\ g^{-1} = \begin{bmatrix} A' & B' \\ C' & D' \end{bmatrix} .$$

Applying this to the special case of the spin representation of the Poincaré group gives the transformation properties of the corresponding spinor fields. It will suffice here to give those under $\tilde{K}$, which it is convenient to take in the form $R^1 \times SU(2) \times SU(2)$. This acts on M as follows: if $p = (t,W) \in \tilde{M} = R^1 \times SU(2)$,

$$g = s \times U \times V:\ (t,W) \longrightarrow (t + s,\ UWV^{-1}) = g(p).$$

The action of g on the indicated parallelized spinor bundle then takes the form

$$\Gamma(g):\ \tilde{\phi}(p) \longrightarrow \begin{bmatrix} V & 0 \\ 0 & V \end{bmatrix} \ \phi(g^{-1}(p)),$$

independently of the 'conformal weight', which determines the action (assumed scalar) of R on the scale transformations.

Parity

In principle, the foregoing covers the treatment of space and time reversal, provided the representation from which one induces is one of the extended Poincaré group, inclusive of space and time reversal. If however one induces from the connected group, then a separate treatment of these symmetries must be made with reference to the connected chronometric group. It is in fact the case that they do extend to this group; i.e. every automorphism of the Poincaré group, outer as well as inner, extends (uniquely) to one of $\tilde{\tilde{SU}}(2,2)$. In abstract principle the automorphism of SU(2,2) extending space reversal on $\tilde{P}^e$ does not need to be implementable by a point transformation on the underlying geometrical space $\tilde{M}$, but in fact it is, taking the form $(t,W) \longrightarrow (t,W^{-1})$, directly extending space reversal in M.

Whether we induce from the connected or extended (by transformations not connected to the identity) Poincaré group, the operator P representing space reversal is, in accordance with normal physical theory, conditioned by the property that $P\,\Gamma(g)P^{-1} = \Gamma(g^{\pi})$, $g \in G$, where π is the automorphism of G, written exponentially, corresponding to space reversal. If one wants to take a maximally conservative (general) position with regard to the parity operator, the minimum that can reasonably be required is this relation, together with analogous relations regarding commutation relations with the operators T and C representing time reversal and charge conjugation. If one now assumes that T and C are given the direct extensions of the conventional ones, as is possible because the curvature of space does not directly affect them; normalizes by requiring that $P^2 = 1$, $PT = -TP$, $CP = -PC$; uses the covariance condition on P only for $g \in \tilde{K}$, the full invariance group of massive particles; and assumes that P takes the form: $\tilde{\phi}(t,W) \longrightarrow M\tilde{\phi}(t,W^{-1})$, where M is a matrix that is *now necessarily space-dependent*, one finds that M must have the form $i\begin{bmatrix} 0 & W^{-1} \\ -W^{-1} & 0 \end{bmatrix} N$, where N is a matrix of the form $\begin{bmatrix} a & b \\ -b & a \end{bmatrix}$, a and b being real scalar matrices. It will be no essential restriction for present purposes to suppose that $N = I$ or $N = \begin{bmatrix} 0 & 1 \\ -1 & 0 \end{bmatrix}$, two possibilities that also arise by induction from the extended Poincaré group.

The generalized Dirac equation

We must now specify the chronometric spinor particles of definite mass. What are the principal physical desiderata concerning the (wave) equations on M providing such specification? The equation will presumably take the form $(\underline{D} + m)\phi = 0$, where $\underline{D}$ is a differential operator (which in the limit of infinite radius of the universe - analogous to the limit $c \longrightarrow \infty$ which goes from the Lorentz to the Galilean group - should coincide with the usual Dirac operator), and m is a constant, the mass, or a function thereof. But what physically appropriate and mathematically efficient constraints uniquely determine $\underline{D}$ (hopefully)? The basic ones are well known, and may be succinctly described as follows:

1. Positivity of the energy (spectral condition);
2. Finiteness of the propagation velocity (causality);
3. Invariance under the essential isometry group;
4. Invariance under the mass-conserving subgroup (here the essential isometry group $\tilde{K}$ of $\tilde{M}$ - 'essential' meaning that $\tilde{K}$ is the universal cover of the isometry group);
5. Covariance under the full causal group of the cosmos (hopefully);
6. Huygens' principle (the existence of the familiar lacuna, transformed into the new cosmos) in the case of vanishing m.

The mathematical interpretation of these physical constraints admits of some flexibility, but it is reassuring that in practice, the ultimate implications are unaffected. In the case of 'positivity of the energy', the most fundamental and cogent constraint, it should be understood that this is no contra-indication for negative-frequency solutions such as are well-known for the Dirac equation in Minkowski space, but should be taken to refer to the associated quantized field, for which a unique appropriate form satisfying this constraint exists. Strictly speaking, it is necessary to carry out the entire treatment of the Dirac equation on the second-quantized level to eliminate ambiguities involved in a single-particle treatment, but here we can only make reference to Weinless (1969).

In the present context, any tenable physical interpretation will exclude the naive wave equation $\Box \phi = 0$, where $\Box = \partial^2/\partial t^2 - \Delta$, Δ designating the Laplace-Beltrami operator on S^3, and require the replacement of $-\Delta$ by $1 - \Delta$ (the 1 scaling as curvature). With this one change, the consequent Klein-Gordon type equation automatically satisfies all six constraints (cf. Jakobsen et al., 1978, and Ørsted, 1980). The decomposition of a general solution into positive and negative frequencies is invariant under G; the equation defines a unitary representation of the G, which, most importantly, is positive-energy in the sense that the generators of temporal evolution, whether in M or in $\tilde{M}$, correspond to positive self-adjoint operators. None of this is true without the additive constant 1, or with any other constant. Of the enumerated constraints, only 2 and 3 are satisfied with other constants.

The case of spinor fields is considerably more complicated. In terms of the parallelization employed earlier, the natural basis for the vector fields on $\tilde{M}$ consists of the right translations, which has no 'internal' action. The simplest natural candidate for a physically appropriate generalization of the Dirac equation takes the form (particularly in the light of the situation described for scalar fields)

$$\underline{D} = \gamma_\mu X_\mu + C,$$

where the γ_μ are the usual constant Dirac matrices, the X_μ are an orthonormal basis for the left-invariant vector fields, relative to the $\tilde{K}$-invariant metric on $\tilde{M}$, and C is a constant matrix. It is useful to normalize the X_μ so that at the point of observation, (0,I), they coincide with the Poincaré generators $\partial/\partial x_\mu$ ($\mu = 0,1,2,3$).

The choice $C = -(3/2)\gamma_4\gamma_5$ turns out to give an analog to the additive constant 1 for the wave equation, yielding what appears to be the physically as well as mathematically appropriate generalization of the Dirac operator in Minkowski space. I thank S.M. Paneitz and H.R. Petry for communicating the results of independent determinations of the Dirac operator. It is perhaps worth noting that the operator with $C = -\gamma_4\gamma_5$ has the property that its square is the wave operator, in analogy with the situation in Minkowski space; however, it is covariant only modulo zero-order terms. Concerning covariance cf. also Kosmann (1976).

The chronometric treatment of parity involves parity doublets that perforce arise because of the non-invariance under space inversion of the generator, say ζ, of the infinite cyclic constituent of the center of the universal cover G of SU(2,2). Because of this ζ and parity cannot both be represented as scalars in the action of G on spinor fields. Parity doublets were proposed by Lee and Yang within the Minkowski framework at a preliminary stage in their work, and indeed arise naturally from Minkowskian harmonic analysis considerations, but the central symmetry ζ is non-existent (or may be said to act only trivially) in Minkowski space.

The main point here however is the treatment of the Fermi and Yukawa interactions in relation to parity invariance, in the chronometric cosmos. Before this, we should however note the parallel between the present treatment and parity doublet theories of various types that have been proposed earlier in Minkowski space (cf. Lee and Yang (1956), Pati (1977), and references given in the latter work). Basically, the theoretical physical role of the Dirac equation is to pick out a class of positive-energy (or negative energy, but then a conjugation changes the sign) subspaces of the general spin field that have the following properties: (a) dynamically, within each subspace, a wave function is determined by its values at any given time, which may be given arbitrarily, and which propagate in accordance with the geometrically given causal structure; (b) group-theoretically, the decomposition should be invariant under the physical symmetry group leaving the putative vacuum invariant (here $\tilde{K}$, in Minkowski space, $\tilde{P}^e$).

From this point of view, the two subspaces defining the parity doublet are equally valid. The suppression of one member of the parity doublet while seemingly plausible from the standpoint of theoretical economy, is arbitrary from the standpoint either of mathematical harmonic analysis or the usual assumption that central quantum members should be absolutely conserved.

Interactions and space inversion

In order to have a parity-invariant free particle formalism, it is natural in the chronometric cosmos for a spinor wave function to have two components, each satisfying the Dirac equation, which transform into each other under parity, as indicated. The Fermi interaction, bilinear in the bilinear invariant local forms in the respective wave functions, correspondingly involves several types of these bilinear local forms, known as 'currents'. Calling the two equations 'left' and 'right', and denoting the corresponding wave function components by superscripts L and R, there will be two types of P-invariant currents between two particles, say 1 and 2:

$$j_\mu^{12} = \bar{\psi}_1^L \gamma_\mu \psi_2^L + \bar{\psi}_1^R \gamma_\mu \psi_2^R \; ; \quad k_\mu^{12} = \bar{\psi}_1^L \gamma_\mu \psi_2^R + \bar{\psi}_1^R \gamma_\mu \psi_2^L \, ,$$

in terms of the usual Dirac notations. The general P-invariant Fermi interaction takes the corresponding form

$$H_I = a_1 j_\mu^{12} j_\mu^{34} + a_2 k_\mu^{12} k_\mu^{34} + a_3 j_\mu^{12} k_\mu^{34} + a_4 k_\mu^{12} j_\mu^{34} \, .$$

The numerical coefficients a_i are undetermined. However, if in partial analogy with charge independence of strong interactions it is proposed that the interaction under consideration, say beta decay, is insensitive to the 'chirality' (i.e. left-ness or right-ness of the particle), the simplest form with equal coefficients results. It is now a straightforward deformation to the limit of vanishing curvature to verify that the resulting interaction in this case appears locally, within the conventional special relativistic framework, identical with the chiral-invariant interactions that is found experimentally for beta decay and many other weak interactions, and interpreted in that framework as maximal parity-violating.

A natural question is why if chiral-invariant Fermi interactions appear as maximal parity-violating in the Minkowski space framework, interactions like quantum electrodynamics do not likewise appear such. The answer is that the bosons involved may have wave functions that exhibit symmetry or anti-symmetry relative to the central element ζ of G (the photon exhibits the former) and so may be defined on the 2-fold cover $S^1 \times SU(2)$ of U(2) (cf. Jakobsen et al., 1978, and Speh, 1979). By conservation of energy, the total energy operator integrated over the time period 2π (and the transition from time $-\pi$ to $+\pi$ chronometrically corresponds to the transition from $-\infty$ to $+\infty$ in Minkowski space) is simply a multiple of the energy at any one time. On the other hand, space inversion acts nontrivially on the center of G, from which follows the symmetry or antisymmetry for the j or k currents, or the reverse. As a consequence, either the integrated j- or k-term in the interaction hamiltonian must vanish. The result then has the conventional limiting form, apart from a possible γ_5 factor that does not affect parity conservation.

In summary, then, the Fermi and Yukawa interactions on chronometric space are compatible with the Minkowski space interpretation of experiments fitted by these theoretical interactions, including the apparent maximal parity violation in the case of weak interactions.

ACKNOWLEDGMENT

I thank S.M. Paneitz and H.R. Petry for private communications, and K. Bleuler and L. O'Raifeartaigh for stimulating discussions. Thanks are also due the Humboldt Foundation for facilitating the present contribution.

REFERENCES

Jakobsen, H.P., Ørsted, B., Segal, I.E., Speh, B., and Vergne, M. (1978). Symmetry and causality properties of physical fields. Proc. Nat. Acad. Sci. USA 75, 1609.

Kosmann, Y. (1976). On Lie transformation groups and the covariance of differential operators. In *Differential geometry and relativity*, ed. M. Cahen and M. Flato, Reidel, Dordrecht.

Lee, T.D. and Yang, C.N. (1956). Mass degeneracy of the heavy mesons. Phys. Rev. 102, 290.

Lichnerowicz, A. (1964). Champ de Dirac, champ du neutrino et transformations C, P, T sur un espace-temps courbe. Ann. Inst. H. Poincaré Sect. A(N.S.) 1, 233.

Ørsted, B. (1979). The conformal invariance of Huygens' principle. Jour. Diff. Geom., in press.

Ørsted, B. (1980). Conformally invariant differential equations and projective geometry. Jour. Funct. Anal., in press.

Pati, J.C. (1977). In *Unification of elementary forces and gauge theories*, report of B.W. Lee Mem. Conf., Batavia, publ. Harwood, London.

Segal, I.E. (1951). A class of operator algebras determined by groups. Duke Math. Jour. 18, 221.

Segal, I.E. (1976). *Mathematical cosmology and extragalactic astronomy*. Academic Press, N.Y.

Segal, I.E. (1980). Time energy, relativity, and cosmology. In *Symmetries in Science*, ed. B. Gruber and R.S. Millman, Plenum Publ., N.Y., 385.

Segal, I.E., Loncaric, J. and Segal, W. (1980a). Uniformity of the radial distribution of quasars in the chronometric cosmology and the X-ray background. Astrophys. Jour. 238, 38.

Segal, I.E., Jakobsen, H.P., Ørsted, B., Paneitz, S.M. and Speh, B. (1981). Covariant chronogeometry and extreme distances. II: Elementary particles. Proc. Nat. Acad. Sci. USA, in press.

Speh, B. (1979). Degenerate series representations of the universal covering group of SU(2,2). Jour. Funct. Anal. 33, 95.

Weinless, M. (1969). Existence and uniqueness of the vacuum for linear quantized fields. Jour. Funct. Anal. 4, 350.

GROUP THEORETICAL ASPECTS OF THE CHRONOMETRIC THEORY.*

H.P. Jakobsen

Mathematics Institute, University of Copenhagen,
DK-2100 Copenhagen Ø, Denmark.

Below is given a brief outline of one particular point, namely that of chronometric quantum-numbers for mass-zero equations. A complete description of these and related matters, which results from the joint investigations of I.E. Segal, B. Speh, B. Ørsted and myself, will appear elsewhere. In this connection I would like to thank Segal, Speh, Ørsted and M. Harris for valuable discussions. The other aspects touched upon in the talk are described in [1], and [2]. The chronometric theory is described in [5]. For additional background and further results, c.f. [3] and literature cited there.

1. Notation

$\tau = \tau_1$ denotes the defining representation of $Gl(2,\mathbb{C})$, i.e. $\tau\begin{pmatrix}\alpha & \beta\\ \gamma & \delta\end{pmatrix} = \begin{pmatrix}\alpha & \beta\\ \gamma & \delta\end{pmatrix}$. τ_n denotes the n-th fold symmetrized tensor product of τ, acting on $V_n = \overset{n}{\underset{s}{\otimes}} \mathbb{C}^2$, and τ_0 denotes the trivial representation. The letters a,b,c,d,h,k,u,v,z,w denote 2 x 2 matrices.

2. Basics

The unitary representations of SU(2,2) corresponding to spin $\frac{n}{2}$ and mass o are

$$(U_n^+(g)f)(z) = \tau_n(cz+d)^{-1}\det(cz+d)^{-1} f(g^{-1}z), \text{ and}$$
$$(U_n^-(g)f)(z) = \tau_n(a-(g^{-1}z)c)^{-1}\det(a-(g^{-1}z)c)^{n+1} f(g^{-1}z), \tag{2.1}$$

where $g^{-1} = \begin{pmatrix}a & b\\ c & d\end{pmatrix} \in SU(2,2)$, $g^{-1}z = \frac{az+b}{cz+d}$, and $f(z) \in V_n$.

* This research was supported in part by a grant from the N.S.F.

Let

$$SU(2,2)_D = \left\{ g = \begin{pmatrix} a & b \\ c & d \end{pmatrix} \;\middle|\; g^*\begin{pmatrix} 0 & 1 \\ -1 & 0 \end{pmatrix} g = \begin{pmatrix} 0 & 1 \\ -1 & 0 \end{pmatrix} \right\},$$

$$SU(2,2)_B = \left\{ g = \begin{pmatrix} a & b \\ c & d \end{pmatrix} \;\middle|\; g^*\begin{pmatrix} 1 & 0 \\ 0 & -1 \end{pmatrix} g = \begin{pmatrix} 1 & 0 \\ 0 & -1 \end{pmatrix} \right\},$$

$$D = \left\{ z \;\middle|\; \frac{z-z^*}{2i} \text{ is strictly positive definite} \right\}, \quad \text{and}$$

$$B = \{ z \mid z^* z < 1 \}.$$

Of course, the groups $SU(2,2)_D$ and $SU(2,2)_B$ are isomorphic. D is the generalized upper half-plane and B is the generalized unit disk. The Shilov boundary of D is $H(2) = \{$ 2 x 2 hermitian matrices $\}$ = Minkowski space, whereas the Shilov boundary of B is $U(2) = \{$ 2 x 2 unitary matrices $\}$. $\widetilde{U(2)}$ = the universal covering group of $U(2) = \mathbb{R} \times SU(2) = \mathbb{R} \times S^3$ is the Segal cosmos.

The formulas (2.1) define representations of $SU(2,2)_D$ as well as $SU(2,2)_B$ provided that for $SU(2,2)_D$, f is a holomorphic function on D, and that for $SU(2,2)_B$, f is a holomorphic function on B. These representations are equivalent. In both versions one can pass to the Shilov boundary and thus realize the representations on the respective space-times. One reason that the following claims are true is that the transition from $SU(2,2)_D$ to $SU(2,2)_B$ (and from D to B) can be obtained through the action of an element in the complexification of SU(2,2), and that all the expressions are analytic in the variables g and z. Observe that once a choice of domain has been made the term $(a - (g^{-1}z)c)^{-1}$ may be simplified: For $g^{-1} \in SU(2,2)_D$ and $z \in D$ it is equal to $zc^* + d^*$ and in the B- situation it is equal to $(a^* + zb^*)$.

3. Chronometric quantum numbers

The problem may be handled by means of the "ladder representations" [4] (see also [1]). In contrast, the approach presented here is intrinsic in the sense that it deals directly with the relevant function spaces and differential equations. In the following we shall only consider the representations U_n^+ since the treatment of the U_n^-'s follows from this by obvious modifications.

The domain D is particularly nice for describing the Hilbert space H_n^+ that carries the unitary irreducible representation U_n^+: H_n^+ is a reproducing kernel Hilbert space and in the D-version the kernel is given as

$$K_n^+(z,w) = \tau_n\left(\frac{z-w^*}{2i}\right)^{-1} \det\left(\frac{z-w^*}{2i}\right)^{-1} = \int_{b(C^+)} \tau_n(k) e^{i\, \mathrm{tr}(z-w^*)k}\, dm(k), \qquad (3.1)$$

where $b(C^+) = \{ k = k^* \mid \mathrm{tr}\, k \geq 0 \text{ and } \det k = 0\}$ is the boundary of the solid forward light-cone C^+, and $dm(k)$ is the usual Lorentz-invariant measure.

We wish to determine a complete set of quantum numbers for the representations U_n^+. From the point of view of Segal's cosmological theory as well as from that of group theory, the natural way to do this is to use the maximal compact subgroup K of SU(2,2). K coincides with the maximal subgroup of isometries of SU(2,2) under its action on U(2). In terms of group theory we shall thus determine the decomposition of the restriction of U_n^+ to K as a direct sum of finite-dimensional representations of K; the so-called K-types.

In $SU(2,2)_B$ we thus choose K as

$$K = \left\{ \begin{pmatrix} u & 0 \\ 0 & v \end{pmatrix} \mid u,v \in U(2),\ \det(u\ v) = 1\right\} \qquad (3.2)$$

In particular, K = U(1) x SU(2) x SU(2).

$T = \left\{ \begin{pmatrix} e^{i\tau/2} & 0 \\ 0 & e^{-i\tau/2} \end{pmatrix} \mid \tau \in \mathbb{R}\right\}$ is Segal's time-translation subgroup of K.

The natural domain to use for the determination of the K-types is B. On B,

$$(U_n^+(u,v)f)(z) = \left(U_n^+\begin{pmatrix} u & 0 \\ 0 & v \end{pmatrix} f\right)(z) = \tau_n(v) \det v\, f(u^{-1}z\, v), \qquad (3.3)$$

hence the finite-dimensional K-irreducible subspaces consist of V_n-valued polynomials. (3.3) clearly defines an action of K on the set of all V_n-valued polynomials on B. We denote this action by $\widetilde{U}_n^+$.

Consider the representation P of K on the space of all $\mathbb{C}$-valued polynomials defined by

$$(P(u,v)\,p)(z) = p(u^{-1}z\, v) \qquad (3.4)$$

<u>Proposition 3.1</u> $P(u,v) = \bigoplus_{l=0}^{\infty} \bigoplus_{r=0}^{\infty} \frac{\tau_l(v) \otimes \tau_l(\bar{u})}{\det u^{2r}}$

For later use we observe that P leaves the ideal I = I(det z) of polynomials proportional to det z, invariant. Clearly then

<u>Proposition 3.2</u> $P(u,v)\Big|_I = \bigoplus_{l=0}^{\infty}\bigoplus_{r=1}^{\infty}\dfrac{\tau_l(v)\otimes\tau_l(\bar{u})}{\det u^{2r}}$

The K-types of U_n^+ are, of course, to be found among the summands of $\widetilde{U}_n^+(u,v) = \tau_n(v)\det v \otimes P(u,v)$.

<u>Proposition 3.3</u>

$$\tau_n(v)\det v\otimes P(u,v) = \bigoplus_{l=0}^{\infty}\bigoplus_{r=0}^{\infty}\bigoplus_{\gamma=0}^{\min\{n,l\}}\frac{\tau_{n+l-2\gamma}(v)\otimes\tau_l(\bar{u})}{\det u^{\gamma+l+2r}}$$

However, since the Hilbert space of U_n^+ consists of solutions to certain differential equations, some of the summands must be excluded. To see exactly which do not occur we return to the domain D. By expanding the functions on D around the point $i = \begin{pmatrix} i & o\\ o & i\end{pmatrix}$ and by choosing K in this version as the subgroup of $SU(2,2)_D$ that leaves i fixed, it follows (e.g. by a Gram-Schmidt argument) that the finite-dimensional subspaces of functions on D that carry the irreducible K-representations are built up from functions of the form

$$F_{\underline{q}}(z) = \int_{b(C^+)}\tau_n(k)\, e^{i\,\mathrm{tr}(z+i)k}\,\underline{q}(k)\,dm\,(k), \tag{3.5}$$

where $\underline{q}$ is a V_n-valued polynomial. However, $\underline{q}$ determines a non-zero function $F_{\underline{q}}$ if and only if there is an x in V_n such that $\langle\tau_n(k)\underline{q}(k),x\rangle$ is not identically zero on $b(C^+)$. That is: Exactly those $\underline{q}$'s for which for all x in V_n, $\langle\tau_n(k)\underline{q}(k),x\rangle = 0$ on $b(C^+)$ do not contribute to the K-types in U_n^+. Now observe that $b(C^+)$ is a sufficiently big subset of the variety of the prime ideal I(det z) that we may conclude:

$$\forall x\in V_n\ :\ \langle\tau_n(k)\underline{q}(k),x\rangle = 0 \quad \text{on } b(C^+) \iff$$

$$\forall x\in V_n\ :\ \langle\tau_n(z)\underline{q}(z),x\rangle \in I\ (\det z).$$

Let $A = \{\underline{q}\mid \forall x\in V_n\ :\ \langle\tau_n(z)\underline{q}(z),x\rangle\in I\ (\det z)\}$

A is clearly invariant under the action of $\widetilde{U}_n^+$, and even though it originally was determined from the domain D, it follows by analyticity that

<u>Proposition 3.4</u> The K-types in A are exactly those in $\widetilde{U}_n^+$ that do not occur in U_n^+.

Let $\mathcal{S}$ denote the linear map of $A \otimes V_n$ into the space of $\mathbb{C}$-valued polynomials,

$$(\mathcal{S}(\underline{q} \otimes x))(z) = \langle \tau_n(z)\underline{q}(z), \bar{x} \rangle \tag{3.7}$$

Then

$$P(u,v)\mathcal{S} = \mathcal{S}(\widetilde{U}_n^+(u,v) \otimes \det \bar{v}\, \tau_n(\bar{u})) \tag{3.8}$$

Now choose a summand $\dfrac{\tau_{n+1-2\gamma}(v) \otimes \tau_1(\bar{u})}{\det u^{\gamma+1+2r}}$

from $\widetilde{U}_n^+$ and assume that $\underline{q} \in A$ transforms according to this representation. It follows from (3.8) and Proposition 3.1 that $\langle \tau_n(z)\underline{q}(z), \bar{x} \rangle$ transforms according to

$\dfrac{\tau_{n+1-2\gamma}(v) \otimes \tau_{n+1-2\gamma}(\bar{u})}{\det u^{2\gamma+2r}}$.

Since $\underline{q} \in A$ it follows from Proposition 3.2 that either $\gamma > 0$ or $r > 0$. Thus

Proposition 3.5

$$U_n^+(u,v) = \bigoplus_{l=0}^{\infty} \frac{\tau_{n+l}(v) \otimes \tau_l(\bar{u})}{\det u}$$

By analogous reasoning it follows that

Proposition 3.6

$$U_n^-(u,v) = \bigoplus_{l=0}^{\infty} \frac{\tau_l(v) \otimes \tau_{l+n}(\bar{u})}{\det u}$$

REFERENCES

[1] Jakobsen, H.P. and Vergne, M., J. Functional Analysis 24, 52 (1977); 34, 29 (1979).

[2] Jakobsen, H.P., in "Non-Commutative Harmonic Analysis", Lecture Notes in Math. 728, Berlin-Heidelberg-New York: Springer Verlag 1979.

[3] Jakobsen, H.P., Ørsted, B., Segal, I.E., Speh B., and Vergne, M., Proc. Nat. Acad. Sci. USA 75, 1609 (1978).

[4] Mack, G. and Todorov, I., J. Math. Phys. 10, 2078 (1969).

[5] Segal, I.E., "Mathematical Cosmology and Extragalactic Astronomy", Academic Press, New York, 1976.

SPINOR STRUCTURES

W. Greub

Department of Mathematics,
University of Toronto, Canada.

and

H.R. Petry

Institut für Theoretische Kernphysik,
Universität Bonn, West Germany

The propose of this note is to clarify the mathematical role of Weyl spinors in relativity. We show that the necessary and sufficient conditions for spinor structures on space-times can be easily derived from the basic algebraic properties of the spinor algebra.

Chapter I: Spinor algebra

1. Minkowski spaces

Let E denote a 4-dimensional vector space with an inner product of type (+,-,-,-). The inner product g(x,y) of two vectors $x,y \in E$ will be simply denoted by $\langle x,y\rangle$. $x \in E$ will be called *time-like* if $\langle x,x\rangle > 0$. The space of time-like vectors consists of two components; a *time-orientation* in E is a choice of one of them. Thus there are exactly two time-orientations in E. Having chosen one, say E_+, we shall say that a time-like vector x is *positive* if $x \in E_+$.

The inner product in E induces an inner product $\langle\, ,\, \rangle$ in the dual space E^* and hence in $\wedge^p E^*$. $\langle\, ,\, \rangle$ is determined by the equation

$$\langle x_1^*\wedge \ldots \wedge x_p^*,\ y_1^*\wedge \ldots \wedge y_p^*\rangle = \det \langle x_i^*, x_j^*\rangle \quad , \qquad (1)$$

valid for all $x_i^*, y_i^* \in E^*$.

Since $\dim \wedge^4 E^* = 1$, there are precisely two determinent functions $\omega \in \wedge^4 E^*$ normed such that

$$\langle \omega, \omega \rangle = -1 \tag{2}$$

(The minus sign arises from the signature of g). Choosing one of them means to fix an *orientation* in E.

A collection (E, g, E_+, ω) will be called a *Minkowski space*. An isomorphism between Minkowski spaces (E, g, E_+, ω) and (E', g', E'_+, ω') is an isometry $\varphi: E \to E'$ such that

$$\varphi(E_+) = E'_+$$

and

$$\varphi^* \omega' = \omega .$$

The group of automorphisms of a Minkowski space is called the *proper orthochronons Lorentz group*. It is a connected 6-dimensional Lie group and will be denoted by $L_+(E)$. Its Lie algebra, denoted by Sk(E), consists of the linear maps $\sigma: E \to E$ which satisfy the equation

$$\langle \sigma(x), y \rangle + \langle x, \sigma(y) \rangle = 0 \tag{3}$$

for all $x, y \in E$. It follows immediately that $\operatorname{tr} \sigma = 0$, and that $\dim Sk(E) = 6$.

2. The Pfaffian

Let $\Gamma: Sk(E) \to \wedge^2 E$ denote the linear isomorphism defined by the equation

$$\Gamma(\varphi)(x,y) = \langle x, \varphi y \rangle ; \qquad x, y \in E. \tag{4}$$

A bilinear function pf, called the Pfaffian, is determined in Sk(E) by the formula [1]

$$\mathrm{pf}(\varphi, \psi) = \frac{1}{2} \langle \omega, \Gamma(\varphi) \wedge \Gamma(\psi) \rangle . \tag{5}$$

It is obviously symmetric, i.e.

$$\mathrm{pf}(\varphi, \psi) = \mathrm{pf}(\psi, \varphi); \qquad \varphi, \psi \in Sk(E) ;$$

and satisfies the relation

$$\mathrm{pf}(\tau \varphi \tau^{-1}, \tau \psi \tau^{-1}) = \mathrm{pf}(\varphi, \psi) \tag{6}$$

for all $\tau \in L_+(E)$. If we set $\tau = \exp + t\mathfrak{z}$, $(t \in R,\ \mathfrak{z} \in S_k(E))$ we obtain by differentiation with respect to t :

$$pf([\mathfrak{z}, \varphi], \psi) + pf(\varphi, [\mathfrak{z}, \psi]) = 0 \qquad (7)$$

The pfaffian of a single skew transformation $\varphi \in Sk(E)$ is defined by the equation

$$pf(\varphi) = pf(\varphi, \varphi) . \qquad (8)$$

It can be shown that

$$pf(\varphi)^2 = - \det \varphi . \qquad (9)$$

3. The complex structure in Sk(E)

Let (;) denote the inner product in Sk(E) given by

$$(\varphi, \psi) = \frac{1}{4}\,\mathrm{tr}\,\varphi\psi \ ; \qquad \varphi, \psi \in Sk(E) \qquad (10)$$

Obviously, we have the identity, valid for all $\varphi, \psi, \alpha \in Sk(E)$:

$$([\alpha, \varphi], \psi) + (\varphi, [\alpha, \psi]) = \frac{1}{4}\,\mathrm{tr}\,[\alpha, \varphi\psi] = 0 .$$

Fix $\psi \in Sk(E)$; then there is a unique element $j_E(\psi) \in Sk(E)$ such that

$$pf(\varphi, \psi) = -(\varphi, j_E(\psi)) \qquad (11)$$

for all $\varphi \in Sk(E)$. The assignment $\psi \rightarrow j_E(\psi)$ defines a linear transformation of Sk(E) which satisfies

$$(\varphi, j_E(\psi)) = (j_E(\varphi), \psi), \qquad (12)$$

as is easily deducted from the symmetry of pf. Moreover, one can show that

$$- j_E^{\,2} = \mathrm{id} . \qquad (13)$$

Thus j_E is a complex structure in Sk(E). From the equations (7), (11), we conclude that

$$[\varphi, j_E(\psi)] = j_E([\varphi, \psi]) = [j_E(\varphi), \psi] , \qquad (14)$$

which shows that j_E makes Sk(E) into a complex Lie algebra, which has obviously complex dimension 3.

Observe that, if the orientation of E is reversed, then the Pfaffian changes sign and hence so does j_E. Using j_E we may define a complex bilinear nondegenerate inner product $\langle\!\langle\ ,\ \rangle\!\rangle$ in Sk(E) :

$$\langle\!\langle \varphi,\psi \rangle\!\rangle = (\varphi,\psi) - i(\varphi, j_E(\psi)) . \tag{15}$$

From the equation (11) we conclude that

$$\langle\!\langle \varphi,\psi \rangle\!\rangle = (\varphi,\psi) + i\ \mathrm{pf}(\varphi,\psi) \tag{16}$$

and that for all $\varphi,\psi,\alpha \in \mathrm{Sk}(E)$, we have the identity

$$\langle\!\langle [\alpha,\varphi] ,\psi \rangle\!\rangle + \langle\!\langle \varphi, [\alpha,\psi] \rangle\!\rangle = 0 \tag{17}$$

4. The space of hermitean functions

Let F be a complex 2-dimensional vector space. A function $\Phi\colon F\times F \longrightarrow C$ which is complex linear in the first and antilinear in the second argument is called sesquilinear. It is called hermitean, if in addition

$$\Phi(x,y) = \overline{\Phi(y,x)}$$

holds for all $x,y \in F$.

The set of such hermitean functions forms a real 4-dimensional vector space which we denote by H(F).

Now fix a non-zero determinant function ε in F. Let Φ, Ψ H(F) and set, for all $x_1, x_2, y_1, y_2 \in F$

$$\begin{aligned}\Omega(x_1,x_2,y_1,y_2) = {} & \Phi(x_1,y_1)\,\Psi(x_2,y_2) - \Phi(x_2,y_1)\,\Psi(x_1,y_2) - \\ & \Phi(x_1,y_2)\,\Psi(x_2,y_1) + \Phi(x_2,y_2)\,\Psi(x_1,y_1)\end{aligned} \tag{18}$$

Then Ω is skew symmetric in x_1,x_2 and y_1,y_2 , respectively. Hence there is a unique complex number $\langle \Phi , \Psi \rangle$ such that

$$\Omega(x_1,x_2,y_1,y_2) = \langle\Phi , \Psi\rangle \cdot \varepsilon(x_1,x_2)\cdot\overline{\varepsilon(y_1,y_2)} \tag{19}$$

From this definition follows that $\langle \Phi , \Psi \rangle$ is bilinear and symmetric in both arguments. Moreover, using the fact that Φ and Ψ are hermitean, one can easily show that $\langle \Phi , \Psi \rangle$ is in fact real. We want to prove now that $\langle \ , \ \rangle$ is an inner product of type (+,-,-,-).
For this purpose we fix a base $\{a,b\}$ of F such that

$$\varepsilon(a,b) = 1 . \tag{20}$$

Set
$$x_1 = y_1 = a ,$$
$$x_2 = y_2 = b , \tag{21}$$

in the equations (18) and (19). Equation (20) implies that

$$\langle \Phi , \Psi \rangle = \Phi(a,a)\Psi(b,b) + \Phi(b,b)\Psi(a,a) - \Phi(a,b)\overline{\Psi(a,b)} - \overline{\Phi(a,b)}\Psi(a,b) \tag{22}$$

Observe now that Φ is uniquely determined by the values of

$$U_0(\Phi) = (\Phi(a,a) + \Phi(b,b)) / \sqrt{2} \tag{23}$$
$$U_1(\Phi) = (\Phi(a,a) - \Phi(b,b)) / \sqrt{2}$$
$$U_2(\Phi) = \mathrm{Re}\, \Phi(a,b) \cdot \sqrt{2}$$
$$U_3(\Phi) = \mathrm{Im}\, \Phi(a,b) \cdot \sqrt{2}$$

It follows that the collection $\{U_0,\ldots U_3\}$ forms a basis of $H(F)^*$. Comparing (23) with equation (22) then yields the identity:

$$\langle \Phi , \Psi \rangle = U_0(\Phi)U_0(\Psi) - \sum_{i=1}^{3} U_i(\Phi)U_i(\Psi) \tag{24}$$

which clearly shows that $\langle \ , \ \rangle$ is indeed an inner product in H(F) of type (+,-,-,-). Moreover, one can show that the time-like vectors in H(F) are precisely those hermitean functions which are either positive or negative definite. Hence by distinguishing the positive definite elements we have a canonical prescription to give H(F) a time-orientation. In the next sections we show that there exists also a canonical orientation in H(F).

5. The double covering of the Lorentz group

Let SL(F) denote the group of complex linear transformations of F with determinant one. SL(F) is simply connected and its Lie-algebra consists of the trace-free linear transformations of F_1 which we denote by $L_0(F)$.

Let τ denote the representation of SL(F) in H(F), given by

$$\tau(\alpha)(\Phi) = (\alpha^{-1})^{*}\Phi \tag{25}$$

for all $\alpha \in SL(F)$ and $\Phi \in H(F)$.
A straightforward computation shows that $\tau(\alpha)$ preserves the inner product in H(F) and hence is an isometry of H(F). The determinant of $\tau(\alpha)$ is then equal to ± 1, and since SL(F) is simply connected, only the plus sign can hold. Evidently, $\tau(\alpha)$ transforms the positive definite hermitean functions into themselves and hence $\tau(\alpha)$ is a proper orthochronons Lorentz transformation.

One can show that $\tau: SL(F) \longrightarrow L_+(H(F))$ is in fact a universal covering of $L_+(H(F))$ and that the kernel of F consists of the identity and its negative. This implies that τ induces an isomorphism $\theta: L_0(F) \longrightarrow Sk(H(F))$ of the corresponding Lie algebras. θ is explicitely given by the formula

$$\theta(\varphi)\Phi(x,y) = -\Phi(\varphi x, y) - \Phi(x, \varphi y) \tag{26}$$

which is valid for all $\Phi \in H(F)$, $\varphi \in L_0(F)$ and all $x,y \in F$.

Lemma 1: For all $\varphi \in L_0(F)$ we have the identities:

$$\operatorname{tr} \theta(\varphi)^2 = 2(\operatorname{tr}\varphi^2 + \overline{\operatorname{tr}\varphi^2}) \tag{27}$$

$$\det \theta(\varphi) = \frac{1}{4}(\operatorname{tr}\varphi^2 - \overline{\operatorname{tr}\varphi^2})^2 \tag{28}$$

Proof: Fix $\varphi \in L_0(F)$ and observe that $\operatorname{tr}\varphi = 0$. The Cayley-Hamilton theorem yields for φ :

$$\varphi^2 + \det\varphi \cdot \mathrm{id} = 0 ,$$

and so

$$\varphi^2 = \lambda\, \mathrm{id} \tag{29}$$

with

$$\lambda = -\det\varphi = \frac{1}{2}\operatorname{tr}\varphi^2$$

Observe that $\theta(\varphi)$ is skew; the same theorem applied to $\theta(\varphi)$ yields

$$\theta(\varphi)^4 - \frac{1}{2}(\operatorname{tr}\theta(\varphi))^2 \cdot \theta(\varphi)^2 + \det\theta(\varphi)\cdot \mathrm{id} = 0$$

or equivalently

$$(\theta(\varphi)^2 - \frac{1}{4}\operatorname{tr}\theta(\varphi)^2\cdot \mathrm{id})^2 + (-\frac{1}{16}(\operatorname{tr}\theta(\varphi)^2)^2 + \det\theta(\varphi))\cdot \mathrm{id} = 0 \qquad (30)$$

Using (29) we find from equation (26):

$$\theta(\varphi)^2 = (\lambda+\bar{\lambda})\,\mathrm{id} + 2\varphi^*$$

The second term does not contribute to the trace, whence

$$\operatorname{tr}\theta(\varphi)^2 = 4(\lambda+\bar{\lambda}) = 2(\operatorname{tr}\varphi^2 + \overline{\operatorname{tr}\varphi^2}) \qquad (31)$$

which proves (27). Moreover, this shows that

$$\theta(\varphi)^2 - \frac{1}{4}\operatorname{tr}\theta(\varphi)^2\mathrm{id} = 2\varphi^*, \qquad (32)$$

whence according to (29):

$$(\theta(\varphi)^2 - \frac{1}{4}\operatorname{tr}\theta(\varphi)^2\mathrm{id})^2 = 4(\varphi^2)^* = 4\lambda\bar{\lambda}.$$

Inserting this into (30) we find that

$$\det\theta(\varphi) = (\lambda+\bar{\lambda})^2 - 4\lambda\bar{\lambda} = (\lambda-\bar{\lambda})^2 = \frac{1}{4}(\operatorname{tr}\varphi^2 - \overline{\operatorname{tr}\varphi^2})^2$$

6. The canonical orientation of H(F)

We want to prove in this section

Lemma 2: There is a unique orientation in H(F) such that for all $\varphi \in L_0(F)$

$$\operatorname{pf}(\theta(\varphi)) = \frac{1}{2i}(\operatorname{tr}\varphi^2 - \overline{\operatorname{tr}\varphi^2}) \qquad (33)$$

Proof: Formula (9) shows that for any choice of the normed determinant function ω in H(F)

$$\operatorname{pf}(\theta(\varphi))^2 = -\det\theta(\varphi)$$

Hence, by lemma 1,

$$\mathrm{pf}(\theta(\varphi)) = \frac{\varepsilon}{2i}(\mathrm{tr}\,\varphi^2 - \overline{\mathrm{tr}\,\varphi^2})$$

where $\varepsilon = \pm 1$. Both sides of these equations are continuous functions of φ. Hence ε cannot depend on φ. Recall now that the Pfaffian changes sign when the orientation is reversed. Thus we can always find a suitable ω such that $\varepsilon = 1$, and ω is indeed unique. q.e.d.

Lemma 3: Let the orientation in H(F) be fixed by lemma (1). Let j denote the corresponding complex structure in Sk(H(F)), and let $\langle\!\langle \ , \ \rangle\!\rangle$ be the complex inner product in Sk(h(F)), as defined in section 3. Then we have for all $\varphi, \psi \in L_0(F)$ the identities:

$$\theta(i\varphi) = j(\theta(\varphi)) \tag{34}$$

$$\mathrm{tr}\,\varphi\psi = \langle\!\langle \theta(\varphi), \theta(\psi) \rangle\!\rangle \tag{35}$$

Proof: Equations (27) and (33) yield

$$\frac{1}{4}\,\mathrm{tr}\,\theta(\varphi)^2 + i\,\mathrm{pf}(\theta(\varphi)) = \mathrm{tr}\,\varphi^2$$

and hence we obtain by polarizing both sides of this equation

$$\mathrm{tr}\,\varphi\psi = \frac{1}{4}\,\mathrm{tr}\,\theta(\varphi)\,\theta(\psi) + i\,\mathrm{pf}(\theta(\varphi), \theta(\psi)) .$$

Equation (16) shows that equation (35) is indeed true. Hence, we also have for all $\varphi, \psi \in L_0(F)$

$$\langle\!\langle \theta(\varphi), j\,\theta(\psi) \rangle\!\rangle = i\,\langle \theta(\varphi), \theta(\psi) \rangle = i\,\mathrm{tr}\,\varphi\psi = \langle\!\langle \theta(\varphi), \theta(i\psi) \rangle\!\rangle$$

which implies (34) since $\langle\!\langle \ , \ \rangle\!\rangle$ is non-degenerate.

Remark: We have seen in section 4, that the real 4-dimensional vector space H(F), of hermitean functions on a 2-dimensional complex vector space F with determinant function ε, carries a canonical inner product g of type (+,-,-,-) and a canonical time orientation $H(F)_+$. Moreover, we have a canonical isomosphism θ, of the real 6-dimensional Lie algebras $L_0(F)$ and Sk(H(F)). Evidently, $L_0(F)$ is also a complex Lie algebra, and, according to section 3, any choice of a normed determinant function ω, makes Sk(H(F)) into a complex Lie algebra.

Lemma 2 and 3 show that ω is fixed uniquely by the requirement that θ is a complex isomorphism. By adopting this choice of ω we give H(F) a canonical orientation. Hence, the collection $(H(F), g, H(F)_+, \omega)$ is a Minkowski space. It is uniquely determined by the pair (F, ε) alone and, therefore, we call it the canonical Minkowski space associated to (F, ε).

Chapter II:

1. Minkowski bundles

Let M be a connected smooth manifold and let (V, π, M, E) be a real vector bundle of rank 4, where $\pi : V \longrightarrow E$ denotes the projection and E is the typical fibre. Assume that, for each $x \in M$, the fibre $E_e = \pi^{-1}(x)$ is a Minkowski space, i.e. each E_x carries an inner product g_x of type (+,-,-,-), a normed determinant function ω_x, and a distinguished subspace E_x^+ of positive time-like vectors. We assume that these additional structures are smoothly defined, which means that

a) the subbundle of V having E_x^+ as fibres is smooth;

b) for all sections $\sigma_1, \ldots, \sigma_4$ in E the functions

$$g(\sigma_1, \sigma_2)(x) = g_x(\sigma_1(x), \sigma_2(x))$$

$$\omega(\sigma_1, \ldots, \sigma_4) = \omega_x(\sigma_1(x), \ldots, \sigma_4(x))$$

are smooth.

If these assumptions hold, then (V, π, M, E) is called a Minkowski bundle. A smooth bundle map γ between Minkowski bundles is called Lorentzian if γ induces an isomorphism of Minkowski spaces on each fibre.

Since g is smoothly defined on V, we can associate to V the smooth bundle $(Sk(V), \tilde{\pi}, M, Sk(E))$ whose fibres at the points $x \in M$ are precisely the vector spaces $Sk(E_x)$.

Now let ∇ be a covariant derivative in V. We shall say that ∇ is metric preserving if

$$Xg(\sigma_1, \sigma_2) - g(\nabla_X \sigma_1, \sigma_2) - g(\sigma_1, \nabla_X \sigma_2) = 0 \qquad (1)$$

holds for all sections σ_1, σ_2 in E and all vector fields X on M.

If ∇ has this property then the curvature form R is a two-form taking values in the bundle Sk(V). More precisely R is defined by the equation [2].

$$R_x(X,Y)\,\sigma(x) = (\nabla_X \nabla_Y \sigma)(x) - (\nabla_Y \nabla_X \sigma)(x) - (\nabla_{[X,Y]} \sigma)(x) \qquad (2)$$

which holds for all sections σ and vector fields X and Y.

2. Bundles of hermitean functions

Let (ξ, π_ξ, M, F) denote a complex vector bundle of rank two with projection π_ξ and typical fibre F. Assume that each fibre $F_x = \pi_\xi^{-1}(x)$ carries a non-vanishing determinant function ε_x which is smooth; i.e. for any two sections α, β in ξ, the function $\varepsilon(\alpha, \beta)$ defined by the equation

$$\varepsilon(\alpha, \beta)(x) = \varepsilon_x(\alpha(x), \beta(x))$$

is smooth.
We can associate to ξ the smooth bundles $(H(\xi), \pi_\xi, M, H(F))$ and $(L_0(\xi), \pi_\xi, M, L_0(F))$ whose fibres at any point x consist of the space of hermitean functions $H(F_x)$, and the space of trace-free linear mappings $L_0(F_x)$, respectively.

A section in H(F) assigns to any two sections α, β in ξ the smooth function $\delta(\alpha, \beta)(x) = \delta_x(\alpha(x), \beta(x))$. Recall now that $H(F_x)$ is a Minkowski space for any point x M. In particular it carries a smooth fibre metric g which is defined by the formulas (19) and (20). These formulas show that for any two sections δ_1, δ_2 in $H(\xi)$ the function $g(\delta_1, \delta_2)$ satisfies

$$g(\delta_1, \delta_2)\varepsilon(\alpha_1, \alpha_2)\varepsilon(\beta_1, \beta_2) = \delta_1(\alpha_1, \beta_1)\delta_2(\alpha_2, \beta_2) + \delta_1(\alpha_2, \beta_2)\delta_2(\alpha_1, \beta_1)$$
$$- \delta_1(\alpha_1, \beta_2)\delta_2(\alpha_2, \beta_1) - \delta_1(\alpha_2, \beta_1)\delta_2(\alpha_1, \beta_2) \tag{3}$$

for all sections $\alpha_1, \alpha_2, \beta_1, \beta_2$ in ξ. It follows that $H(\xi)$ is a Minkowski bundle, and hence the associated bundle $Sk(H(\xi))$, is well-defined according to the last section. The fibres of this bundle at any point $x \in M$ consist of the space $Sk(H(F_x))$ which, according to section 5, is canonically isomorphic to $L_0(F_x)$; i.e. we have a Lie algebra isomorphism $\theta_x : L_0(F_x) \longrightarrow Sk(H(F_x))$ at each point. Hence a strong bundle map

$$\theta : L_0(\xi) \longrightarrow Sk(H(\xi))$$

can be defined, whose restriction to the fibres is precisely θ_x.

Let now $\tilde{\nabla}$ be a covariant derivative in ξ; $\tilde{\nabla}$ induces a covariant derivative in $H(\xi)$ by the formula

$$(\nabla_X \delta)(\alpha, \beta) = X(\delta(\alpha, \beta)) - \delta(\tilde{\nabla}_X \alpha, \beta) - \quad (\alpha, \tilde{\nabla}_X \beta), \tag{4}$$

valid for all sections α, β in ξ, δ in $H(\xi)$ and all vector fields X on M. Moreover, we define the covariant derivatives $\tilde{\nabla}_X \varepsilon$ and $\nabla_X g$, of ε and g, respectively, by

$$(\tilde{\nabla}_X \varepsilon)(\alpha, \beta) = X(\varepsilon(\alpha, \beta)) - \varepsilon(\tilde{\nabla}_X \alpha, \beta) - \varepsilon(\alpha, \tilde{\nabla}_X \beta) \tag{5}$$

$$(\nabla_X g)(\delta_1, \delta_2) = X(g(\delta_1, \delta_2)) - g(\nabla_X \delta_1, \delta_2) - g(\delta_1, \nabla_X \delta_2) \tag{6}$$

where δ_1, δ_2 are arbitrary sections in $H(\xi)$.

<u>Lemma 4:</u>

$$\tilde{\nabla}_X \varepsilon = 0 \qquad (a)$$

implies for all vector fields X :

$$\nabla_X g = 0 \qquad (b)$$

<u>Proof:</u> Assume that ∇^0 is a fixed covariant derivative in ξ with

$$\nabla^0_X \varepsilon = 0$$

Any covariant derivative with property (a) can then be written in the form

$$(\tilde{\nabla}_X \alpha)(x) = (\nabla^0_X \alpha)(x) + A_x(X) \cdot \alpha(x)$$

when A_X is a 1-form with values in $L_0(\xi)$.
Hence for any section δ in $H(\xi)$ we find from (4)

$$(\nabla_X \delta)(\alpha, \beta)(x) = (\nabla^0_X \delta)(\alpha, \beta)(x) - \delta(A_x(X)\alpha(x), \beta(x))(x)$$
$$- \delta(\alpha(x), A_x(X)\beta(x))(x)$$

or, equivalently,

$$\nabla_X \delta = \nabla^0_X \delta + \theta(A(X))\delta .$$

It follows that

$$(\nabla_X g)(\delta_1, \delta_2) = (\nabla^0_X g)(\delta_1, \delta_2) - (g(\theta(A(X))\delta_1, \delta_2) + g(\delta_1, \theta(A(X))\delta_2))$$

Now we know that $\theta(A(X))$ is a section in $Sk(H(\xi))$; hence the last terms in brackets must vanish. Therefore, we have for any covariant derivative $\tilde{\nabla}$ the identity

$$\nabla_X g = \nabla^o_X g.$$

It follows that (b) holds if we can find a single ∇^o with the property (b). Note that for our purpose ∇^o is needed only locally. A suitable choice for ∇^o is then given as follows: In a sufficiently small neighbourhood U of any point x, there are sections γ_1, γ_2 in ξ such that

$$\varepsilon(\gamma_1, \gamma_2) = 1 ;$$

we fix now ∇^o by the requirement that

$$\nabla^o_X \gamma_1 = \nabla^o_X \gamma_2 = 0$$

for every vector field X. It follows that, locally

$$\nabla^o_X \varepsilon = 0$$

Equ. (4) leads locally for any section δ in ξ, to

$$(\nabla_X \delta)(\gamma_i, \gamma_j) = X(\delta(\gamma_i, \gamma_j)). \tag{7}$$

Setting $\alpha_1 = \beta_1 = \gamma_1$, $\alpha_2 = \beta_2 = \gamma_2$ in equation (3) we find for all sections δ_1, δ_2 in $H(\xi)$ the identity

$$\begin{aligned} g(\delta_1, \delta_2) = {} & \delta_1(\gamma_1, \gamma_1)\,\delta_2(\gamma_2, \gamma_2) + \delta_1(\gamma_2, \gamma_2)\,\delta_2(\gamma_1, \gamma_1) \\ & - \delta_1(\gamma_1, \gamma_2)\,\overline{\delta_2(\gamma_1, \gamma_2)} - \overline{\delta_1(\gamma_1, \gamma_2)}\,\delta_2(\gamma_1, \gamma_2) \end{aligned} \tag{8}$$

valid in U.
From (7) and (8) we conclude that

$$X(g(\delta_1, \delta_2))(x) = g(\nabla^o_X \delta_1, \delta_2)(x) + g(\delta_1, \nabla^o_X \delta_2)(x)$$

which is the desired result.

3. Curvature and characteristic classes.

Assume now that $\tilde{\nabla}$ is a covariant derivative in ξ with $\tilde{\nabla}_X \varepsilon = 0$ for all vector fields X on M. Hence the curvature $\tilde{R}$ of $\tilde{\nabla}$ is a two-form with values in $L_o(\xi)$; more precisely $\tilde{R}$ satisfies the equation

$$\tilde{R}(X,Y)\,\alpha \;=\; \tilde{\nabla}_X \tilde{\nabla}_Y \alpha \;-\; \tilde{\nabla}_Y \tilde{\nabla}_X \alpha \;-\; \tilde{\nabla}_{[X,Y]}\alpha \tag{9}$$

for all vector fields X,Y and all sections α in ξ. Let ∇ denote the induced covariant derivative in H(ξ). From the last section we know that ∇ is metric preserving; hence its curvature is the two-form R, with values in Sk(H(ξ)), given by equation (2). Moreover, from the definition of ∇, given in equation (4), we derive that for all vector fields X and Y, all sections α and β in ξ, and all sections $\eth$ in H(ξ), the identity

$$(R(X,Y)\,\eth)\,(\alpha,\beta) \;=\; -\,\eth(\tilde{R}(X,Y)\,\alpha,\beta) - \eth(\alpha,\tilde{R}(X,Y)\beta) \tag{10}$$

holds. This shows that

$$R(X,Y) \;=\; \theta(\tilde{R}(X,Y)) \tag{11}$$

Consider now the 4-form c, which assigns to any tangent vectors $y_1,\dots,y_4$ at $x \in M$ the complex number

$$c_x(y_1,\dots,y_4) \;=\; \frac{1}{8\pi^2}\;\frac{1}{4}\sum_{\tau\in S_4} \operatorname{sig}(\tau)\;\operatorname{tr}\tilde{R}_x(y_{\tau(1)},y_{\tau(2)})\cdot \tilde{R}_x(y_{\tau(3)},y_{\tau(4)})\;. \tag{12}$$

c is known to be closed and to represent the second Chern class of ξ [2]. (Note that the first Chern class is zero, since ξ has a smooth nonzero determinant function on each fibre). In addition we consider the 4 forms p and c, defined by

$$p_x(y_1,\dots,y_4) \;=\; \frac{1}{8\pi^2}\;\frac{1}{4}\sum_{\tau\in S_4} \operatorname{sig}(\tau)\;\operatorname{tr}R_x(y_{\tau(1)},y_{\tau(2)})\cdot R(y_{\tau(3)},y_{\tau(4)}) \tag{13}$$

and

$$e_x(y_1,\dots,y_4) \;=\; \frac{1}{4\pi^2}\;\frac{1}{4}\sum_{\tau\in S_4} \operatorname{sig}(\tau)\;\operatorname{pf}\,(R_x(y_{\tau(1)},y_{\tau(2)})\cdot R_x(y_{\tau(3)},y_{\tau(4)})) \tag{14}$$

where pf denotes the pfaffian defined in chapter 1 (section 2). e and p are known to be closed and to represent the Euler and the first Pontriagin class of H(ξ), respectively [2].

<u>Lemma 5:</u> The 4 forms c, e and p which represent the second Chern class of ξ, the

Euler class and the first Pontriagin class of $H(\xi)$, respectively, are related by the formula

$$c = \frac{1}{4}p + \frac{i}{2}e \qquad (15)$$

Proof: From formula (11) we know that

$$R_x(y_i,y_k) = \theta_x \cdot \tilde{R}_x(y_i,y_k)$$

holds for all y_i, y_k at the tangent space at $x \in M$. Hence we can apply formula (35) of chapter I which yields then

$$\operatorname{tr} \tilde{R}_x(y_{\tau(1)},y_{\tau(2)})\, R_x(y_{\tau(3)},y_{\tau(4)}) + i\operatorname{pf}(R_x(y_{\tau(1)},y_{\tau(2)}), \cdot R_x(y_{\tau(3)},y_{\tau(4)})).$$

Equation (15) is then immediately obtained by comparing (12) with (13) and (14).

4. Spinor-structures

Let (V,π,M,E) be a Minkowski bundle with fibre metric g_x. We shall say that V has a spinor structure if there is a complex rank-two vector bundle (ξ,π_ξ,M,F) with smooth non-zero determinant function ε_x on each fibre, such that the associated Minkowski bundle $H(\xi)$ is isomorphic to (V,π,M,E); i.e. there is a vector bundle isomorphism γ which makes the diagram

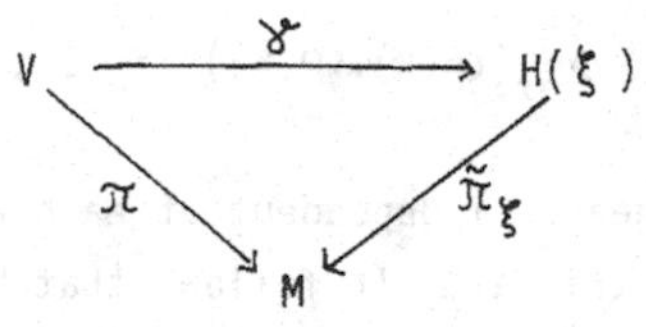

commute. Moreover, we require that γ is Lorentzian (compare section II, 1.).

Lemma 6: The second Chern class $[c]$ of ξ and the Euler and Pontriagin classes of V, denoted $[e]$ and $[p]$ respectively, are related by the formula

$$[c] = \frac{1}{4}[p] + i[e] \qquad (16)$$

Proof: Choose a covariant derivative $\tilde{\nabla}$ in ξ, with $\tilde{\nabla}\varepsilon = 0$. Let ∇ be the induced covariant derivative in $H(\xi)$; then, there is a unique covariant derivative ∇' in V such that

$$\gamma \cdot \nabla' = \nabla \cdot \gamma$$

It follows that the corresponding curvature forms R and R' are related by the formula $\gamma \cdot R' = \tilde{R} \cdot \gamma$. Since the cohomology classes [c], [e] and [p] are independent of the particular choices of the covariant derivatives which lead to their definitions [2], we can apply lemma 5; passing from equation (15) to cohomology yields the desired equation (16).

Since [c] is an integer class we have as a trivial *corollary*:
The Euler class of a Minkowski bundle with a spinor structure vanishes.
Physically, the most important case arises, when M is a space-time; i.e.

a) M is 4-dimensional, and
b) V is the tangent bundle of M.

Lemma 7: Let (V, π, M, E) be a Minkowski bundle with a spinor structure, and assume that M is 4-dimensional and non-compact; then V is trivial.

Proof: Let (ξ, π_ξ, M, F) denote the spinor structure. We choose a hermitean, positive fibre metric δ_0 on ξ and consider the subbundle $\xi_0 \subset \xi$ whose fibres F_x consist of all vectors $y_x \in F_x$ with $\delta_0(y_x, y_x) = 1$. Hence ξ_0 is a 3-sphere bundle over a 4-dimensional non-compact manifold M. By a general theorem every (n-1)-sphere bundle over an n-dimensional non-compact manifold admits a global section [2]. Hence ξ admits a non-vanishing global section φ. Let ψ denote the unique section which for all $x \in M$ and all $y_x \in F_x$ satisfies the equation

$$\varepsilon_x(\psi(x), y_x) = \delta_{ox}(\varphi(x), y_x) .$$

It follows that

$$\varepsilon_x(\psi(x), \varphi(x)) = \delta_{ox}(\varphi(x), \varphi(x)) = 1 ,$$

whence $\psi(x)$ and $\varphi(x)$ are linearly independent at each point $x \in M$. Thus we have a global frame in ξ, i.e. ξ is trivial. It follows that $H(\xi)$ is trivial, and, since $H(\xi)$ and V are isomorphic by assumption, V has to be trivial, too.

Obviously, V has a spinor structure when V is trivial. We, therefore, have the two corollaries:

Corollary 1 (Geroch): A non-compact space-time admits a spinor structure, if and only if it is parallelizable [3].

Corollary 2: A compact space-time admits a spinor structure if it is parallelizable everywhere, except at a single point.

In the compact case we have also an interesting corollary to lemma 6:

Corollary 3: Let ξ be a spinor structure for a compact space-time. Let p_1 and c_2 be the Pontriagin number of M and the Chern number of ξ. Then

$$c_2 = p_1 / 4,$$

i.e. the Pontriagin number of M has to be divisible by 4.

References

[1] Greub, W., Multilinear Algebra, Springer (New York) 1978.

[2] Greub, W., Halpern, S. and Vanstone, R., Connection, Curvature and Cohomology, Academic Press (New York) 1976.

[3] Geroch, R., J. Math. Phys. 9, 1739 (1968).

THE COMPLEX CAYLEY ALGEBRA AND THE LORENTZ GROUP

W.H. Greub

Department of Mathematics,
University of Toronto, Canada.

Introduction

Let $\mathbb{R}^4$ be a 4-dimensional Euclidean space and let $\mathbb{H}$ denote the corresponding algebra of quaternions. Since the multiplication preserves inner products, it defines a group structure on the unit sphere S^3. Now fix two unit vectors a, b on S^3 and set

$$\tau_{a,b}(x) = a\,x\,b^{-1}, \quad x \in \mathbb{R}^4.$$

Then $\tau_{a,b}$ is a proper rotation of $\mathbb{R}^4$. Hence a homomorphism

$$\Phi : S^3 \times S^3 \longrightarrow SO(4)$$

is defined by

$$\Phi(a, b) = \tau_{a,b}.$$

It is well known that Φ is surjective and that its kernel consists of the elements e and - e, where e is the unit element in $\mathbb{H}$ cf. [1] Sec. 8, 24. Thus we have the short exact sequence

$$1 \longrightarrow \mathbb{Z}_2 \xrightarrow{\ i\ } S^3 \times S^3 \xrightarrow{\ \Phi\ } SO(4) \longrightarrow 1,$$

which makes $S^3 \times S^3$ into a double (and hence the universal) covering (and hence the universal) group of SO (4).

It is the purpose of this paper to give a similar explicit construction of the universal covering group of SO (1, 3) (the proper orthochronous Lorentz group) in terms of the complex algebra of quaternions. We shall start with the construction of this algebra using the complex cross product.

1. The complex cross-product in $\mathbb{C}^3$.

Let $\mathbb{C}^3$ be a 3-dimensional complex vector space with a non-degenerate complex inner product. This inner product determines (in two ways) a *normalized determinant* function; that is a skew-symmetric 3-linear function Δ, which satisfies the *Lagrange identity*.

$$\Delta(z_1, z_2, z_3)\,\Delta(w_1, w_2, w_3) = \det\left((z_i, w_j) \right) .$$

Now fix vectors a and b in $\mathbb{C}^3$. Then a vector $a \times b$ is determined by the equation

$$\Delta(a, b, z) = (a \times b, z)\,, \quad z \in \mathbb{C}^3.$$

It will be called the *complex cross-product* of a and b.

This cross-product has exactly the same properties as the cross-product in $\mathbb{R}^3$:

1) $(\lambda a_1 + \mu a_2) \times b = \lambda a_1 \times b + \mu a_2 \times b, \quad \lambda, \mu \in \mathbb{C}$

2) $a \times b = - b \times a$

3) $(a \times b, a) = 0$ and $(a \times b, b) = 0$

4) $(a_1 \times b_1, a_2 \times b_2) = (a_1, a_2)(b_1, b_2) - (a_1, b_2)(a_2, b_1)$

5) $a \times (b \times c) = (a, c)\, b - (b, c)\, a$

2. The complex quaternion algebra.

Let E be a complex vector space of dimension 4 with a complex non-degenerate inner product. A *unit vector* a is a vector for which $(a, a) = 1$ and a (complex) *light vector* is a vector a such that $(a, a) = 0$.

Fix a unit vector e in E and let E_1 denote its orthogonal complement. Then E_1 is a 3-dimensional complex vector space and the inner product in E restricts to a non-degenerate inner product in E_1.

Now define a complex multiplication in E as follows:

$$p \cdot q = - (p, q)\, e + p \times q\,, \quad p, q \in E_1$$

$$e \cdot x = x\,, \quad x \cdot e = x\,, \quad x \in E\,.$$

This multiplication makes E into an associative algebra with unit element e, called the *complex quaternion algebra*. It follows from the definition that

$$(a\,b,\ a\,b) = (a,\ a)\,(b,\ b)\ ,\qquad a,\ b \in E. \tag{1}$$

This formula implies that the unit vectors in E form a (complex 3-dimensional) group, which will be denoted by S.

The *conjugate* of an element $a \in E$ is defined by

$$\tilde{a} = \alpha e - p\ ,$$

where

$$a = \alpha\, e + p\ ,\qquad \alpha \in \mathbb{C}\ ,\quad p \in E_1.$$

It follows easily that

$$a\,\tilde{a} = \tilde{a}\,a = (a,\ a\ )\,e\ .$$

Thus, if a is not a light vector, then the element

$$a^{-1} = \frac{\tilde{a}}{(a,\ a)}$$

is the inverse of a. This shows that all elements a with $(a,\ a) \neq 0$ are invertible. On the other hand, if a is invertible, formula (1) implies that $(a,\ a) \neq 0$.

Finally note that the center of the algebra constructed above consists of the elements λe, $\lambda \in \mathbb{C}$.

3. The homomorphism Φ.

A *proper rotation* of E is a complex linear transformation φ of E which satisfies the relations

$$(\varphi x,\ \varphi y) = (x,\ y)\ ,\qquad x,\ y \in E$$

and

$$\det \varphi = 1.$$

These rotations form a group, denoted by SO (E). Now fix two unit vectors a and b and set

$$\tau_{a,\ b}\,(x) = a\,x\,b^{-1}\ ,\qquad x \in E.$$

Then $\tau_{a,b}$ is a proper rotation. Moreover,

$$\tau_{aa', bb'} = \tau_{a,a'} \circ \tau_{b,b'}.$$

Hence a homomorphism

$$\Phi : S \times S \longrightarrow SO(E)$$

is defined by

$$\Phi(a, b) = \tau_{a,b} .$$

<u>Theorem I:</u> The homomorphism Φ is surjective and its kernel consists of the elements (e, e) and (-e, -e). Thus we have a short exact sequence

$$1 \longrightarrow \mathbb{Z}_2 \xrightarrow{\ i\ } S \times S \xrightarrow{\ \Phi\ } SO(E) \longrightarrow 1.$$

<u>Proof:</u> Suppose that $(a, b) \in \ker \Phi$. Then

$$a x b^{-1} = x , \quad x \in E$$

and so

$$a x = x b , \quad x \in E . \tag{2}$$

Setting $x = e$ we obtain $a = b$. Now equation (2) reads

$$a x = x a, \quad x \in E$$

and so a commutes with all vectors $x \in E$. This implies that $a = \pm e$ and so

$$(a, b) = \pm (e, e).$$

To show that Φ is surjective, consider first the rotations

$$\tau_a (x) = a x a^{-1} , \quad a \in E , \quad (a, a) = 1.$$

Then

$$\tau_a (e) = e$$

and so τ_a restricts to a (proper) rotation of E_1.

In particular, let p be a unit vector in E_1. Then a simple calculation shows that

$$\tau_p(y) = -y + 2(p, y)p, \quad y \in E_1$$

Setting $\sigma_p = -\tau_p$ we obtain

$$\sigma_p(y) = y - 2(p, y)p, \quad y \in E_1$$

Thus σ_p is the *reflection* of E_1 at the plane orthogonal to p.

Now, by a classical theorem of linear algebra, every proper rotation α of E_1 is a product of an even number of reflections (cf. [2], Sec. 11, 7).

$$\alpha = \sigma_1 \circ \dots \circ \sigma_k \quad (k \text{ even}).$$

The argument above shows that every σ_j can be written as

$$\sigma_j = \tau_{p_j} = \Phi(p_j),$$

where p_j is some unit vector in E_1. It follows that

$$\alpha = \sigma_1 \dots \sigma_k = \Phi(p_1) \circ \dots \circ \Phi(p_k) = \Phi(p_1 \dots p_k)$$

Thus every proper rotation of E_1 is in the image of Φ. Finally, let α be any proper rotation of E and set

$$\alpha(e) = c.$$

Then $(c, c) = 1$. Define β by

$$\beta(x) = c^{-1}. \alpha(x), \quad x \in E.$$

Then $\beta(e) = e$ and so β restricts to a rotation of E_1. Hence there is a unit vector b such that

$$\beta(x) = b x b^{-1}.$$

It follows that

$$\alpha(x) = c \cdot \beta(x) = c b x b^{-1} = a x b^{-1}, \quad x \in E,$$

where $a = c b$. This shows that Φ is surjective.

4. Real inner product spaces.

Let F be a 4-dimensional real vector space with a non-degenerate inner product $(\, , \,)_F$ of type (p, q) (p + q = 4). Set

$$E = \mathbb{C} \otimes F$$

and define a complex inner product in E by setting

$$(\alpha \otimes x, \beta \otimes y) = \alpha \beta (x, y)_F, \quad \alpha, \beta \in \mathbb{C}.$$

Let Δ_F be a normed determinant function in F; that is Δ_F satisfies

$$\Delta_F(x_1, \ldots, x_4) \Delta_F(y_1, \ldots, y_4) = (-1)^q \det((x_i, y_j))$$

(cf. [1], Sec. 9, 19). Then a normalized determinant function Δ_E in E is given by

$$\Delta_E(\alpha_1 \otimes x_1, \ldots, \alpha_4 \otimes x_4) = i^q \alpha_1 \ldots \alpha_4 \Delta_F(x_1 \ldots x_4). \quad (3)$$

Next, consider the *complex conjugation* $z \longrightarrow \bar{z}$ in E given by

$$\overline{\lambda \otimes x} = \bar{\lambda} \otimes x, \qquad \lambda \in \mathbb{C}, \, x \in F.$$

It satisfies the relations

$$\overline{\lambda \cdot z} = \bar{\lambda} \cdot z$$

$$\bar{\bar{z}} = z$$

and

$$(\bar{z}_1, \bar{z}_2) = \overline{(z_1, z_2)}. \quad (4)$$

Moreover, formula (3) implies that

$$\Delta_E(\bar{z}_1, \ldots \bar{z}_4) = (-1)^q \bar{\Delta}_E(z_1, \ldots, z_4). \quad (5)$$

Next, fix a unit vector e_F in F. Then $e = 1 \otimes e_F$ is a unit vector in E and so the complex cross product is defined in the orthogonal complement E_1 of e (cf. sec. 1). Relations (4) and (5) imply that

$$(\overline{y_1 \times y_2}, \bar{y}) = (\overline{y_1 \times y_2, y}) = \overline{\Delta}_E (y_1, y_2, y)$$

and

$$(\bar{y}_1 \times \bar{y}_2, \bar{y}) = \overline{\Delta}_E (\bar{y}_1, \bar{y}_2, \bar{y}) = (-1)^q \overline{\Delta}_E (y_1, y_2, y)$$

$$y_1, y_2, y \in E_1 ,$$

where

$$\overline{y_1 \times y_2} = (-1)^q \; \bar{y}_1 \times \bar{y}_2 , \quad y_1, y_2 \in E_1 .$$

Thus,

$$\overline{y_1 \times y_2} = \bar{y}_1 \times \bar{y}_2 \quad \text{if } q \text{ is even, } (q = 0, 2, 4) \tag{6}$$

and

$$\overline{y_1 \times y_2} = \bar{y}_2 \times \bar{y}_1 \quad \text{if } q \text{ is odd, } \; (q = 1, 3) . \tag{7}$$

These relations imply that for $z_1 \in E$ and $z_2 \in E$

$$\overline{z_1 \cdot z_2} = \bar{z}_1 \cdot \bar{z}_2 , \text{ if } q \text{ is even,} \tag{8}$$

while

$$\overline{z_1 \cdot z_2} = \bar{z}_2 \cdot \bar{z}_1 , \text{ if } q \text{ is odd .} \tag{9}$$

5. Lorentz transformations.

A Minkowski space is a 4-dimensional vector space with an inner product of type (1, 3). Let F be a Minkowski space and consider the corresponding complex quaternion algebra E (cf. Sec. 4). Choose a unit vector $a \in E$ and set

$$\tau_a (z) = a \, z \, \bar{a} , \quad z \in E .$$

Then τ_a is a proper rotation of E.

Formula (9) implies that

$$\overline{\tau_a (z)} = a \, \bar{z} \, \bar{a} = \tau_a (\bar{z}) , \quad z \in E$$

and so, in particular,

$$\overline{\tau_a (1 \otimes x)} = \tau_a (1 \otimes x) , \quad x \in F .$$

Hence τ_a restricts to a linear transformation of F which will be also denoted by τ_a,

$$\tau_a (x) = \tau_a (1 \otimes x).$$

Clearly,

$$(\tau_a (x), \tau_a (y)) = (x, y) , \quad x, y \in F$$

and

$$\det \tau_a = 1.$$

Thus τ_a is a proper Lorentz transformation.

To show that τ_a is orthochronous, observe that for every unit vector a

$$(e, a \bar{a}) > 0 ,$$

as is easily checked. It follows that

$$(e, \tau_a e) > 0$$

and so τ_a preserves fore-cone and past-cone.

Hence a homomorphism $\Psi : S \longrightarrow SO^+ (1, 3)$ is given by

$$\Psi (a) = \tau_a , \quad a \in S.$$

<u>Theorem II:</u> The homomorphism Ψ is surjective and its kernel consists of the elements e and -e. Thus we have a short exact sequence

$$1 \longrightarrow \mathbb{Z}_2 \longrightarrow S \longrightarrow SO^+ (1, 3) \longrightarrow 1$$

<u>Proof:</u> Since the homomorphisms

$$\Phi : S \times S \longrightarrow SO (E)$$

and

$$\Psi : S \longrightarrow SO^{+}(1, 3)$$

(cf. Sec. 3) are connected by the relation

$$\Psi(a) = \Phi(a, \bar{a}^{-1}), \quad a \in S,$$

the second part of Theorem 1 shows that $\ker \Psi = \{e, -e\}$.

Next, let $\Psi : F \longrightarrow F$ be any proper orthochronous Lorentz transformation and consider the map $\varphi : E \longrightarrow E$ given by

$$\varphi(\alpha \otimes x) = \alpha \otimes \Psi(x), \qquad \alpha \in \mathbb{C}, \; x \in F.$$

In view of Theorem 1, there are unit vectors a and b in E such that

$$\varphi(z) = a\, z\, b^{-1}, \qquad z \in E.$$

We shall show that $b^{-1} = \bar{a}$. In fact, since

$$\varphi(\bar{z}) = \overline{\varphi(z)}, \qquad z \in E,$$

it follows that

$$a\, \bar{z}\, b^{-1} = \bar{b}^{-1}\, \bar{z}\, \bar{a}, \qquad z \in E,$$

whence

$$\bar{b}\, a\, \bar{z} = \bar{z}\, \bar{a}\, b, \qquad z \in E.$$

Setting $\bar{a}\, b = c$, we obtain

$$\bar{c}\, \bar{z} = \bar{z}\, c, \qquad z \in E. \tag{10}$$

In particular, for $z = e$,

$$\bar{c} = c.$$

Now formula (10) reads

$$c\, \bar{z} = \bar{z}\, c, \qquad z \in E.$$

This implies that $c = \varepsilon e$, $\varepsilon = \pm 1$ and so we obtain

$$\bar{a}\, b = \varepsilon e,$$

whence

$$\bar{a} = \varepsilon b^{-1}, \qquad \varepsilon = \pm 1.$$

Thus

$$\Psi(x) = \varepsilon\, a\,(1 \otimes x)\,\bar{a} .$$

Since Ψ is orthochronous, we have

$$(e, \Psi e) > 0 ,$$

whence

$$\varepsilon\,(a\,\bar{a}, e) > 0 .$$

But

$$(a\,\bar{a}, e) > 0 ,$$

and so it follows that $\varepsilon = +1$. This completes the proof of Theorem II.

6. The fundamental group of $SO^+(1, 3)$.

Finally we shall use Theorem II to give a simple proof that the fundamental group of $SO^+(1, 3)$ is $\mathbb{Z}_2$,

$$\pi_1(SO^+(1, 3)) \cong \mathbb{Z}_2 .$$

By Theorem II the group S is a double covering of $SO^+(1, 3)$, and so we have only to show that S is simply connected. Define a family of maps $\varphi_t : S \longrightarrow S$ by

$$\varphi_t(z) = \frac{x + t\,i\,y}{\left[(1-t^2)(x, x) + t^2\right]^{1/2}}, \qquad 0 \le t \le 1,$$

where

$$z = x + i\,y .$$

Then

$$\varphi_0(z) = \frac{x}{(x,x)^{1/2}} \qquad z \in S$$

and

$$\varphi_1(z) = z, \qquad z \in S.$$

Thus S^3 is a deformation retract of S. It follows that

$$\pi_1(S) \cong \pi_1(S^3) = 0 .$$

REFERENCES

[1] Greub, W., Linear Algebra, Fourth Edition, Graduate Texts in Mathematics, Springer New York, 1975.

[2] Greub, W., Multilinear Algebra, Second Edition, Universitext, Springer New York 1978

V. QUANTIZATION METHODS

PREQUANTISATION FROM PATH INTEGRAL VIEWPOINT

P.A. Horváthy

CNRS, Centre de Physique Théorique,
F-13288 Marseille, Cedex 2.

and

Ist. Fis. Mat. dell'Università *
Via Carlo Alberto 10, 10123 Torino, Italy.

ABSTRACT

The quantum mechanically admissible definitions of the factor $\exp\left[i/\hbar \; S(\gamma)\right]$ - needed in Feynman's integral - are put in bijection with the prequantisations of Kostant and Souriau. The different allowed expressions of this factor - the inequivalent prequantisations - are classified in terms of algebraic topology.

1. Introduction

In [1] a first attempt was made to use the *geometric techniques* of Kostant and Souriau [2,3] ("K-S theory") in studying *path integrals*. The method was applied to Dirac's monopole and the Bohm - Aharonov experiment.

Here we intend to develop a more general theory. We show that a general symplectic system is *quantum mechanically admissible* (Q.M.A.S.) iff it is *prequantisable* with transition functions depending on space-time variables. If the configuration space is not simply connected, the different physical situations correspond to different prequantisations. A classification scheme [6] - implicitly recognized already by Kostant [2] and Dowker [5] - is presented.

The basic object of our considerations below is the factor

$$\exp\left[\frac{i}{\hbar} \; S(\gamma)\right] \tag{1}$$

---

* Present address

where $S(\gamma)$ is the classical action along the path γ.

Our results contribute to the *physical interpretation of prequantisation*, and are hoped to provide physicists with a kind of *introduction* to this theory.

2. Quantum Mechanically Admissible Systems (Q.M.A.S.)

Let us restrict ourselves to classical systems (E, σ) with evolution space $E = T^* Q \times \mathbb{R}$ (Q is a configuration space) and presymplectic structure of the form

$$\sigma = d\,\theta_0 + e\,\mathbb{F} \tag{2}$$

$d\,\theta_0$ - where θ_0 is the restriction to the energy surface $H = H_0\,(q,p,t)$ of the canonical 1-form of $T^*(Q \times \mathbb{R})$ - describes a free system; $\mathbb{F}$ - a closed 2-form on space-time $X = Q \times \mathbb{R}$ - represents the external field coupled to our system by the constant $\underline{e}$ (cf. [3]).

If the system admits a Lagrangian function, then $\sigma = d\,\theta$ and it is exactly this "Cartan 1-form" θ which has to be integrated along paths in phase space (whose initial resp. final points project to the same $x = (q,t)$ resp. $x' = (q',t') \in X$) when computing a *path integral in phase space:*

$$S(\gamma) = \int_\gamma \theta \tag{3}$$

Now, by Poincaré's lemma, for any point there exists a contractible neighbourhood U_j and a 1-form θ_j defined here such that $\sigma_{U_j} = d\,\theta_j$. Thus we would be tempted to define $S_j(\gamma)$ by (3) even if no global Lagrangian - and consequently no global θ - exists. It was pointed out in [1], that the different expressions $S_j(\gamma)$ and $S_k(\gamma)$ may be completely different. The following notion will be useful:

<u>Definition</u> (E, σ) is a *quantum mechanically admissible system* (Q.M.A.S.) iff there exists a collection $\{U_j, \theta_j\}$ of pairs of open contractible subsets U_j and 1-forms θ_j defined there - called *local system* in what follows - such that they are *compatible*, i.e. for any $\gamma \subset U_j \cap U_k$ we have

$$\exp\left[\frac{i}{\hbar}\int_\gamma \theta_j\right] = C_{jk}\,(x,x') \cdot \exp\left[\frac{i}{\hbar}\int_\gamma \theta_k\right] \tag{4}$$

where the unitary factors C_{jk} depend only on the projections to space-time of the initial resp. end point of γ, but not on γ itself.

Clearly, in such situations the Feynman propagators corresponding to θ_j resp. θ_k will be related by unobservable phase factors.

In [1] we have shown that this happens iff

$$\frac{1}{2\pi\hbar} \int_S \delta \in \mathbb{Z} \tag{5}$$

for any 2-cycle S in space. Expressed in fiber bundle language we have (Weil's lemma)

Theorem [1]

A.) (E, δ) is Q.M.A. - iff *prequantisable* with transition functions depending on X. Then for *any* γ with end points in U_j we can define

$$\text{" } \exp\left[\frac{i}{\hbar} S(\gamma)\right] \text{ "} \tag{6a}$$

such that there exists phase factors C_{jk} with

$$\text{" } \exp\left[\frac{1}{\hbar} S_j(\gamma)\right] \text{ "} = C_{jk}(x,x') \text{ " } \exp\left[\frac{i}{\hbar} S_k(\gamma)\right] \text{ "} \tag{6b}$$

For $\gamma \subset U_j \cap U_k$, we have

$$\text{" } \exp\left[\frac{i}{\hbar} S_j(\gamma)\right] \text{ "} = \exp\left[\frac{i}{\hbar}\int_\gamma \theta_j\right] \tag{6c}$$

B.) Explicitely, we have the transition function $Z_{jk} : U_j \cap U_k \longrightarrow U(1)$ with

$$\theta_j - \theta_k = \frac{d\, Z_{jk}}{i\, Z_{jk}} \tag{6d}$$

yielding

$$C_{jk}(x,x') = \frac{Z_{jk}(x)}{Z_{jk}(x')} \tag{6e}$$

Let γ be any path in E joining $y = (x,.)$ to $y' = (x',.)$. Denote (Y,ω,π) a prequantisation [3] of (E, δ). Lift γ to Y *horizontally* through a $\xi \in \pi^{-1}(y)$, denote ξ' the end point of this horizontal lift $\bar{\gamma}$. If $y, y' \in U_j$, we can write locally $\xi = (y, Z_j)$, $\xi' = (y', Z_j')$

The expression (6a) is then

$$\text{" } \exp\left[\frac{i}{\hbar} S_j(\gamma)\right] \text{ "} = \frac{Z_j}{Z_j'} \tag{7}$$

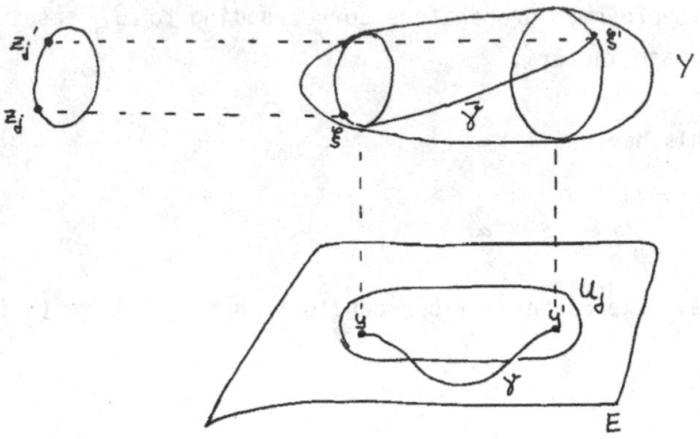

3. Geometric Expression for the Integrand

Now we can give a completely coordinate free form to the integrand in Feynman's expression. Following a suggestion of Friedmann and Sorkin [8] let us consider *any path* $\tilde{\gamma} \subset Y$ projecting to γ. Write

$$\tilde{\gamma}(o) = \xi \simeq (y, Z_j) \qquad \tilde{\gamma}(1) = \xi' \simeq (y', Z_j')$$

Lemma

$$\text{"} \exp\left[\frac{i}{\hbar} S_j(\gamma)\right] \text{"} \frac{Z_j'}{Z_j} = \exp\left[\frac{i}{\hbar} \int_{\tilde{\gamma}} \omega\right] \tag{8}$$

The product of two coordinate-dependent quantities is thus coordinate-independent!

Now all we need is to remember that the wave functions can be represented by complex functions on Y satisfying [3]

$$\psi(\underline{Z}_Y(\xi)) = Z \cdot \psi(\xi) \tag{9}$$

(where $\underline{Z}_Y$ denotes the action of U(1) on Y) rather than merely functions on Q; the usual wave functions are the *local* representants of these objects obtained as

$$\psi(\xi) = Z_j \cdot \psi_j \qquad \xi \in \pi^{-1}(U_j) \tag{10}$$

Thus, we get finally the geometric formula for the time evolution

$$(\mathcal{U}_{t'-t}\psi)(\xi') = \int_Q dq \int_{P_{xx'}} D\gamma \; \exp\left[\frac{i}{\hbar}\int_{\tilde{\gamma}} \omega\right] \psi(\xi) \tag{11}$$

where $\xi = \tilde{\gamma}(o)$, $P_{xx'} = \{ \gamma \subset E \quad \gamma(o) = (x,.) \quad \gamma(1) = (x',.)\}$. Note that

$$\exp\left[\frac{i}{\hbar}\int_{\tilde{\gamma}}\omega\right]\psi(\xi)$$

is, in fact, a function of γ , independently of the choice of $\tilde{\gamma}$ supposing $\tilde{\gamma}(1) = \xi'$ is held fixed.

Remarks.

1. We do not try to give a geometric definition for "D γ". An attempt in this direction was made by Simms [9].
2. The introduction of the bundle (Y,ω,π) allows for developing a generalized variational formalism [8] and makes it easy to study conserved quantities.

4. A Classification Scheme [6]

If the underlying space is not simply connected, we may have *more than one* prequantisation and thus several inequivalent *meanings* of (1) (two local systems are said to be equivalent if their union is again an admissible local system).

The general construction for all the prequantisations are found in Souriau [3]. Denote $(\tilde{E}, \pi_1, q)$ the universal covering of E; define $\tilde{\sigma} = q^* \sigma$. π_1, the first homotopy group of E, acts then on $\tilde{E}$ by symplectomorphisms.

Let us choose a *"reference prequantisation"* (Y_0,ω_0,π_0) of (E, σ). As $(\tilde{E}, \tilde{\sigma})$ is simply connected, it has a unique prequantisation $(\tilde{Y}, \tilde{\omega}, \tilde{\pi})$, which can be obtained from (Y_0,ω_0,π_0) as

$$(\tilde{Y}, \tilde{\omega}, \tilde{\pi}) = q^*(Y_0,\omega_0,\pi_0) \tag{12}$$

If $\chi : \pi_1 \longrightarrow U(1)$ is a *character*, then π_1 admits an *isomorphic lift* to $(\tilde{Y}, \tilde{\omega}, \tilde{\pi})$ of the form

$$\hat{g}^{\chi}(\tilde{x}, \xi) = (g(\tilde{x}), \underline{\chi(g)}_{Y_0}(\xi)) \tag{13}$$

$g \in \pi_1$, $\underline{Z}_{Y_0}$ denoting the action of $Z \in U(1)$ on Y_0.

Now, Souriau has shown that

$$(Y_\chi,\omega_\chi,\pi_\chi) = (\tilde{Y}, \tilde{\omega}, \tilde{\pi}) / \hat{\pi}_1^{\chi} \tag{14}$$

is a prequantisation of (E, σ), and all prequantisations can be obtained in this way. The inequivalent prequantisations are thus in (1-1) correspondence with the characters of the homotopy group.

In [1] we rederived this theorem from our path-integral consideration noting that we are always allowed to add a closed but not exact 1-form α to θ_o, which - due to non simply connectedness - may change the propagator in an inequivalent way. The corresponding character is then

$$\chi(g) = \exp\left[\frac{i}{\hbar} \oint_{[\gamma]=g} \alpha \right] \tag{15}$$

For instance, in the Bohm-Aharonov experiment [4] $\pi_1 = \mathbb{Z}$ and all the characters have this form.

This is, however, *not* the general situation. A physically interesting counter-example is that of *identical particles* [3], [10].

Example

Consider two identical particles moving in 3-space. The appropriate configuration is then [10] $Q = \tilde{Q}/\mathbb{Z}_2$ where

$$\tilde{Q} := \mathbb{R}^3 \times \mathbb{R}^3 \setminus \{q_1 = q_2\}$$

which has the homotopy group $\pi_1 = \mathbb{Z}_2$.

E is then $T^*Q \times \mathbb{R}$ with $\sigma = d\theta_o^{(1)} + d\theta_o^{(2)}$. $\pi_1 = \mathbb{Z}_2$ has two characters:

$$\chi_1(z) = 1 \qquad \text{and} \qquad \chi_2(z) = -1$$

where z is the interchange of two configurations.

Thus we have two prequantum lifts of π_1 and two prequantisations, one of which is *trivial*, while the second is *twisted*. The first corresponds to *bosons*, the second to *fermions*. Now, it is easy to see that χ_2 is *not* of the form (15):

Proposition

If the homotopy group is finite, $|\pi_1| < \infty$, then $H^1(E, \mathbb{R}) = 0$, i.e. every closed 1-form is exact.

Proof. Let α be a closed 1-form on E, define $\tilde{\alpha} = q^*\alpha$; $\tilde{\alpha} = d\tilde{f}$ for $\tilde{E}$ is simply connected; define

$$\tilde{h} := \frac{1}{|\pi_1|} \sum_{g \in \pi_1} g^* f$$

$\tilde{h}$ is invariant under $g \in \pi_1$ and projects thus to a $h : E \longrightarrow R$. On the other hand $\tilde{\alpha} = d\tilde{h} = dq^*h = q^*\alpha$, and thus $\alpha = dh$.

The general situation can be treated by algebraic topological means [11]. Consider the exact sequence of groups

$$0 \longrightarrow \mathbb{Z} \longrightarrow \mathbb{R} \xrightarrow{2\pi_1} U(1) \longrightarrow 0 \tag{16}$$

giving rise to the long exact sequence

$$\longrightarrow H^1(E,\mathbb{Z}) \xrightarrow{1} \underset{\text{closed / exact}}{H^1(E,\mathbb{R})} \xrightarrow{2} \underset{\text{characters}}{H^1(E, U(1))} \xrightarrow{3} \underset{\text{Chern class}}{H^2(E,\mathbb{Z})} \xrightarrow{4} \underset{\text{curv. class}}{H^2(E,\mathbb{R})} \longrightarrow \tag{17}$$

$$[\sigma/2\pi\hbar]$$

We can make the following observations:

1) σ defines, by (5), an integer-valued element of $H^2_{dR}(E,\mathbb{R})$ which, by de Rham's theorem, is just $H^2(E,\mathbb{R})$.

2) The bundle is topologically completely characterized by its *Chern class* which sits in $H^2(E,\mathbb{Z})$. Thus we have as many bundles as elements in the kernel of $\underset{\smile}{4}$.

3) As $U(1)$ is commutative, a character of π_1 depends only on $\pi_1/[\pi_1,\pi_1]$, which is known to be $H_1(E,\mathbb{Z})$. On the other hand, the Theorem in Universal Coefficients [11] p. 76, yields that

$$\text{Hom}\,(H_1(E,\mathbb{Z}), U(1)) \simeq H^1(E, U(1)) \tag{18}$$

Thus $H^1(E, U(1))$ is just the *set of all characters* classifying the different pre-quantisations.

4) Under quite general conditions, we have

$$H^i(E,\mathbb{Z}) \cong \mathbb{Z}^{b_i} \oplus \text{Tors}\, H^i$$
$$H_i(E,\mathbb{Z}) \cong \mathbb{Z}^{b_i} \oplus \text{Tors}\, H_i \tag{19}$$

where Tors H^i and Tors H_i are groups whose elements are all of finite order;

5) The kernel of the map $H^i(E,\mathbb{Z}) \longrightarrow H^i(E,\mathbb{R})$ is just Tors H^i; the image of $\mathbb{Z}^{b_i}$ is a basis in $H^i(E,\mathbb{R})$;

6) Again, by the Theorem on Universal Coefficients,

$$\text{Tors } H^2(E,\mathbb{Z}) \simeq \text{Tors } H_1(E,\mathbb{Z}) \qquad (\simeq \text{Tors } \pi_1/[\pi_1,\pi_1]) \tag{20}$$

Thus 2), 5), 6) give us

Proposition

The topologically distinct prequantum bundles are labelled by the elements of (20).

7) According to 5), the image of $H^1(E,\mathbb{Z})$ in $H^1(E,\mathbb{R})$ under 1) is made up of integer multiples of a basis. Thus $H^1(E,\mathbb{R})/\text{im } H^1(E,\mathbb{Z}) \simeq (S^1)^{b_1}$ and we get the exact sequence

$$0 \longrightarrow (S^1) \xrightarrow{b_1} H^1(E, U(1)) \longrightarrow \text{Tors } H_1(E,\mathbb{Z}) \longrightarrow 0 \tag{21}$$

Now by de Rham's theorem, to any element of $H^1(E, R)$ we can associate a closed 1-form $\alpha/_{2\pi\hbar}$ such that its value on $g \in H_1(E, R)$ is

$$\frac{1}{2\pi\hbar} \oint_\gamma \alpha \tag{22}$$

where the homology class of γ is g.

Next, by (21), the image of $(S^1)^{b_1}$ in H^1&U(1)) is composed of characters of the form

$$\chi(g) = \exp\left[\frac{i}{\hbar}\oint_\gamma \alpha\right] \tag{23}$$

As $(S^1)^{b_1}$ is connected, and Tors $H^2(E,\mathbb{Z})$ is finite, we have

Proposition

The characters of the form (23) make up the *connected component* containing $\chi \equiv 1$ of the group of characters;

8) Let us choose a basis $\alpha_1,\ldots,\alpha_{b_1}$ in $H^1_{dR}(E,\mathbb{R})$ and pick up a $\chi_k \in H^1(E, U(1))$ corresponding to each element of Tors H^2 = Tors H_1.

Proposition

Any character can be written as

$$\chi(g) = \exp\left[\frac{i}{\hbar}\sum_{j=1}^{b_1} a_j \oint_{[\gamma]=g} \alpha_j\right]\chi_k \tag{24}$$

where $a_j \in \mathbb{R}$ (mod 2π). The χ_k's can be chosen in such a way that they form a *subgroup* of the group of characters, however, there is *no canonical choice* for them.

Finally, we get the following refinement of Souriau's construction (14)

Proposition

Y_{χ_1} and Y_{χ_2} are *topologically identical* iff χ_1 and χ_2 belong to the *same component* of the group of characters.

The different *connection forms* on the *same bundle* are labelled by the elements of the *connected component* containing the identity character $\chi \equiv 1$.

Proof: If a character is of the form (23) then, by carrying smoothly the coefficients to 0, the bundle has to change also smoothly. On the other hand, the Chern class has to change discretely. Consequently, it remains constant.

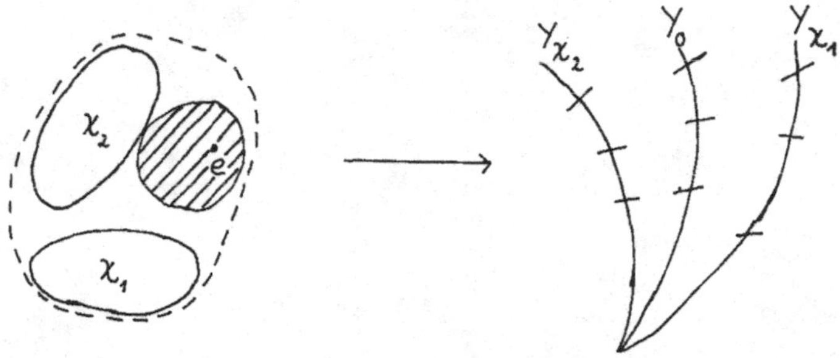

group of characters bundles

ACKNOWLEDGMENTS

I am indebted to Jean-Marie Souriau for hospitality at Marseille. Special thanks are due to János Kollár for his generous help in obtaining the results in Section 4. Discussions are also acknowledged to John Rawnsley.

REFERENCES

[1] Horváthy, P.A., Classical action, the Wu-Yang phase factor and prequantisation, to appear in the Proceedings of the Int. Coll. on Diff. Geom. Meths. in Math. Phys., held in Aix-en-Provence, 1979, Springer Lecture Notes in Mathematics.

[2] Kostant, B., Quantisation and unitary representations in Lecture Notes in Math. 170, Springer (1970).

[3] Souriau, J.M., Structure des systèmes dynamiques, Dunod, Paris (1970), and 2nd Edition to appear at North Holland.

[4] Woodhouse, N.M.J. and Simms, D.J., Geometric quantisation, Springer Lecture Notes in Physics 53, (1976).

[5] Dowker, J.S., Selected Topics in Topology and Quantum Field Theory, Lectures at Austin, Jan.-May 1979.

[6] The results presented here were obtained in collaboration with J. Kollár.

[7] Asorey, M. and Boya, L.J., J. Math. Phys. 20, 10 (1979).

[8] Friedman, J.L. and Sorkin, R., Phys. Rev. D20, 2511 (1979), and Comm. Math. Phys. 73, 161 (1980).

[9] Simms, D.J., talk presented at the Int. Coll. on Diff. Geom. Meth. in Math. Phys. held at Aix-en-Provence (1979), to appear.

[10] De Witt, C. and Laidlaw, M.G.G., Phys. Rev. D3, 6, 1375 (1971).

[11] McLane, S., Homology, Springer (1967).

A GEOMETRICAL PROPERTY OF POV-MEASURES AND SYSTEMS OF COVARIANCE

S. Twareque Ali

Institut für Theoretische Physik, T.U. Clausthal,
3392 Clausthal-Zellerfeld, Fed. Rep. of Germany.

ABSTRACT

A Choquet type of an integral representation is found for a class of normalized positive operator valued (POV) measures on a Hilbert space. An arbitrary POV-measure within this class is thereby represented uniquely as an integral over projection valued (PV) measures. As an application, the case of a commutative system of covariance (representing a generalization of the imprimitivity theorem of Mackey) is discussed. The relevance of these results to the theory of quantum mechanical observables, admitting stochastic value spaces, is pointed out.

1. INTRODUCTION

The subject of positive operator valued (POV) measures acting on a Hilbert space, has received much attention in the recent mathematical and physical literature [1,2,3]. Of central interest here is the situation where the POV-measure is also covariant under the action of some locally compact topological group. In that case it is interesting to study possible generalizations of the imprimitivity theorem of Mackey [4]. In the physical literature, POV-measures have been used to define generalized observables in quantum mechanics [5,6,7], and recently they have been extensively employed in a certain formulation of quantum mechanics on stochastic phase spaces [8,9,10].

In this paper we present some new mathematical results on POV-measures. These results are of a geometrical nature, and are the strongest when the measures are commutative. Accordingly, this report consists of two parts. In the first (Sections 2 and 3), we study certain geometrical properties of POV-measures, defined on the Borel sets of a locally compact space X, and taking values in the set of all bounded operators on a separable Hilbert space $\mathcal{H}$. The set of all such measures a, which satisfy the additional condition $a(X) \leq I$ (= identity operator on $\mathcal{H}$), forms a convex cone. We analyze, in some detail, the extreme points of this cone, with special attention to the case where the ranges of the POV-measures lie in a fixed commutative von Neumann algebra $\mathcal{A}$. This allows for an identification of the set of all extreme points with the set of all projection valued (PV) measures on X which have ranges in $\mathcal{A}$. We then obtain a Choquet type of an integral representation for a POV-measure in terms of PV-measures, and also study the uniqueness of the representing measure. In the second part (Sections 4 and 5) we consider commutative systems of covariance, i.e., commutative POV-measures which satisfy a covariance property (similar to the imprimitivity property in the sense of Mackey [4]), under the action of the group G. We show how the notion of a stochastic value space [10] naturally appears for a quantum mechanical observable defined by a system of covariance (or, for that matter, by a normalized commutative POV-measure). Using the integral representation of Secs. 2 and 3 we derive an explicit embedding of the system of covariance in a system of imprimitivity defined on an extended Hilbert space $\tilde{\mathcal{H}}_a$ which is minimal in the sense of Naimark [11]. We also analyze the associated extended group representation.

Throughout this paper G will be a locally compact, separable and metrizable group, H amd M closed subgroups of G, X and Y separable and metrizable locally compact topological spaces and $\mathcal{H}$ a separable Hilbert space. Furthermore, $g \longmapsto U_g$ will denote a strongly continuous unitary representation of G on $\mathcal{H}$ and $\mathcal{A}$ a von Neumann algebra of operators on $\mathcal{H}$.

Proofs of results have either been given completely, or else been sketched with adequate details.

2. EXISTENCE OF REPRESENTING MEASURES

Notation

$C_\infty(X)$ = C^*-algebra of complex continuous functions on X which vanish at infinity.

$\mathcal{L}[C_\infty(X), \mathcal{A}]^+$ = set of all positive linear maps from $C_\infty(X)$ to $\mathcal{A}$.

$\mathcal{L}_1[C_\infty(X), \mathcal{A}]^+ = \{ \hat{a} \in \mathcal{L}[C_\infty(X), \mathcal{A}]^+ \mid \|\hat{a}\| \leq 1 \}$.

$\mathcal{M}(X, \mathcal{A})$ = set of all bounded, regular POV-measures on X having ranges in $\mathcal{A}$.

$\mathcal{M}_I(X, \mathcal{A}) = \{ a \in \mathcal{M}(X, \mathcal{A}) \mid a(X) \leq I \}$.

In this section we set down preliminary definitions and then obtain a Choquet type of an integral representation for certain positive linear maps

$$\hat{a} \; : \; C_\infty (X) \longrightarrow \mathcal{A}$$

($\hat{a}$ is positive if $\hat{a}$ (f) is a positive operator on $\mathcal{H}$, whenever f is a positive function in C_∞ (X)). Any such map is necessarily bounded (cf. for example [12]), and we restrict ourselves to the case where the (usual Banach space) norm $\|\hat{a}\| \leq 1$. The main results are stated in Theorems 1 and 2, where representing measures are found for $\hat{a}$ and its corresponding POV-measure a.

a) Topology on $\mathcal{L}_1 [C_\infty (X), \mathcal{A}]^+$

Definition 1

A net $\hat{a}_\alpha$ in $\mathcal{L}[C_\infty (X), \mathcal{A}]^+$ converges to $\hat{a} \in \mathcal{L}[C_\infty (X), \mathcal{A}]^+$ in the τ-*topology* if and only if

$$(\phi, \hat{a}_\alpha (f) \psi) \longrightarrow (\phi, \hat{a} (f) \psi),$$

$$\forall \quad \phi, \psi \in \mathcal{H} \qquad \text{and} \qquad f \in C_\infty (X).$$

Definition 2

The τ-topology on $\mathcal{L}_1 [C_\infty (X), \mathcal{A}]^+$ is the topology induced on it by the τ-topology of $\mathcal{L}[C_\infty (X), \mathcal{A}]^+$.

The following result will be crucial for our subsequent analysis.

Lemma 1.

The space $\mathcal{L}_1 [C_\infty (X), \mathcal{A}]^+$ is compact and metrizable under the τ-topology.

Proof: $\mathcal{L}_1 [C_\infty (X), \mathcal{A}]^+$ is the set of positive transformations which map the unit ball $C_\infty (X)_1$ of C_∞ (X) into the unit ball $\mathcal{A}_1$ of $\mathcal{A}$. Since $C_\infty (X)_1$ spans C_∞ (X) linearly, and $\mathcal{A}_1$ is weakly compact [13], and further since the set of functionals

$$\{ A \longmapsto (\phi, A\psi) \mid A \in \mathcal{A}, \; \phi, \psi \in \mathcal{H} \}$$

is a separating family, it follows from [14] that $\mathcal{L}_1 [C_\infty (X), \mathcal{A}]^+$ is τ-compact. On the other hand, since X is separable and metrizable, C_∞ (X) has a countable dense set $\{f_i\}$; while $\mathcal{H}$, being separable, also has a countable dense set $\{\xi_i\}$. Then, it is easily demonstrated that the countable family of seminorms:

$$q_{i,j,k} (\pi, \pi') \;=\; (\xi_i, [\pi(f_j) - \pi' (f_j)] \, \xi_k)$$

$\forall\ \pi, \pi' \in \mathcal{L}_1\,[C_\infty(X), \mathcal{A}]^+$ generates the τ-topology on it. Hence τ is metrizable.

Q.E.D.

b) The extreme points $\hat{\mathfrak{G}}$ of $\mathcal{L}_1\,[C_\infty(X), \mathcal{A}]^+$.

It follows from Lemma 1 that the convex set $\mathcal{L}_1\,[C_\infty(X), \mathcal{A}]^+$ is the closed convex hull of its extreme points $\hat{\mathfrak{G}}$. We characterize in the next lemma these extreme points in the particular case where $\mathcal{A}$ is commutative. Let $\mathcal{N}\,[C_\infty(X), \mathcal{A}]$ be the set of all multiplicative elements in $\mathcal{L}_1\,[C_\infty(X), \mathcal{A}]^+$, i.e. $\hat{P} \in \mathcal{N}\,[C_\infty(X), \mathcal{A}]$ if and only if $\hat{P}(f \cdot g) = \hat{P}(f)\,\hat{P}(g),\ \forall\ f, g \in C_\infty(X)$.

Lemma 2

The extreme points $\hat{\mathfrak{G}}$ of $\mathcal{L}_1\,[C_\infty(X), \mathcal{A}]^+$ form a G_δ-set, and $\mathcal{N}\,[C_\infty(X), \mathcal{A}] \subset \hat{\mathfrak{G}}$. If $\mathcal{A}$ is commutative $\hat{\mathfrak{G}} = \mathcal{N}\,[C_\infty(X), \mathcal{A}]$, i.e., the extreme points are all the mulplicative elements.

Proof: The fact that $\hat{\mathfrak{G}}$ is a G_δ-set follows from Lemma 1 and Proposition 1.3 in [15], while the rest of the lemma follows from [16].

c) The representing measure for $\hat{a}$

To state now our first main result:

Theorem 1

Let $\hat{a} \in \mathcal{L}_1\,[C_\infty(X), \mathcal{A}]^+$; then there exists a probability Baire measure ν on $\hat{\mathfrak{G}}$ such that $\forall\ f \in C_\infty(X)$ and $\phi, \psi \in \mathcal{H}$,

$$(\phi, \hat{a}(f)\psi) = \int_{\hat{\mathfrak{G}}} (\phi, \hat{e}(f)\psi)\, d\nu(\hat{e})\,. \qquad (2.1)$$

Proof: The functionals $\hat{a} \longmapsto (\phi, a(f)\psi)$ on $\mathcal{L}\,[C_\infty(X), \mathcal{A}]^+$ are linear and τ-continuous. Hence by Choquet's theorem [15] the measure ν exists. Further, since $\mathcal{L}_1\,[C_\infty(X), \mathcal{A}]^+$ is metrizable, ν is carried by $\hat{\mathfrak{G}}$ and is in fact a Baire measure on it.

Q.E.D.

We next use this theorem to get, for an arbitrary POV-measure $a \in \mathcal{M}_I(X, \mathcal{A})$, a representing measure which is carried by the extreme points of $\mathcal{M}_I(X, \mathcal{A})$. The result is stated in Theorem 2, which is the analogue of Theorem 1.

d) Correspondence between $\mathcal{L}[C_\infty(X),\mathcal{A}]^+$ and $\mathcal{M}(X,\mathcal{A})$

Let $\mathcal{L}_1[C_\infty(X),\mathcal{A}]^+_n$ denote the subset of elements in $\mathcal{L}_1[C_\infty(X),\mathcal{A}]$ which are normalized in the following sense: if $\hat{a}\in\mathcal{L}_1[C_\infty(X),\mathcal{A}]^+$ and $\{f_j\}$ is a sequence of elements in $C_\infty(X)^+$ such that $f_j\nearrow 1$, pointwise, then $\hat{a}(f_j)\nearrow I$, weakly. Let $\mathcal{M}_I(X,\mathcal{A})_n$ denote the subset of normalized elements a in $\mathcal{M}_I(X,\mathcal{A})$, i.e., elements for which $a(X)=I$. Also, let $\mathcal{P}(X,\mathcal{A})$ denote the set of all projection valued measures in $\mathcal{M}_I(X,\mathcal{A})$. The following useful lemma is then a restatement of Theorems 15, 19 and 21 in [17].

Lemma 3

There exists a one-to-one surjection

$$i\ :\ \mathcal{L}[C_\infty(X),\mathcal{A}]^+\longrightarrow\mathcal{M}(X,\mathcal{A})$$

through the relationship

$$\hat{a}(f)=\int_X f(x)\,da(x)\ ,\qquad i(\hat{a})=a\ , \tag{2.2}$$

for all $f\in C_\infty(X)$, by which $\mathcal{L}_1[C_\infty(X),\mathcal{A}]^+$ is mapped onto $\mathcal{M}_I(X,\mathcal{A})$, $\mathcal{L}_1[C_\infty(X),\mathcal{A}]^+_n$ is mapped onto $\mathcal{M}_I(X,\mathcal{A})_n$ and, $\mathcal{N}[C_\infty(X),\mathcal{A}]$ is mapped onto $\mathcal{P}(X,\mathcal{A})$.

Definition 3

The τ-topology on $\mathcal{M}_I(X,\mathcal{A})$ is the topology it inherits from $\mathcal{L}_1[C_\infty(X),\mathcal{A}]^+$ through the bijection i; i.e., a set $A\subset\mathcal{M}_I(X,\mathcal{A})$ is τ-open if and only if $i^{-1}(A)$ is τ-open in $\mathcal{L}_1[C_\infty(X),\mathcal{A}]^+$.

Let $\mathcal{G}$ denote the set of extreme points of $\mathcal{M}_I(X,\mathcal{A})$. It is then obvious that i maps $\hat{\mathcal{G}}$ onto $\mathcal{G}$, and that Lemmata 1 and 2 imply the following result.

Lemma 4

$\mathcal{M}_I(X,\mathcal{A})$ is a metrizable, compact convex set in the τ-topology. The set of extreme points $\mathcal{G}$ is a G_δ, and if $\mathcal{A}$ is commutative, $\mathcal{G}=\mathcal{P}(X,\mathcal{A})$.

e) The representing measure for $a\in\mathcal{M}_I(X,\mathcal{A})$.

Let $\mathcal{B}(X)$ be the σ-algebra of all Borel sets of X.

Theorem 2

If $a \in \mathcal{M}_I(X, \mathcal{A})$, there exists a probability Baire measure ν on $\mathcal{G}$ such that for all $f \in C_\infty(X)$, $\phi, \psi \in \mathcal{H}$ and $E \in \mathcal{B}(X)$,

$$(\phi, a(E)\psi) = \int_{\mathcal{G}} (\phi, e(E)\psi)\, d\nu(e) \tag{2.3}$$

In particular, if $\mathcal{A}$ is commutative,

$$(\phi, a(E)\psi) = \int_{(X,\mathcal{A})} (\phi, P(E)\psi)\, d\nu(P) . \tag{2.4}$$

Proof: A representation of the type

$$(\phi, a(f)\psi) = \int_{\mathcal{G}} (\phi, e(f)\psi)\, d\nu(e) ,$$

$\forall f \in C_\infty(X)$, where

$$a(f) = \int_X f(x)\, da(x), \qquad e(f) = \int_X f(x)\, de(x)$$

follows from Theorem 1 and Lemma 3. To extend this representation to arbitrary sets $E \in \mathcal{B}(X)$, we note that the real, positive, affine function

$$g_E : \mathcal{M}_I(X, \mathcal{A}) \longrightarrow \mathbb{R}^+$$

defined by

$$g_E(a) = (\phi, a(E)\phi) ,$$

for any $\phi \in \mathcal{H}$, may be obtained as a pointwise limit of a sequence of τ-continuous functions of the type

$$g_f(a) = (\phi, a(f)\phi) ,$$

$f \in C_\infty(X)^+$. Using the extended version of Choquet's theorem (cf. Sec. 12 of [15]), the required result follows.

If $\mathcal{A}$ is commutative, the second part of the theorem follows from Lemma 4.

Q.E.D.

3. UNIQUENESS OF REPRESENTING MEASURES

In this section we deal with the question of the uniqueness of the representing measure

ν obtained in Theorems 1 and 2. The strongest result is in the case where $a(X) = I$ and, either $\mathcal{A}$ is commutative, or the representing measure is supported entirely by $\mathcal{P}(X, \mathcal{A})$. In either of these cases ν is unique (Theorems 4 and 5). The case where $a(X) < I$ will be dealt with elsewhere.

Throughout this section, $\mathcal{G}_I$ will denote a measurable subset of $\mathcal{P}(X, \mathcal{A})$, $\mathcal{A}$ not necessarily commutative, and we shall consider all elements $a \in \mathcal{M}_I(X, \mathcal{A})$ which have representations of the type

$$(\phi, a(E)\psi) = \int_{\mathcal{G}_I} (\phi, P(E)\psi)\, d\nu(P) \tag{3.1}$$

We shall also write (3.1) as

$$a(E) = \int_{\mathcal{G}_I} P(E)\, d\nu(P) . \tag{3.2}$$

In the special case where $\mathcal{A}$ is commutative, we can always take (cf. Theorem 2) $\mathcal{G}_I = \mathcal{G} = \mathcal{P}(X, \mathcal{A})$.

a) The case where $a \in \mathcal{M}_I(X, \mathcal{A})_n$.

Lemma 5

If $a \in \mathcal{M}_I(X, \mathcal{A})_n$, (i.e., $a(X) = I$) $\mathcal{A}$ not necessarily commutative, and if a has an integral representation of the type in (3.1), then in this representation $P(X) = I$, for ν-almost all P in $\mathcal{G}_I$.

Proof: Let ϕ be a unit vector in $\mathcal{H}$. Then from Eq. (3.1)

$$\|\phi\|^2 = 1 = \int_{\mathcal{G}_I} (\phi, P(X)\phi)\, d\nu(P).$$

Let $f(P) = (\phi, P(X)\phi)$, $\forall\, P \in \mathcal{G}_I$. Then $P \longmapsto f(P)$ is a measurable function satisfying

$$f(P) = \|P(X)\phi\|^2 \leq \|\phi\|^2.$$

Hence,

$$0 \leq f(P) \leq 1, \quad \forall\, P \in \mathcal{G}_I,$$

and,

$$1 = \int_{\mathcal{G}_I} f(P)\, d\nu(P) .$$

But since $\nu(\mathcal{G}_I) = 1$, it follows that $f(P) = 1$, for ν- almost all $P \in \mathcal{G}_I$. Thus, for any unit vector $\phi \in \mathcal{H}$,

$$(\phi, P(X)\phi) = 1 ,$$

for ν- almost all P in $\mathcal{G}_I$. But the null set on which $(\phi, P(X)\phi) \neq 1$ could possibly depend on ϕ. However, since $\mathcal{H}$ is separable, we can always choose it to be independent of ϕ. Indeed, let $\{\phi_i\}$ be an orthonormal basis set for $\mathcal{H}$, and let $\mathcal{N}_i$ be the null set on which $(\phi_i, P(X)\phi_i) \neq 1$. Let $\mathcal{N} = \bigcup_i \mathcal{N}_i$, so that (being a countable union of ν-null sets) $\mathcal{N}$ itself is again a ν-null set. Thus for $P \notin \mathcal{N}$ we have,

$$(\phi_i, P(X)\phi_i) = \|P(X)\phi_i\|^2 = 1 ,$$

$\forall\ \phi_i$. Since P(X) is a projector, and $\|\phi_i\|^2 = 1$, it follows that

$$P(X)\phi_i = \phi_i ,$$

$\forall\ \phi_i$. Next, for an arbitrary $\phi \in \mathcal{H}$, such that $\|\phi\|^2 = 1$ and $\phi = \sum_i l_i \phi_i$, l_i = complex coefficients, we have, for any $P \notin \mathcal{N}$,

$$(\phi, P(X)\phi) = \sum_i |l_i|^2 (\phi_i, P(X)\phi_i) + \sum_{i \neq j} \overline{l_i}\, l_j (\phi_i, P(X)\phi_j)$$

$$= \sum_i |l_i|^2 = 1 .$$

which proves our contention.

Hence, for any unit vector ϕ in $\mathcal{H}$,

$$(\phi, P(X)\phi) = 1 ,$$

except at most on $\mathcal{N}$, so that P(X) = I for ν- almost all P in $\mathcal{G}_I$. Q.E.D.

b) The Naimark extension of a

According to a well-known theorem of Naimark [18] any normalized POV-measure $E \longmapsto a(E)$ on $\mathcal{H}$, may be extended to a normalized PV-measure $E \longmapsto \tilde{P}_a(E)$ on an enlarged Hilbert space $\tilde{\mathcal{H}}_a$ in the following minimal sense:

1) if $\mathbb{P}_a$ is the projector for which $\mathbb{P}_a \tilde{\mathcal{H}}_a = \mathcal{H}$, then $a(E) = \mathbb{P}_a \tilde{P}_a(E) \mathbb{P}_a$, $\forall\ E \in \mathcal{B}(X)$;

2) the set of vectors $\{\tilde{P}_a(E)\xi \mid E \in \mathcal{B}(X),\ \xi \in \mathcal{H}\}$ is dense in $\tilde{\mathcal{H}}$;

3) this minimal extension is unique up to an isometric isomorphism.

For any $a \in \mathcal{M}_I(X, \mathcal{A})$, $\mathcal{A}$ not necessarily commutative, which has a representation of the type (3.1), and which satisfies $a(X) = I$, let us make the following extension.

Let

$$\tilde{\mathcal{H}}_a = \mathcal{H} \otimes L^2(\mathfrak{S}_I, \nu), \tag{3.3}$$

where ν is the measure in Eq. (3.1) which represents a. Elements $\tilde{\xi}$ in $\tilde{\mathcal{H}}_a$ are functions

$$\tilde{\xi} : \mathfrak{S}_I \longrightarrow \mathcal{H},$$

such that

$$\|\tilde{\xi}\|^2_{\tilde{\mathcal{H}}} = \int_{\mathfrak{S}_I} \|\tilde{\xi}(P)\|^2_{\mathcal{H}} \, d\nu(P) < \infty .$$

Moreover, elements of the type $\xi \otimes f$, where $\xi \in \mathcal{H}$ and $f \in L^2(\mathfrak{S}_I, \nu)$ form a dense set in $\tilde{\mathcal{H}}$.

Let $\mathbb{1}$ be the element in $L^2(\mathfrak{S}_I, \nu)$ for which

$$\mathbb{1}(P) = 1, \tag{3.4}$$

for ν- almost all P in $\mathfrak{S}_I$, and let us define an isometric injection

$$j : \mathcal{H} \longmapsto \tilde{\mathcal{H}}_a$$

by,

$$j(\xi) = \xi \otimes \mathbb{1} \tag{3.5}$$

$\forall \ \xi \in \mathcal{H}$. Then clearly,

$$(\mathbb{P}_a \tilde{\xi})(P) = \int_{\mathfrak{S}_I} \tilde{\xi}(P') \, d\nu(P') \, \mathbb{1}(P), \tag{3.6}$$

$\forall \ \tilde{\xi} \in \tilde{\mathcal{H}}$ and ν - almost all $P \in \mathfrak{S}_I$ defines a projection $\mathbb{P}_a$ from $\tilde{\mathcal{H}}_a$ to $j(\mathcal{H})$. Also,

$$\mathbb{P}_a(\xi \otimes f) = (f, \mathbb{1})_{L^2(\mathfrak{S}_I, \nu)} \ \xi \otimes \mathbb{1} . \tag{3.7}$$

Finally, on $\tilde{\mathcal{H}}_a$ let us define, $\forall \ E \in \mathcal{B}(X)$, the projection operators $\tilde{P}_a(E)$ by

$$(\tilde{P}_a(E) \tilde{\xi})(P) = P(E) \ \tilde{\xi}(P), \tag{3.8}$$

for ν- almost all $P \in \mathcal{G}_I$.

Theorem 3

The operators $\tilde{P}_a(E)$ constructed in (3.8) define a normalized PV-measure $E \longmapsto \tilde{P}_a(E)$, on $\tilde{\mathcal{H}}_a$ as defined in (3.3), which extends the normalized POV-measure $E \longmapsto a(E)$ on $\mathcal{H}$ in the minimal sense of Naimark.

Proof: That $E \longmapsto \tilde{P}_a(E)$ is a PV-measure is trivial to verify, as well as the relation

$$ja(E)j^{-1} = \mathbb{P}_a P_a(E) \mathbb{P}_a , \tag{3.9}$$

with j as in (3.5) and $\mathbb{P}_a$ as in (3.6). On the other hand, $\tilde{P}_a(X) = I_{\tilde{\mathcal{H}}_a}$ (i.e., the identity operator on $\tilde{\mathcal{H}}$) follows from Lemma 5. We only prove here the minimality of $E \longmapsto \tilde{P}_a(E)$.

For any $E \in \mathcal{B}(X)$ and $\xi \in \mathcal{H}$, let $\tilde{\xi}_E \in \tilde{\mathcal{H}}_a$ be the vector

$$\tilde{\xi}_E(P) = (P(E)\xi \otimes \mathbb{1})(P) = P(E)\xi .$$

We have to show that the closure $\tilde{\mathcal{H}}_o$ of the linear span of the set

$$\{ \tilde{\xi}_E \mid \xi \in \mathcal{H} , E \in \mathcal{B}(X) \}$$

is equal to $\tilde{\mathcal{H}}_a$ itself. Let $\tilde{\xi}_\perp$ be a vector in $\tilde{\mathcal{H}}_a$ which is completely disjoint from $\tilde{\mathcal{H}}_o$, i.e.,

$$(\tilde{\xi}_\perp, \tilde{\xi}_E)_{\tilde{\mathcal{H}}_a} = 0 ,$$

$\forall \tilde{\xi}_E$. Since $\tilde{\mathcal{H}}_a = \mathcal{H} \otimes L^2(\mathcal{G}_I, \nu)$, we have

$$\tilde{\mathcal{H}}_a = \int_{\mathcal{G}_I}^{\oplus} \mathcal{H}(P)\, d\nu(P) \tag{3.10}$$

as a direct integral, where $\mathcal{H}(P)$ is a copy of the Hilbert space $\mathcal{H}$ for ν- almost all $P \in \mathcal{G}_I$. The operators $\tilde{P}_a(E)$ may now be looked upon as decomposable operators

$$\tilde{P}_a(E) = \int_{\mathcal{G}_I}^{\oplus} \Pi_E(P) \tag{3.11}$$

where $\Pi_E(P) = P(E)$, for ν- almost all $P \in \mathcal{G}_I$. On the other hand, since $E \longmapsto \tilde{P}_a(E)$ is a PV-measure, the von Neumann algebra $\mathcal{A}(\tilde{P}_a)$ generated by the $\tilde{P}_a(E)$'s is the weak closure of the set $\{\tilde{P}_a(E) \mid E \in \mathcal{B}(X)\}$. Let $P \longmapsto \gamma(P)$ be a bounded, complex-valued, ν-measurable function on $\mathcal{G}_I$, and let $\tilde{P}_\gamma$ be the operator on $\tilde{\mathcal{H}}_a$ defined by it:

$$(\tilde{P}_\gamma \tilde{\xi})(P) = \gamma(P)\,\tilde{\xi}(P),$$

$\forall\ \tilde{\xi} \in \tilde{\mathcal{H}}$ and ν- almost all P. Then $\tilde{P}_\gamma \in \mathcal{A}(\tilde{P}_a)$ (cf. [13], Chap. II, Sec. 2, Cor. following Theor. 1). Hence, any $\tilde{P}_\gamma$ may be obtained as a weak limit of elements in $\mathcal{A}(\tilde{P}_a)$ of the form $\sum_{i=1}^{n} c_i\, \tilde{P}_a(E_i)$, with n finite. It follows, therefore, that $\forall\ \xi \in \mathcal{H}$,

$$(\tilde{\xi}_\perp, \tilde{P}_\gamma\, \xi \otimes \mathbb{1})_{\tilde{\mathcal{H}}_a} = 0,$$

$\forall\ \nu$ -measurable functions γ, i.e.,

$$\int_{\mathcal{G}_I} \gamma(P)\,(\tilde{\xi}_\perp(P), \xi)_{\mathcal{H}}\, d\nu(P) = 0,$$

$\forall\ \xi \in \mathcal{H}$, and ν-measurable functions γ. Hence, $\tilde{\xi}_\perp(P) = 0$ for ν- almost all $P \in \mathcal{G}_I$, so that $\tilde{\mathcal{H}}_0 = \tilde{\mathcal{H}}_a$. Q.E.D.

Remark 1

Since $\mathcal{H}$ is assumed to be separable and $\mathcal{M}_I(X, \mathcal{A})$ is metrizable, it follows that the Hilbert spaces $L^2(\mathcal{G}_I, \nu)$ and $\tilde{\mathcal{H}}_a = \mathcal{H} \otimes L^2(\mathcal{G}_I, \nu)$ are separable.

c) Uniqueness in the case where $a \in \mathcal{M}_I(X, \mathcal{A})_n$.

We are now in a position to prove the uniqueness of representing measures for normalized POV-measures which satisfy Eq. (3.1).

Theorem 4

If $a \in \mathcal{M}_I(X, \mathcal{A})_n$ has a representation of the type

$$a(E) = \int_{\mathcal{G}_I} P(E)\, d\nu(P),$$

where $\mathcal{G}_I$ is a measurable subset of $\mathcal{P}(X, \mathcal{A})$, then the measure representing a is unique.

Proof: Let ν and ν' be two measures, both representing a in the above sense. Then by Theorem 3, both $\mathcal{H} \otimes L^2(\mathcal{G}_I, \nu)$ and $\mathcal{H} \otimes L^2(\mathcal{G}_I, \nu')$ are minimal extension spaces for embedding $E \longmapsto a(E)$ into a PV-measure. Hence $\mathcal{H} \otimes L^2(\mathcal{G}_I, \nu)$ and $\mathcal{H} \otimes L^2(\mathcal{G}_I, \nu')$ must be isometrically isomorphic [18] in such a way as to preserve (3.6) and (3.8). Hence $\nu = \nu'$.

Q.E.D.

Let $\mathcal{A}(a)$ be the von Neumann algebra generated by the normalized POV-measure $E \longmapsto a(E)$, i.e., $\mathcal{A}(a)$ is the double commutant of the set $\{a\,(E) \mid E \in \mathcal{B}(X)\}$. We then have the following result when $\mathcal{A}(a)$ is commutative.

Theorem 5

If a is a normalized, regular, commutative POV-measure on X, there exists a unique probability Baire measure ν carried by the set $\mathfrak{S} = \mathcal{P}(X, \mathcal{A}(a))$ such that

$$a\,(E) \;=\; \int_{\mathfrak{S}} P\,(E)\; d\,\nu\,(P)\;.$$

The measure ν is unaltered if we replace $\mathcal{A}(a)$ by any other von Neumann algebra which contains $\mathcal{A}(a)$.

Proof: The first part of the theorem is a consequence of Theorems 2 and 4, since if we set $\mathcal{A} = \mathcal{A}(a)$ in Theorem 2, then clearly $a \in \mathcal{M}_I\,(X, \mathcal{A}(a))$.

To prove the second part, let $\mathcal{A}$ be any von Neumann algebra such that $\mathcal{A} \supset \mathcal{A}(a)$, and let ν' be another measure carried by the extreme points $\mathfrak{S}'$ of $\mathcal{M}_I\,(X, \mathcal{A})$, for which we again have

$$a\,(E) \;=\; \int_{\mathfrak{S}'} P\,(E)\; d\,\nu\,(P)\;.$$

But since any PV-measure in $\mathcal{M}_I\,(X, \mathcal{A}(a))$ is also extreme in $\mathcal{M}_I\,(X, \mathcal{A})$, therefore, $\mathfrak{S} \subset \mathfrak{S}'$. Hence ν may be considered to be a measure on $\mathfrak{S}'$ with support contained in $\mathfrak{S}$. By uniqueness therefore, $\nu = \nu'$. Q.E.D.

4. COMMUTATIVE SYSTEMS OF COVARIANCE

In this section we begin an analysis of a commutative POV-measure a and its representing measure ν (in the sense of Eq. (3.1)) with the further assumption of group covariance.

As introduced before, let G be a locally compact, separable and metrizable group, H a closed subgroup of G, and let

$$X \;=\; G/H \tag{4.1}$$

be the corresponding left coset space. Then X is a separable, locally compact topological space and the action of G on X, denoted by

$$x \longmapsto g\,[x]\,, \qquad x \in X\,, \quad g \in G$$

is transitive. Let $g \longmapsto U_g$ be a strongly continuous unitary representation of G on $\mathcal{H}$, and let the commutative POV-measure $E \longmapsto a(E)$ be defined on the Borel sets $\mathcal{B}(X)$ of X.

Definition 4

A commutative system of covariance is a pair $\{a, U\}$, where $E \longmapsto a(E)$ is a normalized commutative POV-measure, and $g \longmapsto U_g$ is a strongly continuous unitary representation of G, satisfying

$$U_g \, a(E) \, U_g^* = a(g[E]) , \tag{4.2}$$

$\forall$ $E \in \mathcal{B}(X)$ and $g \in G$, it being assumed that X = G/H, for some closed subgroup H of G.

Remark 2

Systems of covariance (not necessarily commutative) have also been termed generalized systems of imprimitivity in the literature [2,6,7,8]. In the case where a is a PV-measure, the definition above is the same as that of the usual Mackey system of imprimitivity [4].

a) The $*$-algebras $\mathcal{A}_c(a)$ and $\mathcal{A}(a)$

Let K(X) denote the set of all complex valued functions on X which have compact supports, $\mathcal{A}_c(a)$ the C^*-algebra generated by the set $\{a(f) \mid f \in C_\infty(X)\}$ and $\mathcal{A}(a)$, as before, the von Neumann algebra generated by the set $\{a(E) \mid E \in \mathcal{B}(X)\}$. Let $\mathcal{A}_c(a)''$ be the von Neumann algebra generated by $\mathcal{A}_c(a)$.

Lemma 6

$\mathcal{A}(a) = \mathcal{A}_c(a)''$, and $\mathcal{A}_c(a)$ is contained densely (in the weak operator topology) in $\mathcal{A}(a)$.

Proof: From the properties of a POV-measure, it follows [17] that, $\forall$ $E \in \mathcal{B}(X)$, a(E) is contained in the weak closure of the set $\{a(f) \mid f \in K(X)\}$. Hence if $\overline{\mathcal{A}_c(a)}^{weak}$ denotes the weak closure of $\mathcal{A}_c(a)$, we have

$$a(E) \in \overline{\mathcal{A}_c(a)}^{weak} \tag{4.3}$$

On the other hand, by von Neumann's density theorem [13],

$$\overline{\mathcal{A}_c(a)}^{weak} = \mathcal{A}_c(a)'' , \tag{4.4}$$

implying, therefore,

$$a(E) \in \mathcal{A}_c(a)'' , \quad \forall E \in \mathcal{B}(X) .$$

Hence,

$$\mathcal{A}(a) \subset \mathcal{A}_c(a)'' \tag{4.5}$$

Next, since $\forall f \in C_\infty(X)$,

$$a(f) = \int f(x)\, da(x) ,$$

it follows that everything which commutes with $a(E)$, $\forall E \in \mathcal{B}(X)$, also commutes with $a(f)$, $\forall f \in C_\infty(X)$. Hence,

$$\mathcal{A}_c(a)'' \subset \mathcal{A}(a) \tag{4.6}$$

The proof of the lemma is completed upon combining (4.5) and (4.6) with (4.4).

Q.E.D.

<u>Remark 3</u>

Equation (4.2) implies that the algebras $\mathcal{A}_c(a)$ and $\mathcal{A}(a)$ are stable under G, i.e.,

$$\left.\begin{aligned} U_g \mathcal{A}_c(a) U_g^* &= \mathcal{A}_c(a) \\ U_g \mathcal{A}(a) U_g^* &= \mathcal{A}(a) \end{aligned}\right\} \tag{4.7}$$

Let Y denote the spectrum of the C^*-algebra $\mathcal{A}_c(a)$. Then the following statements are valid (cf., for example, [13,19]).

1) There exists an algebraic isometry

$$i : \mathcal{A}_c(a) \longmapsto C_\infty(Y)$$

which is bijective, and which can be extended to

$$i : \mathcal{A}(a) \longmapsto L^\infty(Y, \lambda) ,$$

where λ is a basic measure on Y. This latter algebraic isometry is also bijective, and if $u \in L^\infty(Y, \lambda)$ is the image of $A \in \mathcal{A}(a)$ under i , then

$$\|A\| = \operatorname*{Ess.\,Sup}_{y \in Y} |u(y)|$$

2) The action (4.2) of the group G induces an action on the space Y, making it into a

homogeneous space. Let us denote this action by

$$y \longmapsto g\,[y]\ , \quad g \in G\ , \quad y \in Y\ ,$$

and assume that it is *transitive*. Then there exists a closed subgroup M of G for which

$$Y = G/M \tag{4.8}$$

and the measure λ is quasi-invariant under the action of G. Hence λ satisfies

$$\int_Y u\,(g\,[y]\,)\,d\,\lambda\,(y) = \int_Y u(y)\ \xi\,(g,\,y)\,d\,\lambda\,(y)\ , \tag{4.9}$$

for all integrable Borel functions u on Y. Here ξ is a multiplier satisfying

$$\begin{aligned} \xi\,(e,\,y) &= 1 \\ \xi(g_1\,g_2,\,y) &= \xi(g_1,\,y)\ \xi\,(g_2,\,g_1^{-1}\,[y]\,)\ , \end{aligned} \tag{4.10}$$

for almost all $y \in Y$ and all $g_1, g_2 \in G$. (e = unit element of G). It is well-known that

$$\xi : G \times Y \longrightarrow \mathbb{R}^+$$

is a Borel function which is equal almost everywhere to the Radon-Nikodym derivative $d\,[{}_g\lambda]/d\lambda$, ${}_g\lambda$ being the measure defined as; $({}_g\lambda)\,(F) = \lambda(g^{-1}\,[F]\,)$, $\forall\ F \in \mathcal{B}(Y)$.

3) The representation $g \longmapsto U_g$ is induced from a unitary representation $m \longmapsto V(m)$ of M, in the sense of Mackey [4].

Definition 5

The commutative system of covariance $\{a,\,U\}$ is said to be transitive if the action $y \longmapsto g\,[y]$ of G on Y = Spectrum $[\mathcal{A}_c(a)]$ is transitive, so that, in particular, Eq. (4.8) holds [19].

b) Stochastic value spaces.

For any $g \in G$, let

$$g = k_g\ m_g \tag{4.11}$$

be its (unique) Mackey decomposition corresponding to the subgroup M, so that $m_g \in M$, and $k_e = e$. Let

$$\beta \ : \ Y \longmapsto G$$

be the corresponding Borel section, i.e.,

$$\beta(y) = k_{\beta(y)} , \quad \forall y \in Y \tag{4.12}$$

and let y_0 denote the element in Y for which

$$\beta(y_0) = e , \tag{4.13}$$

so that

$$\beta(y) [y_0] = y . \tag{4.14}$$

Let $\mathcal{H}_0$ be the Hilbert space which carries the representation $m \longmapsto V(m)$ of M, from which the representation $g \longmapsto U_g$ of G is induced. It follows then from the Mackey theory, that

(1) $$\mathcal{H}' = \mathcal{H}_0 \otimes L^2 (Y, \lambda) , \tag{4.15}$$

so that if $\phi \in \mathcal{H}$, then ϕ is a $\mathcal{H}_0$ valued function $y \longmapsto \phi(y)$ on Y.

(2) For all $\phi \in \mathcal{H}$

$$(U_g \phi)(y) = B(g, y) \ \phi(g^{-1} [y]) , \tag{4.16}$$

where the operator valued function $B : G \times Y \longmapsto \mathcal{L}(\mathcal{H}_0)$ is defined as

$$B(g, y) = [\xi(g, y)]^{\frac{1}{2}} \ V(m_{g^{-1} \beta(y)})^{-1} . \tag{4.17}$$

(3) The projection valued measure $F \longmapsto \Pi_a(F)$, $F \in \mathcal{B}(Y)$, defined as

$$(\Pi_a(F)\phi)(y) = \chi_F(y) \ \phi(y) \tag{4.18}$$

satisfies the system of imprimitivity

$$U_g \Pi_a(F) U_g^* = \Pi_a (g [F]) , \tag{4.19}$$

(compare Eq. (4.2)), and furthermore,

$$\mathcal{A}(a) = \mathcal{A}(\Pi_a) \tag{4.20}$$

Theorem 6

Let $\{a, U\}$ be a commutative and transitive system of covariance. Then there exists a unique probability measure ν on X, such that

$$(a(E)\phi)(y) = (\chi_E * \nu)(y)\,\phi(y), \tag{4.21}$$

$\forall\, E \in \mathcal{B}(X)$, and $\forall\, \phi \in \mathcal{H}$, where $*$ denotes the convolution

$$(\chi_E * \nu)(y) = \int_X \chi_E(\beta(y)[x])\, d\nu(x). \tag{4.22}$$

For fixed E, the function $y \longmapsto \nu_y(E) = (\chi_E * \nu)(y)$ is a bounded Borel function on Y.

Proof: (We only sketch the proof of this theorem.)
From Lemma 6, and the resulting isometry

$$i : \mathcal{A}_c(a) \longmapsto C_\infty(Y),$$

it follows that the mapping:

$$\nu_y : C_\infty(X) \longmapsto \mathbb{C}$$

defined by

$$\nu_y(f) = (i[a(f)])(y), \tag{4.23}$$

$\forall\; y \in Y$, is positive, linear and satisfies

$$\nu_y(f_n) \nearrow 1,$$

for a sequence of functions f_n in $C_\infty(X)$ such that $f_n \nearrow 1$, pointwise. Hence, each ν_y is a normalized positive Radon measure on X, and furthermore, for fixed $f \in C_\infty(X)$, $y \longmapsto \nu_y(f)$ is a bounded continuous function. Using standard dominated convergence arguments and the fact that the isometry i can be extended to

$$i : \mathcal{A}(a) \longmapsto L^\infty(Y, \lambda),$$

it is possible to prove that $\forall\, E \in \mathcal{B}(X)$

$$(a(E)\phi)(y) = \nu_y(E)\,\phi(y), \tag{4.24}$$

where $\nu_y(E)$ is now a probability measure, obtained from the Radon measure $\nu_y(f)$. Furthermore, for fixed E, $y \longmapsto \nu_y(E)$ is a bounded Borel function. Let

$$\vartheta \equiv \vartheta_{y_0} \quad , \tag{4.25}$$

with y_0 as in (4.13). Then, from (4.14) and (4.2) it follows that

$$\vartheta_y (E) = \vartheta(\beta(y)^{-1} [E]) \ , \tag{4.26}$$

$\forall$ $E \in \mathcal{B}(X)$. On the other hand,

$$\vartheta(\beta(y)^{-1} [E]) = \int_X \chi_{\beta(y)^{-1}[E]} (x) \, d \vartheta (x) = \int_X \chi_E (\beta(y) [x]) \, d\vartheta(x) \tag{4.27}$$

The theorem follows upon combining (4.26) and (4.27) with (4.24). Q.E.D.

Remark 4

Equation (4.24) shows that the commutative POV-measure a associates to each point $y \in Y$ a probability measure ϑ_y. In physics one often encounters a situation, in which, when a point x (for example the position of a particle) is measured, the outcome is in fact only describable with the help of a probability distribution f_x 'centred' at x. In other words, when a reading x is registered by the measuring apparatus, the probability that the observed point actually lies between x' and x' + dx' ($x \neq x'$, in general) is $f_x(x')\,dx'$. Thus to each point x is associated the probability measure ϑ_x with $d\vartheta_x(x') = f_x(x')\,dx'$. In the limiting case of absolutely accurate measurements, ϑ_x is the delta measure δ_x at x (cf. [10] for a fuller discussion and relationship to quantum mechanics).

There is a canonical procedure for associating an observable A, in quantum mechanics, to a POV-measure a [5]. The set of all possible values of an observable is referred to as its *value space*. In the context of imprecise measurements, of the type just mentioned, it is useful to consider, for each observable A and its possible 'precise' value space X, also a *stochastic value space*, X.

Definition 6

A stochastic value space for an observable A (having precise value space X), is a set of ordered pairs

$$\hat{X} = \{ (x, \vartheta_x) \mid x \in X \}$$

such that $\forall$ x, ϑ_x is a probability measure on X, and $\forall$ $E \in \mathcal{B}(X)$, $x \longmapsto \vartheta_x (E)$ is a Borel function.

From Eq. (4.24) the following result is easily seen to follow:

Theorem 7

Let $E \longmapsto a(E)$, $E \in \mathcal{B}(X)$, be a normalized, regular, commutative POV-measure. Let $\mathcal{A}(a) \cong L^{\infty}(Y, \lambda)$, and suppose that there exists a Borel isomorphism

$$j : Y \longmapsto X .$$

Then a determines a stochastic value space $\hat{X} = \{(x, \nu_x) \mid x \in X\}$, with ν_x given by

$$\nu_x = \nu_{j^{-1}(x)} ,$$

$$(a(E)\phi)(j^{-1}(x)) = \nu_{j^{-1}(x)}(E)\,\phi(j^{-1}(x)) , \tag{4.28}$$

$\forall\, \phi \in \mathcal{H}$.

The question remains of course, under what conditions the isomorphism j exists, but this can be settled under certain physical assumptions [8].

5. THE REPRESENTING MEASURE IN THE PRESENCE OF COVARIANCE

In this section we first combine the results of the last two to show that the representing measure ν of Eq. (3.2) is the same as the measure ν in Eq. (4.21) which arises when group covariance prevails. Secondly we use the construction of Section 3.b to embed the whole system of covariance $\{a, U\}$ on $\mathcal{H}$ into a system of imprimitivity $\{\tilde{P}_a, \tilde{U}\}$ on $\tilde{\mathcal{H}}_a$. We also analyze the resulting representation $g \longmapsto \tilde{U}_g$ as an induced representation.

Consider once again the system of covariance $\{a, U\}$, so that Eqs. (4.21) and (4.22) hold. On $\mathcal{H}$, for each fixed $x \in X$ and $E \in \mathcal{B}(X)$, consider the operator $P_x(E)$ such that,

$$(P_x(E)\phi)(y) = \chi_E(\beta(y)[x])\,\phi(y) , \tag{5.1}$$

$\forall\, \phi \in \mathcal{H}$. Then, it is easy to see that $P_x(E)$ is a projector, and that for fixed x, $E \longmapsto P_x(E)$ is a PV-measure on X. Furthermore, it is clear that $P_x(E) \in \mathcal{A}(a)$ $\forall$ E and x. Hence, the set of PV-measures $\{P_x \mid x \in X\}$ is a subset $\mathcal{G}_X$ of $\mathcal{G}$.

The map $x \in X \longmapsto P_x \in \mathcal{M}(X, \mathcal{A}(a))$ defined by Eq. (5.1) is easily proved to be continuous, so that $\mathcal{G}_X$ being the image of X under this map is a Borel set. On the other hand, combining (4.21) and (4.22) with (5.1) we have,

$$(\Phi , a (E)\Phi) = \int_X (\Phi, P_x (E)\Phi) d \nu (x) \tag{5.2}$$

In view of the continuity of the map $x \longmapsto P_x$, the measure ν in (5.2) may actually be thought of as being a measure ν on $\mathcal{G}_X \subset \mathcal{G}$, and we get back to Eq. (3.2). Thus we have proved

Lemma 7

Under the Borel isomorphism

$$\theta : X \longrightarrow \mathcal{G}_X$$

defined by the continuous function $x \longmapsto P_x$, the measure ν in Eq. (4.21) goes over to the measure ν in Eq. (3.2).

Using now the methods of Sec. 3.b, it is a routine matter to extend $\{a, U\}$ to $\{\tilde{P}_a, \tilde{U}\}$. In fact, from (3.3) and (4.15) we get

$$\tilde{\mathcal{H}}_a = \mathcal{H}_0 \otimes L^2 (X, \nu) \otimes L^2 (Y, \lambda) , \tag{5.3}$$

while (3.11) and (5.1) yield

$$(\tilde{P}_a (E) \tilde{\Phi}) (x, y) = \chi_E (\beta(y) [x]) \tilde{\Phi}(x, y) , \tag{5.4}$$

$\forall$ $E \in \mathcal{B}(X)$ and all vectors $\tilde{\Phi} \in \tilde{\mathcal{H}}_a$. Similarly, from (3.6),

$$(\mathbb{P}_a \tilde{\Phi}) (x, y) = \int \tilde{\Phi}(x', y) d \nu (x') \mathbb{1}(x) . \tag{5.5}$$

The representation $g \longmapsto U_g$ now extends to $g \longmapsto \tilde{U}_g$ on $\tilde{\mathcal{H}}_a$ via:

$$(\tilde{U}_g \tilde{\Phi}) (x, y) = B(g, y) \tilde{\Phi}(m_{g^{-1}\beta(y)} [x], g^{-1} [y]) , \tag{5.6}$$

with m and B as in (4.11) and (4.17), respectively.

It is then easily verified that $\{\tilde{P}_a, \tilde{U}\}$ is indeed a system of imprimitivity. To analyze $g \longmapsto \tilde{U}_g$ also as an induced representation, we note first, that, in view of (4.12) and (4.25), the measure ν is invariant under the subgroup M of G. Consider now the unitary representation $m \longmapsto W(m)$ of M on $L^2 (X, \nu)$

$$(W (m) f) (x) = f (m^{-1} [x]) , \tag{5.7}$$

$\forall\ f \in L^2(X, \nu)$. Next, an element $\psi \in \tilde{\mathcal{H}}_a$ can be looked upon as being a function

$$\psi : Y \longrightarrow \mathcal{H}_0 \otimes L^2(X, \nu),$$

which is square integrable w.r.t. λ. On $\mathcal{H}_0 \otimes L^2(X, \nu)$ consider the unitary representation $m \longmapsto \tilde{W}(m)$ of M

$$\tilde{W}(m) = V(m) \otimes W(m), \tag{5.8}$$

with V (m) as in (4.17). Then it is easily seen, that in terms of vectors $y \longmapsto \psi(y)$, Eq. (5.6) can be rewritten as

$$(\tilde{U}_g \psi)(y) = \tilde{B}(g, y) \quad \psi(g^{-1}[y]), \tag{5.9}$$

where

$$\tilde{B}(g, y) = \left[\xi(g, y)\right]^{\frac{1}{2}} \quad \tilde{W}(m_{g^{-1}\beta(y)})^{-1} \tag{5.10}$$

(cf. Eq. (4.17)). Hence, $\tilde{U}_g$ is the representation of G induced from the representation $\tilde{W}(m)$ of M.

ACKNOWLEDGMENTS

Part of this work was done while the author held a fellowship granted by the Alexander von Humboldt-Stiftung. He would like to thank the Stiftung for its support, and Prof. H.D. Doebner for his constant encouragement, and for having made his stay in Clausthal possible in the first place.

REFERENCES

[1] Cattaneo, U., Comment. Math. Helvetici 54, 629 (1979).
[2] Scutaru, H., Letters Math. Phys. 2, 101 (1979).
[3] Castrigiano, D.P.L. and Heinrichs, R.W., Letters Math. Phys. 4, 169 (1980).
[4] Mackey, G.W., Proc. Natl. Acad. Sci. U.S.A. 35, 537 (1949).
[5] Davies, E.B. and Lewis, J.T., Commun. Math. Phys. 17, 239 (1969).
[6] Neumann, H., Helv. Phys. Acta 45, 811 (1972).
[7] Ali, S.T. and Emch, G.G., J. Math. Phys. 15, 176 (1974).
[8] Ali, S.T. and Prugovečki, E., J. Math. Phys. 18, 219 (1977).

[9] Ali, S.T. and Prugovečki, E., 'Consistent models of spin o and 1/2 extended particles scattering in external fields', in *Mathematical Methods and Applications of Scattering Theory*, Eds., J.A. De Santo, A.W. Saenz and W.W. Zachary; Series: Lecture Notes in Physics; Springer-Verlag, Berlin-Heidelberg-New York, Vol. 130 (1980), p. 197.

[10] Ali, S.T., 'Aspects of relativistic quantum mechanics on phase space', in *Differential Geometric Methods in Mathematical Physics*, Ed., H.D. Doebner, Series: Lecture Notes in Physics; Springer-Verlag, Berlin-Heidelberg-New York, Vol. 139 (1981), p. 49.

[11] Naimark, M.A., C.R. (Doklady) Acad. Sci. URSS 41, 359 (1943).

[12] Nachbin, L., *Topology and Order*, Van Nostrand Co. Inc., Princeton, N.J. (1965).

[13] Dixmier, J., *Les Algèbres d'operateurs dans l'espace Hilbertien*, Gauthier-Villars, Paris (1957).

[14] Kadison, R.V., Proc. Amer. Math. Soc. 12, 973 (1961).

[15] Phelps, R.R., *Lectures on Choquet's Theorem*, Van Nostrand Co., Inc., Princeton, N.J. (1966).

[16] Espelie, M.S., Pacific J. Math. 48, 57 (1973).

[17] Berberian, S.K., *Notes on Spectral Theory*, Van Nostrand Co., Inc., Princeton, N.J. (1966).

[18] See, for example, B. SZ-Nagy, *Extensions of Linear Transformations in Hilbert Space which Extend Beyond this Space*, Appendix to F. Riesz and B. SZ-Nagy, *Functional Analysis*, Frederick Ungar, New York (1960).

[19] Takesaki, M., Acta Math. 119, 273 (1967).

PATH INTEGRALS OVER MANIFOLDS *

Jan Tarski

Institut für Theoretische Physik, T.U. Clausthal
3392 Clausthal-Zellerfeld, Fed. Rep. of Germany.

ABSTRACT

Path integrals are considered for the cases where the underlying manifold is multiply connected or non-flat. In case of multiple connectivity, the contributions of different homotopy classes of paths are analyzed with the help of covering spaces. In case of a non-flat manifold, it is pointed out that a judicious choice of the free Hamiltonian operator and of normalizing factors can eliminate the explicit occurrence of the scalar curvature. (A heuristic approach to path integrals is adopted in case of non-flat spaces.)

1. Introduction

In quantizing physical systems on manifolds, two new features arise. First, the manifold may be multiply connected, and then there arise problems of homotopy and of associated phases. Second, the scalar curvature may appear explicitly, e.g. in equations of motion. We should like to elaborate on these features in turn, from the point of view of the path-integral approach to quantization.

In case of a multiply connected manifold, with the fundamental group Γ, the propagator was expressed in the following way in [1]:

$$\check{G}(t;v,u) = e^{i\alpha} \sum_{\delta\in\Gamma} \chi(\delta)\; G_\delta(t;v,u) . \tag{1.1}$$

Here the group elements δ also label the homotopy classes of paths from u to v, and

---

* The present article constitutes a revised text of a seminar which the author gave at a summer school at T.U. Clausthal in 1979. This article is included in the present proceedings by a special arrangement with the editors.

the partial propagator G_{δ} is to be given by a path integral, with the paths restricted to the homotopy class labeled by δ. The $\chi(\delta)$ define a one-dimensional representation of Γ, and $e^{i\alpha}$ is a phase factor which $= 1$ if suitable conventions are made.

We will describe (in sec. 2) the construction of the G_{δ}, by employing a covering space of the manifold in question. Our construction is in fact equivalent to that of Dowker in [2]. However, we emphasize homotopy classes of paths and path integrals, while such integrals were hardly mentioned in *loc. cit.* (Cf. also the discussion of systems on multiply connected manifolds in [3].)

In the considerations of multiple connectivity, we include the example of indistinguishable particles, following [1]. For this particular example the underlying manifold is flat, and the usual mathematical constructions of path integrals may be applicable. A slight complication is caused by the removal of the diagonal set E, cf. eqs. (2.1) below, and we discuss (in the appendix) the irrelevance of this set.

With regard to the scalar curvature R, path integrals and equations of motion with the correction terms R/12 or R/6 were derived in the pioneering work of B. DeWitt [4]. Subsequently a number of authors commented on these correction terms, and we refer to a recent review [5]. (The term R/8 has also been encountered, cf. [6,7]. We will not be concerned with this possibility, or with the approaches of the works just cited.)

We indicate (in sec. 3) how the correction terms can be eliminated, by choosing appropriately the free Hamiltonian operator and the normalization of path integrals. To this end, we present equations of motion and (heuristic) path integrals which are equivalent to those of [4,5], but which are expressed differently. One could say that we reinterpret the results of *loc. cit.*

We remark also that there exists an elementary example, the rotating top, where the effects of curvature and of multiple connectivity are combined ([8]; cf. also [9]). This example will not be considered here further.

The author thanks Professors H.D. Doebner and J. Tolar for useful discussions. He gratefully acknowledges the hospitality at T.U. Clausthal.

2. Multiply connected manifolds.

We first recall the example of k indistinguishable particles on R^n $(k \geqslant 2, n \geqslant 3)$, following Laidlaw and C. DeWitt [1]. We start with the space R^{nk} of points

$\underset{\sim}{x} = (\underset{\sim}{x}_1,\ldots,\underset{\sim}{x}_k)$ where each $\underset{\sim}{x}_j \in R^n$, we let

$$E = \{ \underset{\sim}{x} \in R^{nk} : \underset{\sim}{x}_i = \underset{\sim}{x}_j \text{ for some } i \neq j \}, \tag{2.1a}$$

and we take for the configuration space

$$C = (R^{nk} \setminus E) / S_k , \tag{2.1b}$$

where S_k is the permutation group on k symbols. It follows that C has S_k as the fundamental group, and that the homotopy classes of paths correspond to permutations $\delta \in S_k$.

In order to motivate the construction of the partial propagators G_δ in eq. (1.1), let us consider also an alternate, more elementary approach to indistinguishable particles. In this approach R^{nk} is the configuration space, and symmetry properties of a propagator are ensured by the initial conditions and by the symmetry of the Hamiltonian.

We let $\underset{\sim}{x} \in R^{nk}$ as before, let $\varrho \in S_k$, let $\underset{\sim}{x}_\varrho = (\underset{\sim}{x}_{\varrho(1)},\ldots,\underset{\sim}{x}_{\varrho(k)})$, let $y \in R^{nk}$, and we set

$$G_\chi(t;y,x) = (n!)^{-1/2} \sum_\varrho \chi(\varrho)\; G(t;\underset{\sim}{y},\underset{\sim}{x}_\varrho), \tag{2.2a}$$

where $\chi(\varrho) = 1$ or $\operatorname{sgn}\varrho$, and where each G is the usual propagator, satisfying the Schrödinger equation, and the initial conditions

$$\lim_{t \searrow 0} G(t;y,x) = \delta(\underset{\sim}{y}_1 - \underset{\sim}{x}_{\varrho(1)}) \ldots \delta(\underset{\sim}{y}_k - \underset{\sim}{x}_{\varrho(k)}). \tag{2.2b}$$

The function G_χ then has the desired symmetry. Moreover, there is a bijective correspondence between the partial propagators G in (2.2) and the G_δ of (1.1) (assuming (1.1) is specialized to the case at hand). This correspondence depends on the following:

<u>Lemma 1.</u> Let W_0 be a topological space with the fundamental group Γ, and let W_1 be a simply connected covering space of W_0. Let $v_0 \in W_0$, let $v \in W_1$ be in the inverse image of v_0 under the natural projection, and let the set $\{u^\delta\}_{\delta \in \Gamma}$ constitute the (complete) inverse image of $u_0 \in W_0$. Let η_0 be a path on W_0 from u_0 to v_0. Then η_0 is the image (under the natural projection) of a unique path in W_1 from a point u^ϱ, $\varrho \in \Gamma$, to v. Two paths η_1, η_2 on W_0 from u_0 to v_0 are homotopic if and only if the corresponding points $u^{\delta_1}, u^{\delta_2} \in W_1$ coincide (i.e., if and only if $\delta_1 = \delta_2$).

This lemma summarizes some elementary results in algebraic topology [1,2,10].

Let us now write a path-integral representation of one G in (2.2), with the set E eliminated as shown:

$$G(t;\underset{\sim}{y},\underset{\sim}{x}_{\varrho}) = \int_{\substack{\eta(0) = \underset{\sim}{x}_{\varrho} \\ \eta(\tau) \notin E,\ \forall \tau}} \mathcal{D}(\eta)\ e^{iS(\eta)}\ \delta(\eta(t) - \underset{\sim}{y}) . \qquad (2.3)$$

This representation then defines by projection a path-integral representation over C of a unique $G_{\mathcal{E}}$ in (1.1) (aside from the factor $(n!)^{-\frac{1}{2}}$).

We close with two remarks relating to rigorous studies. First, the mathematical definitions of path integrals generally presuppose paths over a space R^N, cf. e.g. [11,12]. We discuss therefore the problem of the diagonal set E in (2.3) in the appendix. (We remark that such definitions are sometimes given in a form which requires $\eta(0) = 0$, but the shift of the initial point can be easily handled.)

Second, the assignment of paths η_0 to homotopy classes assumes continuity of the η_0, and is not meaningful otherwise. It is therefore worth noting that some of the mathematical definitions of path integrals over the coordinate space indeed exploit spaces of continuous functions (cf. *loc. cit.*). For contrast, it appears that path integrals which utilize coherent states require discontinuous paths (or functions) [13,14].

3. The effects of curvature.

As references to quantization in curved spaces, and to the possible effects of curvature, we cite in particular [4,5,9,15,16]. We will not repeat here the calculations of *loc. cit.*, which can be somewhat tedious. Rather, we will concentrate on discussing the results.

We assume a configuration manifold M homeomorphic to R^n, in order to avoid problems such as that of the definition of the coordinates u^j. Let M be equipped with the metric $(g_{jk}(u))$ and let $g = \det(g_{jk})$. Since we do not consider relativistic situations here, we assume (g_{jk}) to be positive-definite. We also assume (g_{jk}) to be time-independent.

An invariant volume element is given by $d^n u\, g^{1/2}$. We have the option of using wave functions $\varphi(u)$ which are scalar, or those which are densities of weight $\frac{1}{4}$, i.e. $\varphi g^{\frac{1}{4}}$. We will ordinarily use scalar functions. (Spinless particles are assumed here.)

The following operators are symmetric in their action on scalar functions (cf. [4,5, 15]; here $\partial_j = \partial/\partial u^j$):

$$\hat{p}_j = i^{-1}(\partial_j + \frac{1}{4}\partial_j \log g) = i^{-1}g^{\frac{1}{4}}\partial_j g^{-\frac{1}{4}} \tag{3.1}$$

$$\Delta_{LB} = -g^{-\frac{1}{2}}\partial_j(g^{jk}g^{\frac{1}{2}}\partial_k) = g^{jk}\nabla_j\nabla_k , \tag{3.2}$$

Δ_{LB} being the Laplace-Beltrami operator. We now define the free Hamiltonian operator as follows,

$$\hat{H}_0 := \frac{1}{2}\hat{p}_j g^{jk}\hat{p}_k = \frac{1}{2}\Delta_{LB} + \frac{1}{2}g^{\frac{1}{4}}(\Delta_{LB}g^{-\frac{1}{4}}) . \tag{3.3}$$

The last equality is given in [5], and it can be established by a direct computation. For the last term in (3.3) we obtain

$$\frac{1}{2}g^{\frac{1}{4}}(\Delta_{LB}g^{-\frac{1}{4}}) = R/12 , \tag{3.4}$$

where R is the scalar curvature, and we use the convention $R_{jk} = \Gamma^i_{jk},_i - \Gamma^i_{ji},_k +- \ldots$ Equation (3.4) was derived in [16] with the help of normal coordinates. We might note in this connection that $\partial_j \log g^{\frac{1}{2}} = \Gamma^k_{jk}$, cf. [17], p. 18.

We next consider the classical time evolution, or the Hamiltonian flow, generated by a Hamiltonian of the form $\frac{1}{2}(p_j - A_j)g^{jk}(p_k - A_k) + V$, as in [4]. To fix the notation, we suppose that this flow maps (p,u) at $\tau = 0$ into (p',v) at $\tau = t$. Associated to this flow is the Van Vleck determinant D defined as follows,

$$D_{jk} = \partial p_j/\partial v^k = -\partial^2 S/\partial u^j \partial v^k , \qquad D = \det(D_{jk}) , \tag{3.5}$$

where S is the associated classical action. The calculations of B. DeWitt [4] show that for small t and small geodesic distance r(v,u),

$$D(t;v,u) = t^{-n}g^{\frac{1}{2}}(v)g^{\frac{1}{2}}(u)\Big[1 + \frac{1}{6}R_{jk}(u)(v^j - u^j)(v^k - u^k)$$
$$+ \mathcal{O}(r(v,u)^3) + \mathcal{O}(r(v,u)t) + \mathcal{O}(t^2)\Big] . \tag{3.6}$$

We utilize D in Pauli's ansatz for Green's function [18] (see also [19]),

$$G_P(t;v,u) = (2\pi i)^{-\frac{1}{2}n}g^{-\frac{1}{4}}(v)g^{-\frac{1}{4}}(u)\, D^{\frac{1}{2}}(t;v,u)\exp[iS(t;v,u)] \tag{3.7}$$

We now specialize to the free Hamiltonian:

$$H_0 = \tfrac{1}{2}p_j g^{jk} p_k = \tfrac{1}{2}\dot{u}^j g_{jk}\dot{u}^k . \qquad (3.8)$$

(It appears that scalar or vector potential terms would not affect the curvature terms of interest, nor would a time dependence of (g_{jk}).) The function S now becomes tH_0, but the change of variables $(p,u) \longrightarrow (v,u)$ has to be effected, yielding $S = r(v,u)^2/2t$ [9]. The function G_P satisfies

$$(-i^{-1}\partial_t - \hat{H}_0(v))G_P(t;v,u) \sim \mathcal{O}(r(v,u)) + \mathcal{O}(t) , \qquad (3.9a)$$

$$\lim_{t \searrow 0} G_P(t;v,u) = \delta(v,u) , \qquad (3.9b)$$

where the δ-function refers to the volume element $d^n u\, g^{\frac{1}{2}}$, and v in $H_0(v)$ indicates the argument of g^{jk} and of g.

We next partition the interval $[0,t]$ by the points t_j, where

$$0 = t_0 < t_1 < \ldots < t_N < t_{N+1} = t , \qquad (3.10)$$

and we iterate the G_P, to obtain the exact Green's function in the limit (here $\Delta t = \max(t_{j+1} - t_j)$):

$$\begin{aligned} G(t;v,u) &= g^{-\frac{1}{4}}(v) g^{-\frac{1}{4}}(u) \lim_{\Delta t \searrow 0} \int d^n w_N \ldots d^n w_1 \\ &\times \left[D(t - t_N;v,w_N)\, D(t_N - t_{N-1};w_N,w_{N-1}) \ldots D(t_1;w_1,u)\right]^{\frac{1}{2}} \\ &\times \exp\left\{ i\left[S(t - t_N;v,w_N) + \ldots + S(t_1;w_1,u)\right]\right\}. \end{aligned} \qquad (3.11)$$

We identify the last limit as a Feynman integral, and we write the last equation as

$$G(t;v,u) = g^{-\frac{1}{4}}(v) g^{-\frac{1}{4}}(u) \int_{\substack{\eta(0)=u \\ \eta(t)=v}} \mathcal{D}_R(\eta) \exp\left[\tfrac{1}{2}\, i \int_0^t d\tau\, \dot{\eta}^j(\tau) g_{jk}(\eta(\tau))\, \dot{\eta}^k(\tau)\right]. \qquad (3.12a)$$

The function G satisfies

$$(-i^{-1}\partial_t - \hat{H}_0(v))\, G(t;v,u) = 0 \qquad (3.12b)$$

(also, $G \longrightarrow \delta$ when $t \searrow 0$), as follows from (3.9), (3.11). The index R in $\mathcal{D}_R$ means that the Van Vleck determinants in the approximations of (3.11) are those for the Riemannian space in question, i.e. as given in (3.6).

Equations (3.12a-b) may be compared with the following equations, implied in [4,5]:

$$G_{DW}(t;v,u) = g^{-\frac{1}{4}}(v)g^{-\frac{1}{4}}(u) \int_{\substack{\eta(0)=u \\ \eta(t)=v}} \mathcal{D}_F(\eta) \exp\left\{ i \int_0^t d\tau \right.$$

$$\left. \times \left[\tfrac{1}{2}\dot{\eta}^j(\tau) g_{jk}(\eta(\tau))\dot{\eta}^k(\tau) + \tfrac{1}{6}R(\eta(\tau))\right] \right\} \tag{3.13a}$$

$$(-i^{-1}\partial_t - \tfrac{1}{2}\Delta_{LB}(v))\; G_{DW}(t;v,u) = 0 \tag{3.13b}$$

(also, $G_{DW} \longrightarrow \delta$ when $t \searrow 0$). The index F in $\mathcal{D}_F$ means that the Van Vleck determinants in the approximations of (3.11) are those for a Euclidean or flat space, i.e. that the terms $\frac{1}{6}R_{jk}\ldots$ of (3.6) are to be dropped. The term $\frac{1}{6}R$ in (3.13a), on the other hand, could be treated as a contribution to the action.

Each of the two systems (3.12) and (3.13) can be obtained from the other by some straightforward transformations (at the heuristic level; cf. *loc. cit.*). It might therefore be a matter of taste whether (3.12) or (3.13) is preferred. Some relevant points, however, can be noted.

The two systems (3.12) and (3.13) differ with respect to the differential operators (in (3.12b), (3.13b)) and with respect to the normalizations implicit in $\mathcal{D}_R$, $\mathcal{D}_F$. Each difference amounts to R/12, and the net result is the term $\frac{1}{6}R$ included in (3.13a).

Let us consider the differential operators. It seems natural to take $\hat{H}_0$ as a possible quantized form of H_0. On the other hand, when the operator Δ_{LB} is expressed in terms of the $\hat{p}_j$, g^{jk}, and g, its significance becomes unclear, in the following sense.

From (3.1) and (3.2) we obtain

$$\Delta_{LB} = -g^{-\frac{1}{4}}\,\hat{p}_j g^{jk} g^{\frac{1}{2}} \hat{p}_k g^{-\frac{1}{4}} . \tag{3.14}$$

Recall that Δ_{LB} is symmetric when acting on scalar functions, but in general is not symmetric when acting on densities. The same holds for the $\hat{p}_k$. Therefore, for the action of $\hat{p}_k$ to be significant, one must apply Δ_{LB} to a density ψ, so that $g^{-\frac{1}{4}}\psi$ would be a scalar function. We have here a kind of conceptual contradiction.

With regard to the normalizations, consider first the case of a flat space. One encounters there the normalizing factors $t^{-\frac{1}{2}n}$ in free-particle Green's functions, and these functions dominate short-time propagation. The determinant D and the fact that $D^{\frac{1}{2}} \sim t^{-\frac{1}{2}n}$ often seem to be of secondary importance.

When the space is curved, on the other hand, and the $D^{\frac{1}{2}}$ are introduced into the construction of the path integral from the beginning, two alternative ways to exploit the relation $D^{\frac{1}{2}} \sim t^{-\frac{1}{2}n}$ suggest themselves. First, one can factorize $D^{\frac{1}{2}}$, absorb only the factor $t^{-\frac{1}{2}n}$ into the $\mathcal{D}$-symbol (so as to obtain $\mathcal{D}_F$), and let the remaining factor contribute to the integrand in the equivalent form $\exp(itR/12)$. The second alternative depends on absorbing the whole of $D^{\frac{1}{2}}$ into the $\mathcal{D}$-symbol, so as to obtain $\mathcal{D}_R$. In either case, the $g^{\frac{1}{4}}$ in $D^{\frac{1}{2}}$ cancel out.

We are tempted to say that the second alternative is at least as reasonable as the first.

Appendix. The diagonal set for identical particles.

We should like to offer some rigorous considerations on path integrals for identical particles. The integral in (2.3) is in fact (nearly) over a space R^N, and so it could be expected to converge for various potentials, e.g. for those which are Fourier transforms of bounded measures. The elimination of the diagonal set E, however, is a new feature.

We cited in sec. 2 two references with definitions of the path integral. One of these, namely [11], defines an integral over a Hilbert space as a limit of integrals over (linear) subspaces. The restriction of paths to $R^{nk} \setminus E$ could then be rather awkward. We will not consider this approach further in this article.

In the other definition one starts with a measure-theoretic Gaussian integral and continues analytically in variance parameters, so as to obtain a Feynman-type integral as a boundary value [12]. If the Feynman integral of interest is the usual path integral, then for the measure-theoretic integral we may take that of Wiener. Then, in order to justify restricting the paths to $R^{nk} \setminus E$, it is sufficient to show that Brownian paths which start away from E and hit E at a later time have Wiener measure zero. A sufficient condition for this is that E have exterior capacity zero. (See [20] for the necessary background.) We note that the restriction to initial points in $R^{nk} \setminus E$ can be justified in those cases where Green's functions are locally L_1, since they have a distribution-theoretic interpretation and E has Lebesque measure zero.

We now recall a set of sufficient conditions for a closed set $B \subset R^N$, $N \geqslant 3$, to have exterior capacity zero. Consider the function

$$v(y) = \int_{B'} d\mu(x) \, | y - x |^{-N+2} , \tag{A.1}$$

where B' is a compact subset of B, and μ is a finite (positive) measure on Borel subsets of B'. If for every such measure μ, v(y) is an unbounded function on R^N, then B' has zero capacity. If there is a sequence B_1', B_2', ... of such sets which satisfies also $\cup B_j' = B$, then B has exterior capacity zero.

Lemma 2. The set E has exterior capacity zero.

The heuristic basis for this lemma is that $\dim R^{nk} - \dim E = n \geqslant 3$. We expect therefore to find some analogy with the case of a single point in R^n, $n \geqslant 3$, the capacity of the point being zero.

Proof: We consider for definiteness two (identical) particles on R^3, so that E is the set of points satisfying $\underset{\sim}{x}_1 = \underset{\sim}{x}_2$ in $\{(\underset{\sim}{x}_1,\underset{\sim}{x}_2)\} = R^6$. This simplest case illustrates adequately well the general situation. Moreover, for each set B_j' (as above) we may select a unit cube. It suffices therefore to show that a unit cube $E_0 \subset E$ has zero capacity.

Let μ be a measure on E_0 as specified above, with $\mu(E_0) = \mu_0 > 0$. We introduce coordinates so that

$$E_0 = \{ (\xi^1,\xi^2,\xi^3) \in R^3 : 0 \leqslant \xi^j \leqslant 1, \forall j \} . \tag{A.2}$$

We now write E_0 as a union of eight cubical sets $E_1^{(1)},\dots,E_1^{(8)}$ which overlap slightly, as follows. We select some $\varepsilon > 0$ satisfying

$$(\tfrac{1}{2} + \varepsilon)^4 \leqslant \tfrac{2}{3}(\tfrac{1}{2})^3 , \quad \text{or:} \quad (\tfrac{1}{2} + \varepsilon)^{-4} \geqslant \tfrac{3}{2}(2^3) . \tag{A.3}$$

Then we define the sets $E_1^{(i)}$ by restricting each coordinate in the following way:

$$\text{either} \quad 0 \leqslant \xi^j < \tfrac{1}{2} + \varepsilon \quad \text{or} \quad \tfrac{1}{2} - \varepsilon < \xi^j \leqslant 1 . \tag{A.4a}$$

We next subdivide the $E_1^{(i)}$, each into eight cubical sets, in a similar way. I.e., in each of the new sets $E_2^{(i)}$ each coordinate is to be restricted by

$$\begin{aligned} &\text{either} \quad 0 \leqslant \xi^j < (\tfrac{1}{2} + \varepsilon)^2 \quad \text{or} \quad (\tfrac{1}{2} + \varepsilon) - (\tfrac{1}{2} + \varepsilon)^2 < \xi^j \leqslant (\tfrac{1}{2} + \varepsilon) \\ &\text{or} \quad \tfrac{1}{2} - \varepsilon < \xi^j \leqslant \tfrac{1}{2} - \varepsilon + (\tfrac{1}{2} + \varepsilon)^2 \quad \text{or} \quad 1 - (\tfrac{1}{2} + \varepsilon)^2 < \xi^j \leqslant 1. \end{aligned} \tag{A.4b}$$

Clearly, for some k_1, $\mu(E_1^{(k_1)}) \geqslant 2^{-3}\mu_0$. Also, $\exists\, k_2$ such that

$$E_2^{(k_2)} \subset E_1^{(k_1)} \quad \text{and} \quad \mu(E_2^{(k_2)}) \geqslant (2^{-3})^2 \mu_0 . \tag{A.5}$$

In the same way we may find $E_3^{(k_3)}$, etc. We will denote the sets in question simply by $E_1, E_2, E_3, \ldots$.

The corresponding closures have a nonempty intersection, $\cap \bar{E}_j =: \{y_0\}$, and we will estimate the value of $v(y_0)$. Note that in E_0, $|y - x| \leqslant \sqrt{3}$; in E_1, $|y - x| \leqslant (\frac{1}{2} + \varepsilon)\sqrt{3}$; in E_2, $|y - x| \leqslant (\frac{1}{2} + \varepsilon)^2\sqrt{3}$, etc. Therefore (since $-N + 2 = -4$),

$$v(y_0) \geqslant \int_{E_0} d\mu(x) \max_{E_0} (|y - x|^{-4}) = (\sqrt{3})^{-4} \mu_0 = (\mu_0/9). \tag{A.6a}$$

Integration over E_1 shows that, in view of (A.3),

$$v(y_0) \geqslant \left[(\tfrac{1}{2} + \varepsilon)\sqrt{3}\right]^{-4} \mu(E_1) \geqslant \tfrac{3}{2}(2^3)\tfrac{1}{9}(2^{-3}\mu_0) = \tfrac{3}{2}(\mu_0/9) . \tag{A.6b}$$

Similarly, integration over E_2 gives $v(y_0) \geqslant (3/2)^2(\mu_0/9)$, etc. We conclude that $v(y_0)$ is infinite, and the lemma is proved.

REFERENCES

[1] Laidlaw, M.G.G. and DeWitt, C.M., Phys. Rev. D3, 1375 (1971).

[2] Dowker, J.S., J. Phys. A: Gen. Phys. 5, 936 (1972).

[3] Hurt, N.E., in: *Group theoretical methods in physics* (proc., Nijmegen 1975), edited by A. Janner, T. Janssen, and M. Boon (Springer-Verlag, Berlin-Heidelberg-New York, 1976; Lecture notes in physics 50), p. 182.

[4] DeWitt, B.S., Revs. Mod. Phys. 29, 377 (1957).

[5] Dowker, J.S., in: *Functional integration and its applications* (proc., London 1974), edited by A.M. Arthurs (Oxford University Press, London, 1975), p. 34.

[6] Dekker, H., in: *Functional integration, theory and applications* (proc., Louvain-la-Neuve 1979), edited by J-P. Antoine and E. Tirapegui (Plenum Press, New York and London, 1980), p. 207.

[7] Mizrahi, M.M., J. Math. Phys. 16, 2201 (1975).

[8] Schulman, L., Phys. Rev. 176, 1558 (1968).

[9] DeWitt, C.M., Ann. Inst. H. Poincaré A11, 153 (1969).

[10] Spanier, E.H., *Algebraic topology* (McGraw-Hill Book Co., New York etc., 1966), chapters 1 and 2.

[11] Tarski, J., in: *Feynman path integrals - proc., Marseille 1978*, edited by S. Albeverio, Ph. Combe, R. Høegh-Krohn, G. Rideau, M. Sirugue-Collin, M. Sirugue, and R. Stora (Springer, as in ref. 3, 1979; Lecture notes in physics 106), p.254.

[12] Tarski, J., in: *Functional integration, theory and applications* (as in ref. 6), p. 143.

[13] Tarski, J., Acta Phys. Austr. 44, 89 (1976), especially pp. 101-102.

[14] Berg, H.P. and Tarski, J., in: *Functional integration, theory and applications* (as in ref. 6), p. 125, especially sec. 5.

[15] DeWitt, B.S., Phys. Rev. 85, 653 (1952).

[16] Śniatycki, J., *Geometric quantization and quantum mechanics* (Springer, as in ref. 3, 1980; Applied math. sciences, vol. 30), especially sec. 7.2.

[17] Eisenhart, L.P., *Riemannian geometry* (Princeton University Press, Princeton, N.J., 1960).

[18] Pauli, W., *Ausgewählte Kapitel aus der Feldquantisierung*, 2nd edition (lecture notes from ETH, Zürich, 1957; reprinted: Boringhieri, Torino, 1962), Anhang: p. 139 ff.

[19] Choquard, Ph., Helv. Phys. Acta 28, 89 (1955).

[20] Doob, J.L., Trans. Amer. Math. Soc. 77, 86 (1954), especially secs. 5, 7.

VI. QUANTUM FIELD THEORY

GAUGE-THEORY GHOSTS AND GHOST-GAUGE THEORIES

Yuval Ne'eman *+

Tel Aviv University, Tel Aviv, Israel x

and

C.P.T., University of Texas, Austin, Texas, USA.

Abstract

We review the geometric unitarity equations for an Internal gauge theory, for a super-gauge, for the Kalb Ramond field, and for a Non-Internal gauge. We present recent results relating to the role of "antighosts".

1. Prehistory

After the success of QED in 1948, attempts were made to extend the methods of Relativistic Quantum Field Theory to both Strong and Weak interactions. However, pseudo-scalar mesons, which were thought to mediate the basic Strong Interaction, happened to involve couplings of the order of $g^2_{\pi \overline{N}N}/4 = 14.5$, totally unsuitable for a perturbation approach. Weak interactions were either unrenormalizable due to their dimensionality if treated as a four-fermion interaction, or symmetry (and gauge) breaking due to the Intermediate Boson masses. Field Theory was already somewhat discredited anyhow. In particular it was known to involve in most cases a breakdown of Unitarity off-mass-shell, a feature that invalidated much of what could still be achieved with Feynman diagrams.

Happily for the evolution of Physics, we still live in a somewhat heterogeneous world. Although the Western U.S. and much of the rest of the Physics world abandoned it, Field Theory and its Unitarity problem continued to be treated by non-mainstream groups

---

* Supported in part by the U.S.-Israel Binational Science Foundation

+ Supported in part by the U.S. DOE, Grant

x Wolfson Chair Extraordinary of Theoretical Physics

or individuals. Feynman made the first step [1], when treating the Yang Mills gauge (as a pilot-project for Gravity at the suggestion of Gell-Mann). Feynman was, of course, beyond being influenced by "the concensus"... B. de Witt continued the project [2]; as a General Relativist, he had not heard that Field Theory was "out". The Unitarity issue was indeed resolved in the USSR, where Fradkin, Faddeev and Popov [3] perfected the method; and in Holland G. 't Hooft [4] achieved the final breakthrough, by adding Regularization to Unitarity, and doing it even for the case of "Spontaneous Breakdown", i.e. massive vector mesons.

2. The BRS equations

The procedure developed by Feynman, De Witt and perfected by Faddeev and Popov, was based upon the introduction of ghost fields $\chi^a(x)$. These were scalar fields with Fermi statistics. They were set up so as to cancel the unphysical contributions of redundant components of the gauge fields. Fermi statistics provided the necessary minus signs for closed loops. The Yang Mills invariant Lagrangian for a local gauge group G (a,b...: the indices in its adjoint representation)

$$L_{INV} = -\frac{1}{4} F^a_{\mu\nu} F_a^{\mu\nu} \tag{2.1}$$

does not possess an invertible Fourier transform, needed to construct a propagator. To obviate this difficulty, a gauge-fixing term was added (we use here a linear treatment)

$$L_{GF} = -\eth \Sigma \tag{2.2}$$

with $\eth$ a Lagrange multiplier field ensuring $\Sigma \overset{\Omega}{=} 0$. ($\overset{\Omega}{=}$ implies equality resulting from an equation of motion). This *is* the gauge fixed by L_{GF}.

One then added a Lagrangian for the ghosts

$$L_{GH} = \bar{\chi}_a \hat{m} \chi \tag{2.3}$$

where $\hat{m}$ is given by the gauge transformation of the Yang Mills field-potential (D_μ is the covariant derivative)

$$\delta A^a_\mu = (D_\mu \varepsilon)^a \tag{2.4}$$

$$\delta \Sigma (A^a_\mu) = \hat{m} \varepsilon^a \tag{2.5}$$

$\bar{\chi}$ is the "anti-ghost", another Fermi scalar (or χ read from future to past). With

the spin-statistics theorem broken, there is no requirement that $\delta\bar{\chi} = \pm\delta\chi$.
The new Lagrangian

$$L = L_{INV} + L_{GF} + L_{GH} \tag{2.6}$$

has lost local gauge invariance but has retained "global" (or "rigid") invariance. In addition it has gained Slavnor-Taylor invariance [5]. As a result of laborious efforts, this produced Ward-Takahashi identities through which Unitarity can be proven.

After 't Hooft had achieved the final aim, the procedure was greatly simplified [6] by Becchi, Rouet and Stora ("BRS") and by Tyutin. Defining

$$\varepsilon^a := \chi^a \Lambda \qquad\qquad s := -\frac{\partial}{\partial\Lambda}\delta \tag{2.7}$$

with Λ a constant Grassmann element, anticommuting with itself, with χ^a and with $\bar{\chi}^a$, we have (:= is a definition, λ_a are the algebra matrices in the adjoint representation)

$$s\phi = -D\chi \text{ or } sA_\mu = D_\mu\chi, \text{ for } \phi = A^a_\mu\lambda_a \cdot dx^\mu,\ D := dx^\mu D_\mu \tag{2.8}$$

$$s\chi = -\tfrac{1}{2}[\chi,\chi] \tag{2.9}$$

$$s\Sigma = \hat{m}\chi \tag{2.10}$$

$$s\psi = -[\chi,\psi] := \zeta \tag{2.11}$$

Equations (2.9) and (2.11) represent the homogeneous group action under (2.7). Adding

$$s\bar{\chi} = \mathfrak{z} \tag{2.12}$$

$$s\mathfrak{z} = 0 \tag{2.13}$$

ensures that

$$sL = 0 \tag{2.14}$$

since

$$sL_{GF} = -\mathfrak{z}s\Sigma \quad , \qquad sL_{GH} = \mathfrak{z}s\Sigma - \bar{\chi}s^2\Sigma$$

provided that $s^2\Sigma = 0$, which is guaranteed if we impose

$$s^2A^a_\mu = s^2\phi^2 = 0 \tag{2.15}$$

Using the Jacobi identity ("algebraic closure") we also find

$$s^2 \chi = 0 \tag{2.16}$$

and (2.12) - (2.13) imply

$$s^2 \bar{\chi} = 0 \tag{2.17}$$

These equations were shown to guarantee Unitarity [6]. It should be mentioned that in some treatments, equation (2.17) does not hold. This happens for instance when L_{GF} is constructed quadratically as $-\frac{1}{2}C^a C_a$, $\delta C^a = \hat{m}\varepsilon$, yielding $sC^a = \hat{m}\chi^a$, $s\bar{\chi}^a = C^a$, $s^2\bar{\chi}^a = \hat{m}\chi^a$.

3. Groping for a geometrical understanding

Equations (2.15)-(2.17) made several people suspect that the transformation s has a geometrical origin. In particular, R. Stora noted some relationship between χ and the Cartan-Maurer left-invariant forms on the group G. However, it was only in 1978 that Jean Thierry-Mieg found the answer [7]. I had been working with T. Regge on a geometrical interpretation of Supergravity, and we found it useful to work in a manifold with the full dimensionality of the Poincaré group (10 dimensions) for Gravity, or of the Super- (or Graded-) Poincaré supergroup ("supersymmetry") for Supergravity [8]. Flat space-time precisely corresponds to a submanifold of the Poincaré group manifold: the piece constituted by the parameters of translations. We are thus embedding Minkowski space in a 10-dimensional "flat" space, where "flatness" or "rigidity" means that there is no departure from the precise geometry imposed by the Cartan-Maurer equations of the given group (see the second paper in ref. [8] for their derivation). We found that Supergravity and its extensions could be understood geometrically as theories over a "Soft" version of this manifold, departing from the rigid group manifold in the same way that Salvador Dali's Soft Self Portrait (or his Watches) depart from the "rigid" original.

We shall return to this construction later on. However, we just note that we were using actively additional "vertical" dimensions: the parameters of the Lorentz-group, for instance, in the case of the construction of Gravity as a gauge theory over the 10-dimensional Softened Group Manifold (SGM) of the Poincaré group. Every field had ten holonomic components instead of four. The tetrad and Lorentz-connection (to cope with spinors in General Relativity) were $e_M{}^a dz^M$ and $\omega_M^{[ab]} dz^M$, $a,b = 0..3$, $M = 0...3$ and [12], [23] ... , altogether ten values.

Jean Thierry-Mieg had attended Les Houches in 1975 and was studying the BRS equations for a Yang-Mills theory. He looked at the Principal Fiber Bundle, the geometric structure that had been shown by E. Lubkin and others (and especially by Yang C.N. and Wu T.T.) to be equivalent to a gauge theory. This also had "vertical" dimensions sitting over space-time, just as the Lorentz Group was sitting over it in the SGM of the Poincaré group. I gave two talks on the latter subject while on visits to Paris in the spring of 1977 and in early 1978. Thierry-Mieg attended both. From the first one he derived encouragement to look for an interpretation for the extra components of the gauge fields. By the time I gave my second talk, he had conceived the thought that they might just be the "ghosts". I liked the idea: I could see how this fitted with my own picture of the extra-dimensionality and would lead to the correct equations. For a rigid system (a global section over the Bundle) ghosts would indeed reduce to the Left-invariant forms, but in the "softened" version (i.e. a local section) they would become true fields.

The geometrical identification of the ghosts became one of the main elements in Thierry-Mieg's thesis [7]. The approach was very new and some points remained unsolved at that stage (especially the role of antighosts). Jean's first paper on the subject was greeted with skepticism and got into referee-trouble. It had to wait two years before getting published [7].

4. The Geometric Derivation

Using a Principal Bundle P $(P, M, \pi, G, \cdot)$ to represent a Yang-Mills theory, it was thus shown [7], [9] that χ is the "vertical" piece of the connection ω, that s is the vertical piece of the exterior derivative and that the BRS equations are indeed the *Cartan-Maurer structural equations on the Bundle* (reducing to the equations for Left-invariant forms on G for a rigid horizontal section). M is the base manifold, π the projection, G the structure group and the dot is the right action,

$$\pi(p \cdot g) = \pi(p) \qquad \forall\, p \in P;\ \forall g,\, g' \in G \tag{4.1}$$

$$(p \cdot g) \cdot g' = p \cdot (g\, g') \tag{4.2}$$

$$(\cdot) \ :\ P \times G \rightarrow P \tag{4.3}$$

Using (locally) co-ordinates $M(x^{\mu})$ and $G(\alpha^{i})$ we lift them onto a section Σ, now described as

$$\Sigma :\ \alpha^{i} = 0 \tag{4.4}$$

We refer the reader to the 3 different proofs in the J. Math Phys. and Nuovo Cim. articles of reference [8] (and in ref. [10]). Here we shall only write

$$\omega^a = \Xi^a_i \, d\alpha^i + A^a_\mu \, dx^\mu = \Xi^a + \Phi^a \qquad (4.5)$$

and remind the reader that for a flat and trivial section, Ξ is the Cartan left-invariant one form and Φ vanishes. Under a gauge transformation (or upon moving the section), Ξ and Φ become both x and α dependent. Denoting the exterior derivative by $\tilde{d}$, we define

$$\left.\begin{array}{l} \tilde{d}\mathcal{b} = z\mathcal{b} + d\mathcal{b}, \; z\mathcal{b} = d\alpha^i \frac{\partial}{\partial\alpha^i}\mathcal{b}, \; d\mathcal{b} = dx^\mu \partial_\mu \mathcal{b} \\ \tilde{d}^2 = o \longrightarrow d^2 = z^2 = zd + dz = o \end{array}\right\} \qquad (4.6)$$

Defining the curvature 2-form

$$R = \tilde{d}\omega + \frac{1}{2}[\omega,\omega] \qquad (4.7)$$

the Cartan-Maurer equations state that R has only horizontal components $F_{\mu\nu}$ (i.e. in $dx^\mu \wedge dx^\nu$). Setting the $d\alpha^i \wedge d\alpha^j$ and $dx^\mu \wedge d\alpha^i$ components of R to zero yields:

$$z\,\Phi = -D\chi \qquad (4.8)$$

$$z\,\Xi = -\frac{1}{2}[\Xi,\Xi] \qquad (4.9)$$

$$z\,\psi = -[\Xi,\psi] = \zeta \qquad (4.10)$$

Consistency is guaranteed by the Bianchi identity

$$\tilde{D}R = o \qquad (4.11)$$

which, together with R=F yields

$$zF + [\chi, F] = o \quad \text{or} \quad ZF = o \qquad (4.12)$$

Equations (4.8)-(4.10), together with the anticommuting properties of Ξ as a 1-form solve (2.8)-(2.11),

$$s = z, \qquad \chi = \Xi \qquad (4.13)$$

As to the antighost $\bar{\chi}$, it was left without a geometrical interpretation at this stage. Note that since $sL = o$, and $dL = o$ by saturation of the horizontal submanifold, we have a geometric closure [9] of the quantum gauge-Lagrangian,

$$\tilde{d} L = 0 \tag{4.14}$$

To see that the Fermionic properties of χ correspond to the anticommutation of one-forms in the exterior calculus (or to differentials in the measure) we have to display a geometric functional integral. Indeed, using the conventional generating functions Γ, we write the Faddeev-Popov expression ($F = \frac{1}{2}F_{\mu\nu}\, dx^{\mu} \wedge dx^{\nu}$)

$$e^{i\Gamma} = G \int D(\phi)\, e^{i\, S_{INV}(\phi)}, \quad S_{INV} = \int F \wedge {}^{*}F \tag{4.15}$$

Using [8]

$$1 = \int \delta(\Sigma)\tilde{d}\Sigma = \int \delta(\Sigma) s\Sigma, \quad \delta(\Sigma) = \int d\mathfrak{b}\, e^{i\mathfrak{b}\Sigma}$$

and the Berezin integral $\int \bar{d}\theta\, \mathfrak{b}(\theta) = -i\frac{\partial}{\partial\theta}\mathfrak{b}(\theta)\Big|_{\theta=0}$

$$s\Sigma = \int \bar{d}\,\bar{\chi}\; e^{\, i\bar{\chi}s\Sigma}$$

and remembering that the group volume G in front of the integral is the integral over $d\alpha$ or over χ (see ref. [11] for a more formal demonstration) we have,

$$e^{\, i\Gamma} = \int D(\phi, \chi, \bar{\chi}, \mathfrak{b})\, e^{\, iL} \tag{4.16}$$

a geometric integration.

5. Symmetric (or Extended) BRS

The solution to the riddle of the antighost $\bar{\chi}$ is very recent. Right after the publication of the BRS equations [6], Curci and Ferrari [12] developed a "symmetric" version in which there were two transformations, s and $\bar{s}$, and the antighost was the ghost for $\bar{s}$. For example, let us take a Lagrangian with an auxiliary field $\mathfrak{b}(x)$,

$$L = -\frac{1}{4} F_{\mu\nu} F^{\mu\nu} + \frac{1}{2}\mathfrak{b}^2 + \mathfrak{b}\Sigma - \bar{\chi}\hat{m}\chi \tag{5.1}$$

We define the transformations under s and $\bar{s}$:

$$\left.\begin{aligned}
s A_{\mu} &= D_{\mu}\chi & \qquad \bar{s} A_{\mu} &= D_{\mu}\bar{\chi} \\
s \chi &= -\tfrac{1}{2}[\chi,\chi] & \bar{s}\bar{\chi} &= -\tfrac{1}{2}[\bar{\chi},\bar{\chi}] \\
s\bar{\chi} &= \mathfrak{b} & \bar{s}\chi &= -s\bar{\chi} - [\chi,\bar{\chi}] \\
s\mathfrak{b} &= 0 & \bar{s}\mathfrak{b} &= -[\bar{\chi},\mathfrak{b}]
\end{aligned}\right\} \tag{5.2}$$

implying

$$\bar{s}\chi = -\mathcal{B} - [\chi,\bar{\chi}]$$

$$\left.\begin{aligned} s^2 A_\mu &= s^2\chi = s^2\bar{\chi} = 0 \\ \bar{s}^2 A_\mu &= \bar{s}^2\bar{\chi} = \bar{s}^2\chi = 0 \\ (s\bar{s}+\bar{s}s)A_\mu &= (s\bar{s}+\bar{s}s)\chi = (s\bar{s}+\bar{s}s)\bar{\chi} = 0 \end{aligned}\right\} \quad (5.3)$$

Here as before,

$$s\Sigma = \hat{m}\chi \quad (5.4)$$

This Lagrangian is s-BRS invariant. Under $\bar{s}$ (:= "anti-BRS") it becomes invariant after insertion of the $\mathcal{B}$ "equation of motion"

$$\mathcal{B} = -\Sigma \quad (5.5)$$

so that we write

$$\left.\begin{aligned} sL &= 0 \\ \bar{s}L &\stackrel{\varrho}{=} 0 \end{aligned}\right\} \quad (5.6)$$

With this approach, $\bar{\chi}$ seems about as "respectable" as χ. What was missing was a geometrical structure explaining this result, or reproducing it naturally. This has recently been provided by Jean Thierry-Mieg [13].

In the second paper of ref. [3] we displayed the classical derivation of the two sets of one-forms and their orthonormal generator algebras for a Lie group G: the generators of right- and left-translations, and the one-forms invariant under each of these. In the Principal Bundle P, the algebra of G is the algebra of right-translations, and the rigid limit of χ corresponds to the left-invariant one-forms. We can construct a second Bundle $\bar{P}$ ($\bar{P}$, M, G, $\bar{\pi}$, $*$) where the asterisk stands for left-action of G on $\bar{P}$. Here the ghost (which we name $\bar{\chi}$) coincides in the rigid limit with the right-invariant one forms. The two algebras commute [8]. There is an "anti-BRS" geometric algebra $\bar{z}$ for zA_μ and $\bar{z}\bar{\chi}$.

Thierry-Mieg now constructs a third Bundle P_2. This is the pull-back $\bar{\pi}^{-1}(P)$, and also $\pi^{-1}(\bar{P})$. It is thus a Bundle with G x G as its structure group and Fiber. We refer the reader to the original derivations [13, 14]. The connections ω and $\bar{\omega}$ can be pulled-back on P_2,

$$\omega\big|_{P_2} = \bar{\pi}^*(\omega|_P) \quad ; \quad \bar{\omega}|_{P_2} = \pi^*(\bar{\omega}|_{\bar{P}}) \tag{5.7}$$

We now pick a section and apply a reality condition, requiring the horizontal parts to coincide on P_2,

$$\Phi|_{P_2} = \bar{\Phi}|_{P_2} \quad , \quad A_\mu \equiv \bar{A}_\mu \tag{5.8}$$

Together with the compatibility condition requiring π and $\bar{\pi}$ to commute, this means that the connection on P_2 splits into three parts

$$g^a = A^a_\mu \, dx^\mu + \chi^a_i \, d\alpha^i + \bar{\chi}^a_j \, d\bar{\alpha}^j \tag{5.9}$$

The Cartan Maurer equations (i.e. horizontality of R) on P_2 provide the necessary Extended (geometric) BRS algebra:

$$\left.\begin{aligned} zA_\mu &= D_\mu \chi & \bar{z}A_\mu &= D_\mu \bar{\chi} \\ z\chi &= -\tfrac{1}{2}[\chi,\chi] & \bar{z}\bar{\chi} &= -\tfrac{1}{2}[\bar{\chi},\bar{\chi}] \end{aligned}\right\} \tag{5.10}$$

and the mixed contribution

$$z\bar{\chi} + \bar{z}\chi + [\bar{\chi},\chi] = 0 \tag{5.11}$$

Since $z\bar{\chi}$ is not constrained, we introduce the $\mathfrak{b}$ field:

$$\left.\begin{aligned} z\bar{\chi} &= -\tfrac{1}{2}[\bar{\chi},\chi] + \mathfrak{b} \\ \bar{z}\chi &= -\tfrac{1}{2}[\bar{\chi},\chi] - \mathfrak{b} \\ z\mathfrak{b} &= -\tfrac{1}{2}[\chi,\mathfrak{b}] - \tfrac{1}{8}[[\chi,\chi],\bar{\chi}] \\ \bar{z}\mathfrak{b} &= -\tfrac{1}{2}[\bar{\chi},\mathfrak{b}] + \tfrac{1}{8}[[\bar{\chi},\bar{\chi}],\chi] \end{aligned}\right. \tag{5.12}$$

and

$$z^2 = \bar{z}^2 = z\bar{z} + \bar{z}z = 0 \tag{5.13}$$

The Curci-Ferrari Lagrangian (we have written an explicit $\sum$ and m)

$$L = -\tfrac{1}{4} F^a_{\mu\nu} F^{\mu\nu}_a + \tfrac{\lambda}{2}\mathfrak{b}^2 + \mathfrak{b}\,\partial_\mu A_\mu + \tfrac{1}{2}(\partial_\mu \bar{\chi} D_\mu \chi + D_\mu \bar{\chi}\,\partial_\mu \chi) - \tfrac{\lambda}{8}[\bar{\chi},\chi][\bar{\chi},\chi] \tag{5.14}$$

obeys

$$z L \doteq o , \quad \bar{z} L \doteq o \tag{5.15}$$

6. Supergroups

We now take G to be a supergroup [9,10,15,16], with H its even subgroup. The indices a,b,... cover the range of H, while i,j,... cover G/H. The connection is now locally

$$\begin{aligned} \omega^a &= \chi^a + dx^\mu A^a_\mu \\ \omega^i &= \eta^i + dx^\mu \xi^i_\mu \end{aligned} \tag{6.1}$$

Fields and ghosts now appear at cross positions. We have demonstrated [10,15] that η^i is a Goldstone meson multiplet. Replacing G by $G' \supset H$ an even group conjugate to G by analytical continuation or otherwise, the η^i provide for a non-linear realization of G′ over its linear subgroup H. Alternatively, in local gauge theories, the η^i become Higgs fields [15,16] as demonstrated in our suggested super-unification of SU(2) x U(1) of the electroweak interactions in SU(2/1). ξ^i_μ is a set of vector ghost fields, like in gravity. The correct method of gauging an internal super-group is not yet clarified, and it is possible that G should be treated globally only, whereas H is local. Note that in the SU(m/n) or the Q(m) supergroups, supermultiplets come in pairs with inverted statistics. Another example of an internal supergroup is provided by our suggestion of SU(5+k/1) as a super-unification [17], of the electroweak (SU(2/1) with $SU(3)_{color}$ and seriality (2^k or 2^{k+1} "families" or "generations")).

We have recently been able [14] to write an Extended BRS algebra for SU(2/1) or any SU(n/1). We have used P_2(SU(2/1)), with a double reality condition which we have termed "forking". The Yang Mills multiplet is (ξ^i_μ ; A^a_μ ; $\bar{\xi}^i_\mu$) and the "ghost" multiplet is the fork (χ^a; h^i ; $\bar{\chi}^a$). We thus have only one set of Higgs-Goldstone fields (an isodoublet) and of vector mesons (the W_μ mesons and B_μ) and doubled ghosts $\chi, \bar{\chi}$ and $\xi_\mu \bar{\xi}_\mu$.

The ξ^i_μ and χ^a have a unit of "ghost" charge (ectoplasm!) and the $\bar{\xi}^i_\mu$ and $\bar{\chi}^a$ have one negative unit. Expressing the horizontality of $R(P_2)$ through the usual expansion over dx^μ, $d\alpha$, $d\bar{\alpha}$, we get

$$\left.\begin{aligned} zA + D\chi + [\xi, h] &= o \\ \bar{z}A + D\chi + [\xi, h] &= o \end{aligned}\right\} \tag{6.2}$$

$$\left.\begin{aligned} z\,\chi + \tfrac{1}{2}\,[\chi,\chi] &= 0 \\ z\,\bar{\chi} + \tfrac{1}{2}\,[\bar{\chi},\bar{\chi}] &= 0 \end{aligned}\right\} \qquad (6.3)$$

$$\bar{z}\,\chi + z\,\bar{\chi} + [\chi,\bar{\chi}] + \tfrac{1}{2}\,[h,h] = 0 \qquad (6.4)$$

$$\left.\begin{aligned} z\,\xi + [\chi,\xi] &= 0 \\ \bar{z}\,\bar{\xi} + [\bar{\chi},\bar{\xi}] &= 0 \end{aligned}\right\} \qquad (6.5)$$

$$\left.\begin{aligned} zh + [\chi,h] &= 0 \\ \bar{z}h + [\bar{\chi},h] &= 0 \end{aligned}\right\} \qquad (6.6)$$

$$\left.\begin{aligned} z\,\bar{\xi} + [\chi,\bar{\xi}] + \bar{z}\,\xi + [\bar{\chi},\xi] + Dh &= 0 \\ z^2 = \bar{z}^2 = z\bar{z} + \bar{z}z &= 0 \end{aligned}\right\} \qquad (6.7)$$

and the cross-transformations again introduce auxiliary fields σ , β_μ :

$$\left.\begin{aligned} z\,\bar{\chi} &= -\tfrac{1}{2}[\bar{\chi},\chi] - \tfrac{1}{4}\,[h,h] + \sigma \\ \bar{z}\,\chi &= -\tfrac{1}{2}[\bar{\chi},\chi] - \tfrac{1}{4}\,[h,h] - \sigma \end{aligned}\right\} \qquad (6.8)$$

$$\left.\begin{aligned} z\,\bar{\xi} &= -\,[\chi,\bar{\xi}] - \tfrac{1}{2}\,Dh + \beta \\ \bar{z}\,\xi &= -\,[\bar{\chi},\xi] - \tfrac{1}{2}\,Dh - \beta \end{aligned}\right\} \qquad (6.9)$$

$$\left.\begin{aligned} z\,\sigma &= -\tfrac{1}{2}\,[\chi,\sigma] - \tfrac{1}{8}\,\big[[\chi,\chi],\bar{\chi}\big] + \tfrac{1}{8}\,\big[[h,h],\chi\big] \\ \bar{z}\,\sigma &= -\tfrac{1}{2}\,[\bar{\chi},\sigma] + \tfrac{1}{8}\,\big[[\bar{\chi},\bar{\chi}],\chi\big] - \tfrac{1}{8}\,\big[[h,h],\bar{\chi}\big] \end{aligned}\right\} \qquad (6.10)$$

$$\left.\begin{aligned} z\beta &= -\,[\chi,\beta] + \tfrac{1}{4}\big[[h,h],\xi\big] \\ \bar{z}\beta &= -\,[\bar{\chi},\beta] - \tfrac{1}{4}\big[[h,h],\bar{\xi}\big] \end{aligned}\right\} \qquad (6.11)$$

Matter multiplets also fork in the same way. The SU(2/1) multiplets contain states with inverted statistics, and these fulfill the role of ghosts for the matter fields; these are composite fields like the ζ of equation (2.11).

7. The Antisymmetric Tensor Gauge Field

The geometrical treatment has scored a hit [11] in providing Unitarity for the Kalb-Ramond field [18] or its non-Abelian extension [19]. This theory has as Lagrangian,

$$L = B^a \wedge F_a^{A+H} - \frac{1}{2} m^2 \, {}^{*}H^a \wedge H_a - \frac{1}{2} \, {}^{*}F_A^i \wedge F_i^A \tag{7.1}$$

$$\{a\} \in G\,, \quad \{i\} \in G' \subset G\,, \quad \{r\} \in G/G'$$

$$B = \frac{1}{2} B^a_{\mu\nu} \, Y_a \, dx^\mu \wedge dx^\nu \tag{7.2}$$

$$H = H^a_\mu \, Y_a \, dx^\mu \tag{7.3}$$

$$A = A^i_\mu \, Y_i \, dx^\mu \,, \quad A^r = o \tag{7.4}$$

G' is conserved locally, with A as gauge field.

$$\delta A = D\varepsilon = d\varepsilon + [A,\varepsilon] \,, \quad \delta H = [\varepsilon, H] \,, \quad \delta B = [\varepsilon, B] \tag{7.5}$$

For G, A+H serves as an Abelian gauge connection, in gauging locally with a Lorentz 4-vector ξ_μ (or a one-form $\xi = \xi^a_\mu \, Y_a \, dx^\mu$). This is a hybrid system, since the B field is needed in addition, transforming like a gauge field,

$$\delta_G A = o, \quad \delta_G H = o, \quad \delta_G B = D^{A+H} \xi = d\xi + [A+H, \xi] \tag{7.6}$$

The curvatures

$$F^a = dA + \frac{1}{2}[A,A] \qquad (\text{over } G') \tag{7.7}$$

$$F^{A+H} = dA + dH + \frac{1}{2}[A+H, A+H] \qquad (\text{over } G) \tag{7.8}$$

transform homogeneously. $\delta_G (B^a \wedge F_a^{A+H})$ vanishes by integration by parts and the Bianchi identity $D^{A+H} F^{A+H} = o$. Varying B we get the Euler-Lagrange equation $F^{A+H} \stackrel{\Omega}{=} o$, which yields

$$D^2_{A+H} \, (A+H) \stackrel{\Omega}{=} o \tag{7.9}$$

by explicit calculations. This defines a cohomology class on shell, and the physical field is just a corresponding scalar field. Solving the equation of motion by taking A+H as a Left-invariant one form ($\theta(x) \in G$)

$$H = -A + \theta^{-1} d\theta = -\theta(\theta^{-1} A \theta + \theta^{-1} d\theta)\theta^{-1}$$

produces as a 2nd order theory

$$L = \frac{1}{2} m^2 \,\mathrm{Tr}\,(\Theta^{-1} D_\mu \Theta)^2 - \frac{1}{4} (F^A_{\mu\nu})^2 \tag{7.10}$$

which is a generalized sigma model with gauge. In writing the BRS equations for this theory, difficulties were originally encountered. Thierry-Mieg showed [11] that the geometric approach can be applied in a straightforward manner. The ghosts arise in G'

$$\omega = A_\mu dx^\mu + X \tag{7.11}$$

$$\left.\begin{aligned} s'A &= DX \quad , \quad s'X = \frac{1}{2}[X,X] \\ s'H &= [X , H] \end{aligned}\right\} \tag{7.12}$$

and for G (y^i is the group co-ordinate)

$$\left.\begin{aligned} \tilde{B} &= \frac{1}{2} B_{\mu\nu}\, dx^\mu \wedge dx^\nu + \frac{1}{2} B_{\mu i}\, dx^\mu \wedge dy^i + \frac{1}{2} B_{ij}\, dy^i \wedge dy^j \\ B_{\mu i}\, dy^i &= \xi_\mu \quad , \quad \frac{1}{2} B_{ij}\, dy^i \wedge dy^j = \eta \end{aligned}\right\} \tag{7.13}$$

ξ_μ is a 4-vector (Fermi) ghost, η a scalar (Bose) "non-ghostly" ghost. Setting to zero everything but the fully horizontal components of DB, we get the BRS equations,

$$\left.\begin{aligned} S'B + D^{A+H}\xi &= 0 \\ S'\xi + D^{A+H}\eta &= 0 \\ S'\eta &= 0 \end{aligned}\right\} \tag{7.14}$$

With S' the vertical covariant derivative in G'

$$S' = s' + [X, \ldots]$$

$$(s')^2 = 0 \text{ yielding } F = 0$$

8. The Soft Group Manifold (SGM) and Unitarity

Gravity, Supergravity and their extensions involve a geometrical construct more general than a Principal Bundle. As discussed in section 3, this is $\mathcal{G}$, the Soft Group Manifold (SGM) [8,9]. Its dimensionality is that of the gauged Lie group G (10- for Poincaré, 14- for Supersymmetry) and its tangent at any point is the group manifold

G itself. G is a space in which finite motions are specified, up to topological considerations, by the Lie algebra A, with structure constants $f^a{}_{bc}$. The connections ω are the Cartan left-invariant forms, providing a rigid triangulation and a vanishing curvature

$$d\omega + \frac{1}{2}[\omega,\omega] = 0 \tag{8.1}$$

a condition which held only (eq. 4.8-4.1) for the ϕ, $z\chi$ and $d\chi$, but not the $d\phi$ in $\mathcal{P}$. $\mathcal{P}$ was rigid in the fiber G direction, but Soft horizontally, so that the two covariant derivatives $[D_\mu, D_\nu]$ did not commute, yielding curvature terms

$$F^a_{\mu\nu} = \partial_\mu A^a_\nu - \partial_\nu A^a_\mu + [A_\mu, A_\nu]^a \tag{8.2}$$

The SGM is a triplet ($\mathcal{G}$,G,ϱ). $\mathcal{G}$ is the soft manifold, G the rigid group with algebra A. The connection ϱ is an A-valued one-form over $\mathcal{G}$. Besides providing a set of frames, it provides a map from $\mathcal{G}_*$ onto $\mathcal{A}$, an algebra isomorphic to the algebra of Lie-derivatives on $\mathcal{G}_*$.

$$\varrho : \mathcal{G}_* \longrightarrow \mathcal{A} \quad ; \forall \tilde{t} \in \mathcal{G} \quad \tilde{t} \lrcorner \varrho = \varrho(\tilde{t}) = \Lambda, \quad \Lambda \in \mathcal{A} \tag{8.3}$$

Note that whereas the right group action by G still involves A, the connection yields operations in $\mathcal{A} \neq$ A. We define a basis in $\mathcal{P}_*$ or $\mathcal{G}_*$

$$\varrho^A(\tilde{\Lambda}_B) = \delta^A_B, \quad \tilde{\Lambda} \in \mathcal{G}_* \tag{8.4}$$

The structural equations now provide a right-hand-side to (8.1),

$$d\varrho + \frac{1}{2}[\varrho,\varrho] \equiv R \tag{8.5}$$

or in components

$$R^A = d\varrho^A + \frac{1}{2} f^A{}_{BC}\, \varrho^B \wedge \varrho^C \tag{8.6}$$

Projecting R^A onto the ϱ, since they provide a basis in $\mathcal{G}^*$, we can also write

$$d\varrho^A + \frac{1}{2}(f^A{}_{BC} - R^A{}_{BC})\, \varrho^B \wedge \varrho^C = 0 \tag{8.7}$$

The Bianchi identity holds

$$(DR)^A = 0 \tag{8.8}$$

where the covariant derivatives are defined over the entire G,

$$(D\eta)^A = d\eta^A + f^A{}_{BC}\, \varrho^B \wedge \varrho^C \tag{8.9a}$$

$$(D\eta)_A = d\eta_A - f^C{}_{BA}\, \varrho^B \wedge \eta_C \tag{8.9b}$$

$$(D(D\eta))^A = f^A{}_{BC}\, R^B \wedge \eta^C \tag{8.10}$$

From (8.4) and (8.7) we get

$$[\tilde{\Lambda}_B, \tilde{\Lambda}_C] = (f^A{}_{BC} - R^A{}_{BC})\, \tilde{\Lambda}_A \tag{8.11}$$

reducing to A when $R^A = 0$.

We now evaluate $\delta \varrho^A$ for a local transformation. The result is

$$\delta \varrho^A = D\varepsilon^A + \tilde{\varepsilon} \lrcorner R^A = D\varepsilon^A + \varepsilon^B \varrho^C R_{BC}{}^A \tag{8.12}$$

This action on ϱ^A is the *Lie-derivative* with respect to the vector field $\tilde{\varepsilon} = \varepsilon^M(\partial/\partial z^M)$

$$L_{\tilde{\varepsilon}}\, \varrho = d\varepsilon + \tilde{\varepsilon} \lrcorner d\varrho = D\varepsilon + \tilde{\varepsilon} \lrcorner R \equiv \delta\varrho \tag{8.13}$$

and we have

$$L_{\tilde{\varepsilon}}\, R = D(\tilde{\varepsilon} \lrcorner R) - [\varepsilon, R] \tag{8.14}$$

Local AGCT (Anholonomized General Coordinate Transformations) "gauge" transformations on G are thus represented by the Lie derivatives and A or $\tilde{A}$ will have the structure functions $F^A{}_{BC}(A^D_\mu)$

$$F^A{}_{BC}(A_\mu) = f^A{}_{BC} - R^A{}_{BC}$$

as determined by (8.4) and (8.11). The transformations of (8.13) have been used by Hehl and von der Heyde in Gravity, with ϱ^A restricted to space-time. In Supergravity, these are the "local supersymmetry transformations", when $\tilde{\varepsilon}$ is a spinor $\tilde{\varepsilon}^\alpha$.

Space-time (the parameter space of translations P_a in the Poincaré group) is a submanifold of the SGM. Gauging has been understood in those cases in which A, the algebra of G, is Weakly-Reducible. This implies the existence of a WR decomposition

$$A = F + H \; ; \; [F,F] \subset F$$
$$[F,H] \subset H \qquad (8.15)$$

In such a geometry, one can impose the Cartan-Maurer equations for the subgroup F as a constraint. All curvatures are reduced to 2-forms over G/F. The coefficients $R^A_{||}$ and $R^A_{|-}$ of $\rho^f \wedge \rho^{f'}$ and $\rho^f \wedge \rho^h$ ($f,f' \in \{F\}$, $h \in \{H\}$) can be made to vanish, where ρ^{ab} represents an (anholonomic) Lorentz or (F-) connection and ρ^a stands for a horizontal connection. The algebra in (8.11) for A,B,C, $\in \{F\}$ is identical to F itself.

In $\overline{ISO}$ (1,3), $G/F = M^{1,3}$ (Minkowski space), but in $g\overline{ISO}$ (1,3), $G/G = M^{1,3/4}$ known as Superspace. It has four Grassmann (and Majorana spinor component) dimensions besides space-time. The nilpotence of Grassmann elements allows an expansion in powers of the Grassmann generating element θ^α for any "superfield" so that one can still deal unambiguously with fields over space time as coefficients of the $(\theta^\alpha)^r$. For extended supersymmetry the nilpotence appears only at $r > 4n$, n being the dimension of the internal symmetry O(n) vector. Our treatment is limited at this stage to cases in which the bosonic submanifold of G/F is $M^{1,3}$, or S^4 in pre-Wigner-Inönu contraction situations (with G a simple group such as $\overline{SO}$ (1,4) or OSp (4/N)).

In the renormalization of quantum field theories, constraints such as explicit symmetry violations have caused difficulties. These were removed by *spontaneous* symmetry breakdown, i.e. when the breakdown resulted from the equations of motion (and described the on-mass shell situation only). Rather than apply SGM factorization as a constraint, we have described the conditions allowing for its occurring spontaneously [9]. This requires A to obey in addition to (8.15), a (so-called) "symmetric decomposition" WRS,

$$[H,H] \subset F \qquad (8.16)$$

If in addition A is Simple (WRSS), and if the Lagrangian L_{INV} is F-gauge invariant (but not G-gauge invariant) and of the form (A,B span the entire A)

$$L_{INV} = R^A \wedge h_{AB} R^B \; , \quad D^{(F)} h_{AB} = 0 \qquad (8.17)$$

where h_{AB} is a regular (F-invariant) 2-tensor under G one has pseudo-closure (in the geometric sense)

$$d L_{INV} = D L_{INV} = D^{(F)} L_{INV} \stackrel{\rho}{=} 0 \qquad (8.18)$$

In addition,

$$R^h \overset{\Omega}{=} 0, \quad h \in \{H\} \tag{8.19}$$

This vanishing of the torsions (curvatures in the H directions) is always true for $R^h_{||}$ i.e. the $R^h{}_{f'f}$ where $h \in H$; $f, f' \in F$, all anholonomic. It is true for $\overline{O}(1,4)$ and OSp(4/N) for all components, including $R^h_{|-}$ and $R^h_{=}$. Moreover, for WRS in which [H,H] spans the entire F ($R^A_{|-}$ is $R^A{}_{fh}$),

$$R^A_{||} \overset{\Omega}{=} 0, \qquad R^A_{|-} \overset{\Omega}{=} 0, \qquad A \in \{F\} \tag{8.20}$$

i.e. we get factorization, all relevant forms being only over H.

Gravity and Supergravity can be written as in (8.17) in an uncontracted form and arise in their precise form through a Wigner-Inönu contraction. This discussion then fits both cases and one can write on-shell BRS equations as for a Principal Bundle. However, BRS are mainly needed to prove Unitarity and this is an off-mass-shell problem (see for example ref. [20]).

Off-mass-shell, the SGM is not a Principal Bundle and we are not allowed to use the results of sections 2-3 whether or not Spontaneous Fibration has occurred.

We have applied the algebra $\tilde{A}$ (8.16) of Lie derivatives as the off-mass-shell local gauge (for the horizontal directions $H \subset G$, this will remain different from A even after fibration). We treat the connections and curvatures as forms over the full dimensionality of G. The indices U,V and R,S respectively denote holonomic variables over $\mathcal{G}/F$ and over F. $\mathcal{G}/F$ is a submanifold larger or equal to space-time. We expand ρ^A

$$\begin{aligned} \rho^A &= dz^U \tau_U^A + dz^R \psi_R^A = \tau^A + \psi^A \\ \rho^M &= \mathcal{K}_A^M \rho^A = \mathcal{K}_A^M (dz^U \tau_U^A + dz^R \psi_R^A) \end{aligned} \tag{8.21}$$

and since

$$\rho^M = dz^N \mathcal{K}_A^M \rho_N^A = dz^N \delta_N^M = dz^U \delta_U^M + dz^R \delta_R^M$$

we have

$$\mathcal{K}_A^U \tau_V^A = \delta_V^U, \quad \mathcal{K}_A^R \psi_S^A = \delta_S^R$$

so that

$$\left.\begin{aligned} \tau^U &= dz^U, \quad \tau^R = 0 \\ \psi^U &= 0, \quad \psi^R = dz^R \end{aligned}\right\} \tag{8.22}$$

The structural equations for τ^A and ψ^A can be read directly from (8.7)

$$\left.\begin{aligned} s\psi^A &= -\frac{1}{2}[\psi,\psi]^A + R^A_{BC}\,\psi^B\,\psi^C \\ s\tau^A_U &= D_U\,\psi^A - R^A_{BC}\,\psi^B\,\tau^C_U \end{aligned}\right\} \quad (8.23)$$

All indices are anholonomic except for U. Note that just as in an internal gauge, the action of s is of an "alibi" type, i.e. it represents an active vertical group translation.

We have proved [21] that the correspondence we had in an internal gauge theory between structural equations and BRS transformations is preserved: that equations (8.23) reproduce the BRS equations fitting the Lie-derivative (or AGCT) gauge provided we use the identification (2.7).

Note that the curvature term $R^f_{BC}\,\tau^C_U = R^f_{BU}$ can be considered as an auxiliary field over $\mathcal{G}/F$.

We note that for any form

$$L_{\tilde{\varepsilon}}\,\omega = \tilde{\varepsilon} \lrcorner\, d\omega + d(\tilde{\varepsilon} \lrcorner\, \omega)$$

Thus for the Lagrangian

$$L_{\tilde{\varepsilon}}\,\mathcal{L} = \tilde{\varepsilon} \lrcorner\, d\mathcal{L} + d(\tilde{\varepsilon} \lrcorner\, \mathcal{L}) \quad (8.24)$$

The second term vanishes upon integration (as a divergence) but the first term vanishes only after the fibration, when $d\mathcal{L}$ vanishes because it becomes a 5-form over $M^{1,3}$. This thus weakens the applicability of (8.23) in the geometric approach, since it requires fibration anyhow. However, this does not interfere with our first aim, which was to check that the BRS equations on SGM are indeed given again by the structural equations.

Alternatively, one has added to $\mathcal{L}$ a "quartic ghost term" [20,22], apparently cancelling the vertical piece of $\mathcal{L}_{INV}$ over the SGM. We have not yet studied this method geometrically.

REFERENCES

[1] Feynman, R.P., Acta Phys. Polon. 26, 697 (1963).

[2] De Witt, B.S., *"Dynamical Theory of Groups and Fields"*, Gordon & Breach Pub., N.Y./London/Paris, 1965.

[3] Faddeev, L.D. and Popov, V.N., Phys. Lett. B25, 29 (1967).

[4] 't Hooft, G., Nucl. Phys. B33, 436 (1971).

[5] Slavnov, A.A., Teor. Mat. Fiz. 10, 153 (1972).
Taylor, J.C., Nucl. Phys. B33, 436 (1971).

[6] Becchi, C., Rouet, A., and Stora, R., Com. Math. Phys. 42, 127 (1975).
Tyutin, I.V., rep. FIAN 39 (1975).

[7] Thierry-Mieg, J., These de Doctorat d'Etat (Paris-Sud) 1978.

[8] Ne'eman, Y. and Regge, T., Phys. Lett. 74B, 54 (1978).
Ne'eman, Y. and Regge, T., Rivista d. Nuovo Cm. III, 1 n.5 (1978).

[9] Thierry-Mieg, J. and Ne'eman, Y., Ann. of Phys. (N.Y.) 123, 247 (1979).

[10] Ne'eman, Y. and Thierry-Mieg, J., Proc. Nat. Acad. Sci. USA 77, 720 (1980).

[11] Thierry-Mieg, J., Harvard rep. HUTMP 79/B86.

[12] Curci, G. and Ferrari, R., Nuovo Cim. 30A, 155 (1975); 32A, 151 (1976); 35A, 1, 273 (1976).
Ojima, I., Prog. Theoret. Phys. 64, 625 (1980).

[13] Thierry-Mieg, J., to be published.

[14] Thierry-Mieg, J. and Ne'eman, Y., to be published.

[15] Ne'eman, Y., Phys. Lett. B81, 190 (1979).

[16] Ne'eman, Y. and Thierry-Mieg, J., Proc. Salamanca (1979). Int. Conf. Diff. Geom. Methods in Phys., A. Perez-Rendon ed., Springer Verlag L.N. in Math.

[17] Ne'eman, Y. and Sternberg, S., Proc. Nat. Acad. Sci. USA (1980).

[18] Kalb, M. and Ramond, P., Phys. Rev. D9, 2273 (1974).
Cremmer, E. and Scherk, J., Nucl. Phys. B72, 117 (1974).

[19] Freedman, D.Z., rep. CALT 68-624 (1977) unpub.
Townsend, P.K., CERN th. 2753 (1979).

[20] Sterman, G., Townsend, P.K., and van Nieuwenhuizen, P., Phys. Rev. D17, 1501 (1978).

[21] Ne'eman, Y., Takasugi, E., and Thierry-Mieg, J., Phys. Rev. D.

[22] Kallosh, R.E. JETP Lett. 26, 575 (1977).

ON MONOPOLE SYSTEMS WITH WEAK AXIAL SYMMETRY

P. Houston and L. O'Raifeartaigh,

Dublin Institute for Advanced Studies,
10 Burlington Road, Dublin 4, Ireland.

Abstract

Let (Φ, $\vec{A}$) be an SO(3) Yank-Mills-Higgs system which is a real-analytic, static, finite-energy solution of the Bogomolny field equation $\vec{B} = \vec{D}\Phi$. We show that the zero-set of the current $\vec{J} = \Phi \wedge \vec{D}\Phi$ is of dimension at most one. Using this property of $\vec{J}$ we obtain the curious result that if the system is axially symmetric, in the weak sense that all local scalar *gauge-invariants* are axially symmetric, the topological charges must be located on the axis of symmetry and must be of equal magnitude and alternate sign. In particular, if the charges are of uniform sign they must be concentrated at a single point. The fact that the charges of spherically symmetric monopoles are bounded by unity is obtained as a corollary. It is also shown that a master-potential for the invariant fields that was found earlier to exist for systems with additional symmetry, exists as a direct consequence of weak axial symmetry alone.

1. Introduction

Let (Φ, $\vec{A}$) be a static ($\partial_t = 0$) purely magnetic ($A_0 = 0$) finite energy SO(3) Yang-Mills-Higgs system satisfying the first-order Bogomolny [1] field equation in Euclidean 3-space E(3),

$$\vec{B} = \vec{D}\Phi, \quad \text{where} \quad \vec{B} = \vec{\nabla} \times \vec{A} + \frac{1}{2}\vec{A} \overset{\times}{\wedge} \vec{A}, \quad \vec{D}\Phi = \vec{\nabla}\Phi + \vec{A} \wedge \Phi, \qquad (1.1)$$

with boundary conditions

$$\vec{B} \longrightarrow 0, \quad \Phi^2 \longrightarrow c^2 \neq 0, \qquad \text{as} \quad |\vec{x}| \longrightarrow \infty \qquad (1.2)$$

Here x and $\wedge$ denote outer-product in space and isospace respectively, c is a constant, and we suppose that ($\vec{A},\Phi$) are real analytic. The real analyticity is not a strong assumption because it has been shown [2] to hold (in at least one gauge) for solutions of (1.1)(1.2) which satisfy quite mild conditions concerning the Sobolov norms of the fields.

In some previous papers we have shown [3] that if the system ($\vec{A},\Phi$) is axially symmetric in the strong or conventional [4] sense that there exists a local (scalar) isovector ω (x) such that for any local (scalar) isovector $\lambda(x)$ we have

$$D_\psi \lambda(x) = \omega(x) \wedge \lambda(x) \quad \text{where} \quad D_\psi = xD_y - yD_x, \qquad (1.3)$$

then the topological charge distribution must be as stated in the abstract. It has also been shown [5] that if the strong axially symmetric system is mirror-symmetric (symmetric with respect to reflexions in planes through the axis of symmetry) then it admits a (scalar-isoscalar) master-potential W(x) from which invariant fields such as (Φ, Φ) and $(D_\psi\Phi, D_\psi\Phi)$ can be obtained by differentiation. (Bracket denotes inner product in isospace).

The rather surprising nature of the result concerning the charge-distribution raises the question as to whether (1.1) is really the most general definition of axial symmetry and whether the results would still hold under a weaker definition. Accordingly, the purpose of this note is to reconsider the situation under what would seem to be the weakest reasonable definition of axial symmetry, namely, that local *gauge-invariants* such as the inner-products

$$(\Phi, \Phi), \quad (\Phi, B_\alpha), \quad (B_\alpha, B_\psi), \qquad x_\alpha = (z, \rho), \qquad (1.4)$$

be independent of the azimuthal angle ψ. (Here the space indices are expressed in cylindrical coordinates to avoid a spurious ψ-dependence).

It turns out that, for real analytic fields satisfying (1.1) and (1.2), weak axial symmetry actually implies strong axial symmetry so that the previous results still hold. Furthermore, it turns out that the result for the charge-distribution can be obtained more or less directly from weak axial symmetry, and that the master-potential W exists even without the hypothesis of mirror-symmetry. The role played by W also becomes much clearer.

In order to establish these results it is first necessary to establish that the zero-set Z(J) of the current

$$\vec{J} = \Phi \wedge \vec{D}\Phi \qquad (1.5)$$

is located on a manifold which is at most 1-dimensional i.e. consists of at most isolated points and analytic curves. This particular result is independent of axial symmetry, and in the axially symmetric case it implies that Z(J) can lie on at most the axis of symmetry and symmetrical rings around the axis.

Finally for completeness we derive as a simple corollary the known result that a spherically symmetric monopole must have unit charge and derive also the single equation for the master-potential which is sufficient to close the system of field equations in mirror-symmetric case. In the latter derivation we use mirror symmetry only in the weak form

$$(B_{\psi}, B_{\alpha}) = 0 . \qquad (1.6)$$

2. Zero Sets of the Higgs Field and the Current.

We commence with the result that the zero-set Z(J) the current $\vec{J}$ is at most 1-dimensional, and it will be convenient to consider also the zero-set of the Higgs field Φ, although from the definition of $\vec{J}$ the zero-set of Φ is contained in Z(J). Since Φ and $\vec{J}$ are real analytic their zero-sets are analytic submanifolds of E(3), and hence what we have to do is eliminate E(3) itself and 2-dimensional submanifolds. As mentioned before, the results are quite general (independent of axial symmetry) and in the (weak) axially symmetric case they reduce the possibilities for Z(J) to the axis of symmetry and isolated symmetrical rings around the axis.

In the case of the Higgs field we first note that the boundary condition (1.2) excludes E(3) itself, and requires that any 1- and 2-dimensional submanifolds be closed. Next using the Bianchi identity $\vec{D}\cdot\vec{B} = 0$ we obtain from (1.1) the usual second-order field equation

$$D^2 \Phi = 0, \qquad (2.1)$$

for the Higgs field, and from (2.1) we obtain at once the equation

$$\Delta(\Phi, \Phi) = 2(\vec{D}\Phi, \vec{D}\Phi) \geqslant 0 \qquad (2.2)$$

for (Φ, Φ). Since (2.2) shows that (Φ, Φ) is a subharmonic function it follows that (Φ, Φ) cannot vanish on a closed 2-dimensional surface without vanishing in the interior, and hence vanishing throughout E(3), in contradiction to the boundary condition. Thus the zero-set of Φ consists of at most isolated points and analytic closed curves [6]. In particular, in the axially symmetric case it reduces to at most isolated points on the axis of symmetry and isolated rings around the axis. (The whole symmetry axis is excluded by the boundary condition).

In the case of the current $\vec{J}$ the space E(3) itself is excluded for a different reason, namely, that if $\vec{J}$ is identically zero the gauge and Higgs field completely decouple and it is well-known that there are no non-trivial finite-energy solutions for the decoupled system. Now suppose that $\vec{J}$ vanishes on an analytic 2-surface Σ. The results for Φ show that Σ contains finite elements $\delta\Sigma$ on which $(\Phi,\Phi) \neq 0$. But from the Bianchi identity and the field equations (1.1) we obtain after some algebraic manipulations

$$\vec{D}\cdot\vec{J} = 0 \quad \text{and} \quad \vec{D} \times \vec{J} = -\Phi \wedge \vec{J} + (\Phi,\Phi)^{-2}\{ \bar{J} \overset{\times}{\wedge} \bar{J} + 2(\Phi,\vec{B}) \times \vec{J} \} , \tag{2.3}$$

and it is easy to see that this equation implies that on $\delta\Sigma$ the normal derivative to $\vec{J}$ can be expressed as a linear combination of $\vec{J}$ and the tangential derivatives to $\vec{J}$, with coefficients which are smooth on $\delta\Sigma$ (and may be functions of $\vec{J}$). Since, by iteration, the same will be true of the normal derivative of any order, and $\vec{J}$ is real analytic, it follows that $\vec{J}$ cannot vanish on $\delta\Sigma$ without vanishing on a finite 3-volume containing $\delta\Sigma$ and hence vanishing throughout E(3). Thus in the non-trivial case the zero-set Z(J) of $\vec{J}$ can be at most 1-dimensional i.e. can consist only of isolated points and analytic curves (not necessarily closed). In particular, in the (weak) axially symmetric case the zero-set of $\vec{J}$ can consist of at most points on the symmetry-axis and isolated rings around the axis.

3. Orthonormal Triads in the Complement of Z(J).

The reason that we need to locate the zero-set Z(J) of $\vec{J}$ is that in the complement $\tilde{E}(3) = E(3)-Z(J)$ we can construct orthonormal triads of isovectors and use them to implement the weak axial symmetry. In this section, we give the construction. First we note that since Z(J) is at most 1-dimensional $\tilde{E}(3)$ is connected, though not necessarily simply-connected. Now let P be any point of $\tilde{E}(3)$. Then at P, and by analyticity, in a finite neighbourhood N of P, we have $\vec{J} \neq 0$. But then in N we have $\Phi \neq 0$, and for at least one component, J say, of $\vec{J}$ we have $J \neq 0$. Furthermore from the definition of $\vec{J}$ we have $(\Phi, J) = 0$. It follows that in N the isovectors

$$\omega_1 = \Phi / |\Phi| , \quad \omega_2 = J/|J| \quad \text{and} \quad \omega_3 = \omega_1 \wedge \omega_2, \text{ in N} \tag{3.1}$$

form an orthonormal triad. The triad (3.1) formed with fixed[+] component J of $\vec{J}$ may

[+] Here and throughout the components of J are understood to be expressed in cylindrical coordinates.

not be extendable to all of $\tilde{E}(3)$, because J might vanish at finite distances from P. But since $\vec{J} \neq 0$ in $\tilde{E}(3)$ and $\tilde{E}(3)$ is connected, it is clear that $\tilde{E}(3)$ can be covered with overlapping neighbourhoods N, each with at least one triad. As we shall see in the next section this result is sufficient to implement real axial symmetry in $\tilde{E}(3)$, and that is all that we shall need. Note that the ω's are real analytic and single-valued in $\tilde{E}(3)$ since they are quotients of functions which are real analytic in E(3).

4. Implementation of Weak Axial Symmetry on E(3)

Using the triad (3.1) we can construct the isovector

$$\omega = -\frac{1}{2}\varepsilon_{abc}(\omega_a, D_\psi \omega_b)\,\omega_c \quad \text{in N, where } a,b,c = 1,2,3. \qquad (4.1)$$

Now let $\lambda(x)$ be an arbitrary axial-scalar isovector (space-tensor isovector whose space-indices are expressed in cylindrical coordinates). Then $\lambda(x)$ has the expansion

$$\lambda(x) = (\omega_a(x), \lambda(x))\,\omega_a(x) \quad \text{in N,} \qquad (4.2)$$

and since by weak axial symmetry

$$D_\psi(\omega_a, \lambda) \equiv \nabla_\psi(\omega_a, \lambda) = 0 \quad \text{in N,} \qquad (4.3)$$

we obtain at once from (4.1) the relation

$$D_\psi \lambda(x) = \omega(x) \wedge \lambda(x) \quad \text{in N.} \qquad (4.4)$$

Equation (4.4) shows that the vector $\omega(x)$ implements the covariant derivative D_ψ in N. To extend it to $\tilde{E}(3)$ we note that if ω'_a is any alternative basis e.g. for a neighbourhood N', and ω' is the isovector constructed as in (4.1) from ω'_a, then from (4.4) we have

$$(\omega' - \omega) \wedge \lambda = 0 \quad \text{in} \quad N \cap N'. \qquad (4.5)$$

Thus we have

$$(\omega' - \omega) \wedge \omega_a = 0 \quad \text{or} \quad \omega = \omega' \quad \text{in} \quad N \cap N'. \qquad (4.6)$$

Equation (4.6) shows that ω is unique and basis-independent in $N \cap N'$, a result that can also be verified directly from (4.1) using weak axial symmetry. Since $\tilde{E}(3)$ can be covered with overlapping neighbourhoods N, it follows that ω and eq. (4.4) extend to all of $\tilde{E}(3)$ as required. Note that ω will also be real analytic in $\tilde{E}(3)$. In particular ω will be unique or single-valued in $\tilde{E}(3)$.

From (4.4) on $\tilde{E}(3)$ it also follows (for $\lambda = \omega_a$) that

$$F_{i\psi} \wedge \omega_a = (D_i\omega)\wedge\omega_a \text{ on } E(3), \text{ where } F_{ij} = [D_i, D_j] = \varepsilon_{ijk} B_k, \tag{4.7}$$

and since the ω_a are non-degenerate on $\tilde{E}(3)$, we then have

$$D_i\omega = F_{i\psi} = \varepsilon_{i\psi k} B_k \quad \text{on } \tilde{E}(3). \tag{4.8}$$

We shall refer to equations (4.4) on $\tilde{E}(3)$ and (4.8) as the equations of weak axial symmetry. With these equations in hand we turn to the topological charge distribution.

5. The Topological Charge

The general expression for the topological charge contained in a volume $\mathcal{V}$ with smooth surface S on which $\Phi \neq 0$ is well known [7] to be

$$Q_{\mathcal{V}} = \frac{1}{4\pi} \int_S f_{ij} dx^i dx^j \quad \text{where } f_{ij} = (\phi, F_{ij}) - (\phi, D_i\phi\, D_j\phi),\ \phi = \Phi/|\Phi| \tag{5.1}$$

is the Maxwell field projected out of F_{ij} by ϕ. In the axially symmetric case f_{ij} is independent of ψ and hence if we choose $\mathcal{V}$ to be a volume of revolution we have

$$Q_{\mathcal{V}} = \frac{1}{4\pi} \int f_{ij}\, dx^i dx^j = \frac{1}{2} \oint_C f_{i4} dx^i, \tag{5.2}$$

where the line integral is along a curve C in S orthogonal to the azimuthal direction. The precise nature of C depends on the topology of S and will be specified later. Now from (5.1) we have, in particular,

$$\begin{aligned} f_{i\psi} &= (\phi, F_{i\psi}) + (D_i\phi, \phi\wedge(\omega\wedge\phi)), \\ &= (\phi, D_i\omega) + (D_i\phi, \omega) = \nabla_i(\phi, \omega), \qquad \text{on } \tilde{E}(3). \end{aligned} \tag{5.4}$$

Suppose now that the curve C lies in $\tilde{E}(3)$ except possibly for the end-points x_1 and x_2. Then (5.4) can be used in the integral (5.2) and we obtain the closed expression

$$Q_{\mathcal{V}} = \frac{1}{2}\left[b(x_1) - b(x_2)\right] \quad \text{where} \quad b(x_a) = \operatorname*{Lt}_{x\to x_a} (\phi(x), \omega(x)), \tag{5.5}$$

for the topological charge Q_V contained in $\mathcal{V}$.

It is well-known that the topological charge as defined in (5.1) must be located at the zeros of Φ. From our results on these zeros, we see that in the axially symmetric case the charges can be located only at isolated points on the axis of symmetry and on rings around the axis. Our first step will be to show that the rings are not possible, so that the charge must be located only on the axis.

6. Elimination of Rings of Topological Charge.

To show that rings of topological charge are not possible we let R be a ring of zeros of Φ. From the results on Z(J), R is isolated in $\tilde{E}(3)$ and hence can be surrounded by a torus of revolution whose surface lies entirely in $\tilde{E}(3)$. Letting the volume $\mathcal{V}$ of the previous section be such a torus, the curve $\mathcal{C}$ must be a circle in $\mathcal{S}$ which loops the torus in the direction orthogonal to the toroidal axis. Then, since $\mathcal{C}$ is closed and (ω, Φ) is single-valued in $\tilde{E}(3)$, the expression (5.5) for the charge inside the torus yields zero, as required. Thus the rings are eliminated and the charge is located only on the symmetry axis.

7. Limits of $\omega\wedge\Phi$ and ω^2 on the Axis of Symmetry

In order to determine the charge distribution on the symmetry axis (z-axis, say) we shall need the limits of $\omega\wedge\Phi$ and ω^2 as $\rho \longrightarrow 0$, and hence we consider these limits in this section. First from (4.4) we have

$$|(\omega\wedge\Phi)| = |D_\varphi \Phi| \leq \rho\,(|D_x \Phi| + |D_y \Phi|) \qquad \text{in } \tilde{E}(3), \tag{7.1}$$

and since A and Φ are real analytic throughout E(3) we then have

$$\mathop{\mathcal{L}t}_{\rho=0} \; (\omega\wedge\Phi) = 0. \tag{7.2}$$

Next from (4.8) we have

$$|\nabla_i \omega^2| = 2(\omega, F_{i\varphi}) \leq 2\rho|\omega|(|F_{ix}| + |F_{iy}|) \leq \rho\,(\omega^2 + K^2) \text{ in } \tilde{E}(3), \tag{7.3}$$

where K^2 is the maximum of $(\vec{B}, \vec{B})$ in E(3). Since $\vec{B}$ is non-trivial and is real analytic, we have $0 < K < \infty$, and hence on integrating (7.3) along a straight line between any two points x and x_0 in $\tilde{E}(3)$ we have

$$\frac{\omega^2(x) + K^2}{\omega^2(x_o) + K^2} \leqslant e^{R|x-x_o|} \quad \text{where} \quad R = \max(\varrho, \varrho_o). \tag{7.4}$$

Keeping ϱ_o fixed and letting $\varrho \longrightarrow o$ we see from Cauchy convergence that $\omega^2(x)$ has a finite limit as $\varrho \longrightarrow o$. Then, by letting $x = (z, \varrho)$ and $x_o = (z_o, \varrho)$ we see that

$$\underset{\varrho = o}{\mathcal{L}t} \; \omega^2(x) = n^2 \tag{7.5}$$

where the finite value n^2 is *independent of z*. Equations (7.2) and (7.5) give the required limits.

Incidentally, we note that since $\omega \wedge \Phi$ is real analytic in $\tilde{E}(3)$, it must be periodic in ψ, and hence, since from (4.4) we have

$$\omega^2 (\omega\wedge\Phi) = D^2_{\psi} (\omega\wedge\Phi) \longrightarrow \nabla^2_{\psi} (\omega\wedge\Phi), \quad \text{as} \quad \varrho \longrightarrow o, \tag{7.6}$$

the constant n in (7.5) must be an *integer*. In the next section n will be identified with the topological charge.

8. Charge Distribution on the Symmetry-Axis.

To determine the charge distribution on the z-axis, we let the volume of the previous section be any volume of revolution which lies inside all rings of zeros of Φ and cuts the z-axis at just two points z_1 and z_2 where $\Phi \neq o$. Then apart from z_1 and z_2, the surface of V lies entirely in $\tilde{E}(3)$ and the curve C is a curve joining z_1 to z_2 with all its interior points in $\tilde{E}(3)$. From (5.5) we then have for the charge in

$$Q_V = \frac{1}{2}\left[b(z_1) - b(z_2)\right], \quad \text{where } b(z) = \underset{\varrho = o}{\mathcal{L}t} \; b(z, \varrho). \tag{8.1}$$

But from the limits obtained in section 7 we have for $\Phi \neq o$,

$$b^2(z) = \underset{\varrho = o}{\mathcal{L}t} \; (\omega,\Phi)^2 = \underset{\varrho = o}{\mathcal{L}t}\left[\omega^2 - (\omega\wedge\Phi)^2\right] = \underset{\varrho = o}{\mathcal{L}t} \; \omega^2 = n^2 \tag{8.2}$$

where n^2 is independent of z. It follows that

$$Q_V = o, \pm n. \tag{8.3}$$

But since the volume $\mathcal{V}$ may contain any number of charges, and two successive charges of the same sign would yield $Q_v = \pm 2n$, eq. (8.3) implies that the charges must be of alternate sign and of the same magnitude.

In particular, if the charges are required to have the same sign then there can be only a single charge (of arbitrary magnitude).

9. Equivalence of Weak and Strong Axial Symmetry.

The results for the charge-distributions were obtained using only the weak axial symmetry equations (4.4) and (4.8). However, for completeness and for the discussion of the master-potential, we wish to show that for analytic fields satisfying (1.1) and (1.2) weak axial symmetry actually implies strong axial symmetry i.e. eqs. (4.4) and (4.8) can be extended from $\tilde{E}(3)$ to $E(3)$.

For this purpose we note that eq. (4.8) can be integrated along any curve Γ from x_0 to x in $E(3)$ to yield

$$\omega(x) = \omega_\Gamma(\omega(x_0), A) + \oint_\Gamma \varepsilon_{i\varphi k} B_k \, dx^i, \tag{9.1}$$

where ω_Γ denotes the parallel transfer of $\omega(x_0)$ along Γ with respect to the connection $\vec{A}$. The value of $\omega(x)$ is path-independent because the integrability condition for (4.8) is just (4.4) in $\tilde{E}(3)$ in the special case when λ is replaced by the components of $\vec{B}$.

But now since $\vec{A}$ and $\vec{B}$ are real analytic throughout $E(3)$ and the complement $Z(J) = E(3) - \tilde{E}(3)$ consists only of points and curves (so that $\tilde{E}(3)$ is connected) eq. (9.1) defines an analytic extension of $\omega(x)$ as $x \longrightarrow Z(J)$. Furthermore, for any $\lambda(x)$ which is real analytic in $E(3)$ eq. (4.4) then extends analytically to $E(3)$, and this is just the condition of strong axial symmetry.

Note that the result would not necessarily hold if $Z(J)$ contained a 2-dimensional submanifold Σ because Σ would necessarily disconnect $\tilde{E}(3)$. Then $\omega(x)$ would not necessarily be path-independent and the values of $\omega(x)$ obtained coming from the two sides might not agree. This can perhaps be seen more clearly by considering the infinitesimal version of the above proof. First we note from (7.4) that ω^2 remains uniformly bounded as $x \longrightarrow Z(J)$ and that by recycling this result into (4.8) $\omega(x)$ and all its finite derivatives are uniformly bounded as $x \longrightarrow Z(J)$. Thus ω has a smooth (C^∞ limit) as $x \longrightarrow Z(J)$. But if $Z(J)$ contained a 2-dimensional Σ, the smooth limits on either side of Σ might not agree. When $Z(J)$ is at most one-dimensional, however, the values as $x \longrightarrow Z(J)$ are independent of the direction of approach and so ω has a limit which is unique as well as smooth when $x \longrightarrow Z(J)$.

The analyticity of the extension follows by differentiating (4.8) again to obtain the *elliptic* equation

$$D^2 \omega = - J_\psi . \tag{9.2}$$

Since the coefficients in this elliptic equation are analytic the smooth solutions must be analytic as required.

10. The Existence of the Masterpotential $\mathcal{W}$.

We wish to show that the existence of the masterpotential $\mathcal{W}$ found in previous papers [5] is a direct consequence of axial symmetry condition (1.3). Inserting the field equation (1.1) in (1.3) we have

$$D_i \omega = \varepsilon_{i\psi k} D_k \Phi , \tag{10.1}$$

and taking the inner product of this equation with ω we obtain

$$\nabla_i \omega^2 = 2 \,\varepsilon_{i\psi k} \nabla_k(\omega,\Phi) - \rho^2 \nabla_i \Phi^2. \tag{10.2}$$

But eq. (10.2) is just the Cauchy-Riemann-type equation which was used previously to deduce the existence of a masterpotential $\mathcal{W}$ such that

$$\frac{\partial W}{\partial z} = (\omega,\Phi), \qquad \rho\frac{\partial W}{\partial \rho} = \omega^2, \qquad \Delta W = \Phi^2, \tag{10.3}$$

and thus the result is established and the role of $\mathcal{W}$ clarified. Note that (10.1) is itself a type of covariant Cauchy-Riemann equation and hence might be of some use in seeking explicit solutions of the field equations.

11. Field equation for $\mathcal{W}$ in the Mirror-Symmetric Case.

It is known that when the system is mirror-symmetric the field equation for $\mathcal{W}$ contains only the fields ω^2 , $h = |\Phi|$ and $b = (\omega,\Phi)$ which occur in (10.3) and hence this equation and (10.3) form a closed system. We therefore wish to derive the field equation for $\mathcal{W}$ directly from our present results. From the second-order field equation (2.1) for Φ we obtain

$$\Delta h = (\vec{D}\phi)^2 h. \tag{11.1}$$

But from the normalization of ϕ and the mirror-symmetry condition (1.6) we have

$$(\Phi, D_\alpha \Phi) = 0 \quad \text{and} \quad (D_\psi \Phi, D_\alpha \Phi) = 0 \tag{11.2}$$

respectively, where $x_\alpha = (z, \rho)$. Hence $D_\alpha \Phi$ has a component only in the direction of the vector

$$\hat{\omega} = \Phi \wedge D_\psi \Phi = \Phi \wedge (\omega \wedge \Phi) = \omega - (\Phi, \omega) \Phi , \tag{11.3}$$

and

$$(\vec{D}\Phi)^2 = \frac{1}{\rho^2} (D_\psi \Phi)^2 + \frac{(\omega, D_\alpha \Phi)^2}{\omega^2} = \frac{k^2}{\rho^2} + \frac{u^2}{k^2} \tag{11.4}$$

where

$$k^2 = \omega^2 - b^2 \quad \text{and} \quad u_\alpha = \nabla_\alpha b + \rho \, \varepsilon_{\alpha\beta} \nabla_\beta h. \tag{11.5}$$

Hence from (11.1) we have

$$\Delta h = \left(\frac{k^2}{\rho^2} + \frac{u^2}{k^2} \right) h , \tag{11.6}$$

and this is the required equation for $\mathcal{W}$. Of course, we must use (11.5) and (10.3) to make it explicit in $\mathcal{W}$.

12. Corollary for the Spherically Symmetric Case.

We wish to show here that previous results [8], which state that a spherically symmetric charge distribution must be of maximum strength unity can be derived as a simple corollary to our present results. First we note that the results for Z(J) imply that a spherically symmetric charge must be located at the origin. Then since spherical symmetry simply extends axial symmetry to all three axes, we have from (1.3)

$$\vec{L} \lambda + \vec{\varpi} \wedge \lambda = 0, \tag{12.1}$$

where $\vec{L}$ is the angular momentum operator, λ is any scalar isovector and $-\omega_z$ is our previous ω. But from (12.1) we have

$$(\vec{L} \times \vec{L}) \lambda + 2 \vec{\varpi} \wedge \lambda - \frac{1}{2} (\vec{\varpi} \underset{\wedge}{\times} \vec{\varpi}) \wedge \lambda = 0, \tag{12.2}$$

and hence

$$\frac{1}{2}(\vec{\omega} \overset{x}{\wedge} \vec{\omega}) \wedge \lambda = \vec{\omega} \wedge \lambda \quad \text{or} \quad \frac{1}{2}(\vec{\omega} \overset{x}{\wedge} \vec{\omega}) = \vec{\omega}, \tag{12.3}$$

since the λ are non-degenerate in E(3) - 0 . But this means that for each fixed x in E(3) - 0 the ω form a canonical set of generators for the isospin group. Furthermore since they act only on the 3-dimensional space of the λ's they can generate only the trivial or 3-dimensional representation. Thus

$$(\omega, \omega) \leq 2 \quad \text{and} \quad (\omega_z, \omega_z) \leq 1. \tag{12.4}$$

But then from (5.5) we have for a volume V enclosing the origin

$$Q_V \leq |b(z)| = |(\omega_z, \Phi)| \leq |\omega_z| \leq 1, \tag{12.5}$$

which is the required result.

Acknowledgements

We are indebted to E. Seiler, C. Taubes and S. Deser for some very illuminating discussions.

References

[1] E. Bogomolny, Sov. J. Nucl. Phys. 24, 449 (1976).
[2] C. Taubes, Harvard Univ. Preprint HUTMP 79/B94.
[3] P. Houston, L. O'Raifeartaigh, Phys. Lett. 93B, 151 (1980)
94B, 153 (1980).
[4] R. Jackiw, Schladming 1979 Lecture Notes, Acta Physica Austriaca Suppl XXII, 383 (1980).
[5] N. Manton, Nucl. Phys. B135, 319 (1978)
P. Houston, L. O'Raifeartaigh, DIAS Preprints 35 (1979)
28 (1980)
S. Adler, Private Communication.
[6] The result for Φ has been obtained independently by C. Taubes, loc. cit.
[7] G.'t Hooft, Nucl. Phys. B79, 276 (1974)
A. Polyakov, JETP Lett. 20, 194 (1974)
J. Arafune, P. Freund, C. Goebel, J. Math. Phys. 16, 433 (1975).
[8] V. Romanov, A. Schvartz, Y. Tyupkin, Nucl. Phys. B130, 209 (1977)
A. Guth, E. Weinberg, Phys. Rev. D14, 1660 (1976)
L. O'Raifeartaigh, Nuovo Cim. Lett. 18, 205 (1977.

QUANTUM FIELDS IN CURVED SPACE-TIMES AND SCATTERING THEORY

Bernard S. Kay

Institute for Advanced Study, Princeton, N.J., USA.

Blackett Laboratory, Imperial College, London.

Institut für Theoretische Physik, Universität Bern,
Sidlerstraße 5, CH-3012 Bern, Switzerland.*

Table of Contents

---

* Present Address

§§0 INTRODUCTION

In this talk, we discuss certain mathematical aspects of quantum field theory in curved space-time. We have two main aims:

(1) To outline a single consistent framework for the whole subject, and to provide a general introduction to rigorous results of a functional analytic/differential geometric nature.

(2) To present some new results recently obtained - in collaboration with Jonathan Dimock - for scattering theory in curved space-time.

Interest in quantum theory in curved space-time derives from a variety of sources. In astrophysics and cosmology, one expects important quantum effects, such as the creation of particles, in regions of strong and rapidly changing gravitational fields i.e. shortly after the beginning of the universe, and in the vicinity of collapsing stars. In the absence of a complete theory of quantum gravity, the approximate theory of quantized matter "test" fields in a fixed curved background has proved very successful, and in the last few years, many calculations have been performed with many interesting results. In particular, Hawking's startling prediction [1] in 1975 of "black-hole evaporation" has provided a considerable stimulus to the whole subject. These results are, of course, interesting - not only as astrophysical predictions - but also as clues towards the construction of a full theory of quantum gravity.

Quite apart from its special relation to gravity, however, we feel that the study of quantum theory in curved space-time is also of interest as providing a broader perspective on quantum field theory in general. Present day quantum field theory (especially in its more rigorous formulations) is very strongly tied to flat space-time and its associated symmetry group: the Poincaré group. Yet, surely, one (if not *the*) essential feature of a quantum field is that it is the solution of a certain (albeit very special and not yet fully understood!) type of hyperbolic partial differential equation. From this point of view, it is natural to take as our model for space-time a manifold - this being the proper mathematical setting for hyperbolic P.D.E.s. Let us restate this in a slightly less abstract way: Even if one is principally interested in field equations with "constant coefficients"; it may be useful, both conceptually, and as a mathematical tool, to consider transformations of various types which, while preserving hyperbolicity, do not preserve the "constant coefficients" nature of the equations.

It is very much with this philosophy in mind that we have begun a study of rigorous mathematical results for this subject. And while we set as our ultimate goal to improve our understanding of physical results such as the Hawking effect, we feel that our work is also of value for the mathematical techniques developed along the way. A characteristic feature of these techniques is that they combine the traditional functional analysis methods familiar in quantum field theory (a standard reference is [2]) with more differential geometric methods to do with the causal and conformal structure of space-times, the theory of hyperbolic P.D.E.s on manifolds, etc. - Methods which were developed e.g. in connection with classical general relativity (see [3]). There appears to be considerable scope for cross-fertilization between these two fields. And, in view of the title of this conference, this is a point we particularly want to stress in this talk. A couple of examples are:

(1) The intimate relation between the Cauchy problem for hyperbolic P.D.E.s and self-adjointness properties of elliptic operators [4] as exploited in [5] (cf §2c).

(2) The application of Fourier Integral Operators to scattering theory in [10] (- see §2d).

We illustrate the point in this talk with two new results related to scattering theory:

In §3a, we describe work with Jonathan Dimock [13] on the existence of classical wave operators on curved space-time.

And in §3b, we present a new geometrical proof [9] of a classical result on the decay of solutions of the Klein Gordon equation.

In order to provide the necessary background, we begin in §§1, 2 with a statement of the basic structure of the subject*. We have tried to make it self-contained, but it is necessarily brief and we refer to our original papers [5,6,7,8,13] and to related mathematical work ([10,11,12,33,14,15,16,17,18,19,20] is an incomplete sample) for full details, references and credits. Let us also provide an (incomplete) list [21,22,23,24,25,26] of useful review articles and collections on the physics of the subject. We would also like to mention here the early work on the structure of (especially linear - i.e. external field) quantum field theories due to Segal (see e.g. [27]) and also Wightman (see e.g. [28]) and others which has strongly influenced our work.

Finally, we should point out that by no means all interesting mathematical questions in the field are of a functional analytic/differential geometric nature. - See for

---
* The reader who is familiar with this material and mainly interested in scattering theory may turn immediately to §2d.

example the discussion of algebraic topology as applied to quantum field theory in [29].

§§1. GENERAL STRUCTURE OF QUANTUM FIELDS IN CURVED SPACE-TIME

§1a. Algebraic Structure

The familiar structure of flat-space-time quantum field theory may be conveniently divided into two parts. At the most basic level, there is the *algebraic structure* - i.e. local algebras of observables, states, etc. (see below). Superimposed on this is what one might call the *vacuum structure* - i.e. the whole complex of ideas related to the existence of a preferred vacuum state, its associated Hilbert-space representation, the existence of a positive Hamiltonian generating the dynamics, and ultimately, a particle interpretation. In flat space-time, one often ignores the distinction between these two levels of structure, regarding an operator interchangeably as an element of an algebra or as already represented on some fixed physical Hilbert-space. We shall see that, in curved space-time, the distinction becomes much more important.

The algebraic structure generalizes in a straightforward way: We quote a set of axioms due to Dimock [11]. Given a space-time $(\mathcal{M}, g)^{*}$:

(1) *For each open set,* $\mathcal{O} \subset \mathcal{M}$, *there is a* C^**-algebra* $\mathfrak{A}(\mathcal{O})$.

$$\mathcal{O} \subset \mathcal{O}' \implies \mathfrak{A}(\mathcal{O}) \subset \mathfrak{A}(\mathcal{O}');$$

$$\mathfrak{A} = \overline{\bigcup_{\mathcal{O}} \mathfrak{A}(\mathcal{O})}.$$

(2) *There exists a faithful irreducible representation.*

(3) $\mathcal{O}, \mathcal{O}'$ *spacelike separated* $\implies$

$$[\mathfrak{A}(\mathcal{O}), \mathfrak{A}(\mathcal{O}')] = 0$$

(4) $\mathcal{O}$ *is causally dependent on* $\mathcal{O}'$ $\implies$

$$\mathfrak{A}(\mathcal{O}) \subset \mathfrak{A}(\mathcal{O}').$$

(5) *Given an isometry between two space-times* $\varkappa : (\mathcal{M}, g) \longrightarrow (\hat{\mathcal{M}}, \hat{g})$;
there is an automorphism $\alpha_{\varkappa} : \mathfrak{A} \longrightarrow \hat{\mathfrak{A}}$
such that $\alpha_{\varkappa}[\mathfrak{A}(\mathcal{O})] = \hat{\mathfrak{A}}(\varkappa(\mathcal{O}))$, $\alpha_{\mathrm{id}} = \mathrm{id}$, $\alpha_{\varkappa_1 \circ \varkappa_2} = \alpha_{\varkappa_1} \circ \alpha_{\varkappa_2}$

---

* see [3] - we assume $\mathcal{M}$ and g are both C^{∞}

For full details and discussion, see [11]. Suffice it to point out here, that, in the special case of Minkowski space; one recovers - thanks to axiom (5) - the usual action of the Poincaré group as automorphisms of the algebra.

§1b. Vacuum Structure

The construction of an analogue for the vacuum structure of flat space-time is much more problematic. The point being that the essential requirement - the existence of a time-translational symmetry - is in general lacking. A general (say globally hyperbolic - see below) space-time (say, after choosing some system of coordinates) will correspond to a time-dependent external gravitational field. Just like any other time-varying external field (e.g. external electromagnetic field in flat space-time), this will be constantly exchanging energy with our quantum system - thus rendering the concept of a vacuum inappropriate. In general, there is no preferred state which may be called a vacuum and in consequence no preferred Hilbert-space representation*. In particular, one loses the familiar particle interpretation of flat-space-time physics. Failure to recognize this can lead to apparent paradoxes. In general, then, one is forced to be content with the algebraic structure as outlined above (- together with the corresponding generalized algebraic concept of state [30]). It is a rather austere framework, but it is free from paradoxes and in principle provides a complete physical description [30].

In certain (important!) special cases, for restricted classes of space-times, one *can* however, recover certain aspects of vacuum structure. We shall consider two such cases:

(1) For *stationary* space-times; one has a time-translational symmetry and can thus mimic the familiar flat-space-time construction.

(2) For asymptotically flat (or asymptotically stationary) space-times, one may develop a *scattering* theory. By comparing the dynamics with that on a flat reference space-time, one may e.g. view scattering as an automorphism on the familiar (now concretely represented) flat space-time algebra. One can then refer again to flat-space-time concepts such as particles etc. as asymptotic observables.

---

* An example is given in §2c.

§§2. KLEIN GORDON EQUATION ON A CURVED SPACE-TIME

§2a. Reduction to Classical Problem

In the rest of this talk, we illustrate the above discussion by considering the covariant Klein-Gordon equation

$$(g^{\mu\nu} \nabla_\mu \partial_\nu + m^2)\, \hat{\varphi} = 0 \qquad (1)$$

We shall see that for such a simple linear system, one can construct the various structures quite explicitly. The point about linearity is that there is a simple and well-known machinery available for reducing any quantum problem to a corresponding classical problem. This is a point that will be re-made in each of the subsequent sections §2b, c, d *.

In the case of algebraic structure the corresponding classical problem is the Cauchy problem for which a natural setting is the set of *globally-hyperbolic* space-times. These are space-times $(\mathcal{M},g)$ for which there exists a global time-coordinate t such that the equal-time surfaces $\{t\} \times \mathcal{C}$ are Cauchy surfaces. Topologically, such space-times are then necessarily of the form $\mathbb{R} \times \mathcal{C}$. Note that one version of Penrose's "cosmic censorship hypothesis" postulates that the universe is globally hyperbolic [31]. (For some ideas on quantization in non-globally-hyperbolic universes, see the example given in §2c. and also the recent paper [20]).

The nice thing about global-hyperbolicity is that it guarantees global existence and uniqueness for the classical Cauchy problem: One has what we shall call *Leray's Theorem* (see especially [11] for references and further discussion).

Theorem: Given Cauchy data;

$$(f, p) \in C_0^\infty(\mathcal{C}) + C_0^\infty(\mathcal{C})$$

on one Cauchy surface (say $\{0\} \times \mathcal{C}$ *) there is a unique* C^∞ *solution* φ *to* (1) *which has compact support on all other hypersurfaces.*

Here, f is the restriction of the solution $\varphi|_{\mathcal{C}}$ to $\{0\} \times \mathcal{C}$. p is the familiar canonical momentum: given in coordinates as

$$({}^3g)^{\frac{1}{2}} \frac{g^{0\mu}}{(g^{00})^{\frac{1}{2}}} \partial_\mu \varphi|_{\mathcal{C}} \qquad \left(\equiv (-{}^4g)^{\frac{1}{2}} g^{0\mu} \partial_\mu \varphi \right)$$

---

* It also dictates a strategy similar to that needed for treating external field problems other than gravity: For further discussion of this point see §2d.

We can thus view classical dynamics - after a suitable splitting of space-time into space and time such that each constant t surface is a Cauchy surface (for details on space-time splits and exact expressions in coordinates see [6]) - as a family of maps:

$$\mathcal{T}(t_2,t_1) \;:\; C_0^\infty(\mathcal{C}) + C_0^\infty(\mathcal{C}) \longrightarrow C_0^\infty(\mathcal{C}) + C_0^\infty(\mathcal{C})$$

from Cauchy data at time t_1 to that at time t_2. One now equips the space of Cauchy data with the natural linear symplectic form

$$\sigma(f_1,p_1;f_2,p_2) \;=\; \int_{\mathcal{C}} (f_1 p_2 - p_1 f_2)\, d^3x \tag{2}$$

This will be conserved in consequence of the conserved current $\varphi_1 \partial_\mu \varphi_2 - \varphi_2 \partial_\mu \varphi_1$ which exists for any given pair of solutions. Thus $\mathcal{T}(t_2, t_1)$ form a family of *symplectic* maps on the symplectic space $(C_0^\infty(\mathcal{C}) + C_0^\infty(\mathcal{C}), \sigma)$.

§2b Algebraic Structure

It is now a simple step (thanks to by now well-established results on representations of the C.C.R. etc.) to construct the algebraic structure: One constructs the "equal-time algebra" $\mathcal{W}(C_0^\infty(\mathcal{C}) + C_0^\infty(\mathcal{C}), \sigma)$* using the standard construction of the *Weyl algebra* over a symplectic space. - I.e. the algebra generated by elements W(f, p) satisfying:

$$W(f_1,p_1)\, W(f_2,p_2) = \exp\left(\frac{-i\sigma(f_1,p_1;f_2,p_2)}{2}\right) W(f_1 + f_2,\, p_1 + p_2) \tag{3}$$

Note that W is related formally to the usual quantum fields $(\hat{\varphi}, \hat{\pi})$ via

$$W(f, p) \;=\; \exp i\,\sigma(\hat{\varphi}, \hat{\pi}\,;\, f, p) \tag{4}$$

(This may be made rigorous (by Stone's theorem) for representations for which the map $t \longrightarrow W(tf, tp)$ is continuous.)

The correct quantum dynamics is now simply given by the family of automorphisms $\alpha(t_2, t_1)$ defined via

$$\alpha(t_2, t_1)\, W(f, p) \;=\; W(\mathcal{T}(t_2, t_1)(f, p)) \tag{5}$$

---

* There are several candidates for $\mathcal{W}$ discussed in the literature. For almost all our purposes we could in fact equally well work with the free algebra generated by ("unbounded") elements R(f,p) satisfying $[R(f_1,p_1), R(f_2,p_2)] = i\sigma(f_1,p_1;f_2,p_2)$.

It is helpful to see this relation between the classical and quantum theories in terms of the commutative map diagram:

$$\begin{array}{ccc}
(C_0^\infty(\mathcal{C}) + C_0^\infty(\mathcal{C}), \sigma) & \xrightarrow{W(\cdot,\cdot)} & \mathcal{W}(C_0^\infty(\mathcal{C}) + C_0^\infty(\mathcal{C}), \sigma) \\
\Big\uparrow \mathcal{T}(t_2,t_1) & & \Big\uparrow \alpha(t_2,t_1) \\
(C_0^\infty(\mathcal{C}) + C_0^\infty(\mathcal{C}), \sigma) & \xrightarrow{W(\cdot,\cdot)} & \mathcal{W}(C_0^\infty(\mathcal{C}) + C_0^\infty(\mathcal{C}), \sigma)
\end{array} \qquad (6)$$

Note that it is now natural to define the time t quantum field $W_t(f, p)$ by

$$W_t(f, p) = \alpha(0, t)\, W(f, p) = W(\mathcal{T}(0, t)(f, p))$$

One then has the formal correspondence

$$W_t(f, p) = \exp i\, \sigma(\hat{\varphi}_t, \hat{\pi}_t; f, p) \qquad (7)$$

where $(\hat{\varphi}_t, \hat{\pi}_t) = \mathcal{T}(t, 0)(\hat{\varphi}, \hat{\pi})$

This picture - of an equal-time algebra together with a family of automorphisms - provides a complete algebraic description for linear fields, and we shall use it as a starting point in discussing stationary space-times and scattering theory in §2c, d. To make contact with the general framework of §1a. however, we need a more covariant way of expressing things [11,6]: For this purpose, one appeals to another aspect of the classical Cauchy problem: The advanced and retarded fundamental solutions E^+ and E^- satisfying

$$(g^{\mu\nu} \nabla_\mu \partial_\nu + m^2)\, E^{\pm} = E^{\pm}(g^{\mu\nu} \nabla_\mu \partial_\nu + m^2) = \mathrm{id} \qquad (8)$$

Again, thanks to global hyperbolicity, these exist as operators from $C_0^\infty(\mathcal{M})$ to $C^\infty(\mathcal{M})$.

One then obtains a system of local algebras in the following way: First, define $E = E^+ - E^-$. Then, given any function $F \in C_0^\infty(\mathcal{M})$; one can form the solution $\varphi = EF$. One then associates to each such F the Weyl operator W(f, p) for the Cauchy data of that solution (say at time zero). Let us now call this W(F).

$$\text{(Formally, } W(F) = \exp i \int_{\mathcal{M}} \hat{\varphi}(x)\, F(x)\, (-{}^4g)^{1/2}\, d^4x) \qquad (9)$$

One then has for each region $\mathcal{O}$ the subalgebra $\mathfrak{A}(\mathcal{O})$ generated by the W(F)'s for F's with support in $\mathcal{O}$. One can now show that all the axioms of §1a. are satisfied (see [11] for details).

§2c. Vacuum Structure for Stationary Space-Times

A *stationary* space-time is one which admits a space-time split such that - in suitably adapted coordinates - the metric becomes time-independent. Now, the time-evolution operators arise from a one-parameter group $\mathcal{T}(t)$:

$$\mathcal{T}(t_2,t_1) = \mathcal{T}(t_2)\,\mathcal{T}(t_1)^{-1} \qquad (10)$$

$$\mathcal{T}(t_1)\,\mathcal{T}(t_2) = \mathcal{T}(t_1 + t_2)$$

and correspondingly, (cf (5) or (6)) there is a quantum automorphism group $\alpha(t)$ such that

$$\begin{aligned} \alpha(t_2,t_1) &= \alpha(t_2)\,\alpha(t_1)^{-1} \\ \alpha(t_1)\,\alpha(t_2) &= \alpha(t_1 + t_2) \end{aligned} \qquad (11)$$

where $\alpha(t)\,W(f, p) = W(\mathcal{T}(t)\,(f, p))$.

It now becomes appropriate to seek a vacuum state i.e. a state such that in the corresponding representation on a Hilbert space $\mathcal{H}$, the automorphisms are implemented by a unitary group with positive energy:

$$\begin{aligned} \alpha(t)\,\rho\,(W(f,p)) &= U(t)\,\rho\,(W(f,p))\,U(t)^{-1} \\ U(t) &= e^{-iHt} \\ H &\geqslant 0 \\ \exists\, \Omega \in \mathcal{H} \text{ s.t. } \quad H\Omega &= 0 \end{aligned} \qquad (12)$$

We then have the

Theorem: If in a stationary space-time:

(i) *There is a Cauchy surface* $\mathcal{C}$

(ii) *In coordinates suitably adapted to* $\mathcal{C}$*, the metric satisfies*

$$g_{00}\,(g^{00})^{1/2} > \varepsilon > 0.$$

Then a vacuum representation exists and furthermore, H has a mass gap.

Details of the proof are given in [5] (see also [8]). Let us just point out here that the general strategy has one or two features in common with our approach to scattering theory in §2d., i.e. introduction of energy-norm and reduction to a corresponding "classical problem" (in this case, the construction of a "one-particle structure").

Notes (1) This vacuum representation is also in a certain sense "almost unique" (see [5], [7], [8] for details).

(2) A sufficient condition guaranteeing (i) and (ii) is that $\mathcal{C}$ be compact.

(3) In flat space-time, this construction reduces to the familiar free field (Fock) construction.

As we stressed in §1b., the essential requirement here is stationarity. We illustrate the need for the algebraic approach in the generic non-stationary case with a simple example: Consider a universe of form $\mathbb{R} \times S^3$ which is stationary with radius R_1 before some "early" time t_1; and again with a different radius R_2 after some "late" time t_2 and varies smoothly in-between. (One can equally well think in analogy of the Klein Gordon equation in flat space-time with a time-dependent mass: simply read "mass" for "radius"). Clearly, the vacuum representation for radius R_1 will be appropriate for $t < t_1$; and that for radius R_2 will describe appropriately dressed particles at late times. But, these two representations will be unitarily inequivalent and in general, the dynamics between times t_1 and t_2 will not be implemented. So, it makes neither mathematical nor physical sense to have a single representation that will do for all times.

Conditions (i) and (ii) are less essential. (ii) can sometimes be dropped (see [8]) at the expense of losing the mass gap. One can also drop condition (i) that $\mathcal{C}$ be Cauchy at the expense of uniqueness: Consider e.g. the strip $\mathbb{R} \times (0,1)$ in the flat two-dimensional space-time $\mathbb{R}^2$ with metric $\begin{pmatrix} 1 & 0 \\ 0 & -1 \end{pmatrix}$. (This is of course *not* a globally hyperbolic space-time). Here, one can easily construct several different quantizations corresponding to different choices of boundary conditions at the edge of the space-time (and corresponding in turn to different self-adjoint extensions of $-\partial_x^2 + m^2$ on $C_0^\infty(0,1) \subset L^2(0,1)$). This theme has been developed in [20].

§2d. Scattering for Asymptotically Stationary Space-Times

In the rest of this talk, we describe recent work in collaboration with J. Dimock on scattering theory for the covariant Klein Gordon equation . We consider space-times over the manifold $\mathbb{R}^4$ and from now on assume a fixed global coordinate system. We also restrict to metrics satisfying the condition that in our fixed coordinate system: (A) *Each constant* t *surface is a Cauchy surface*

- so that the discussion of §2a. will go through. Our aim is to compare - at large time - the progagation in such a background metric with that in the flat background metric η = diag (1, -1, -1, -1). Thus: on the *same* equal-time Cauchy data, we want to compare the *interacting* dynamics $\mathcal{T}(t, s)$ corresponding to our equation

$$(g^{\mu\nu}\nabla_\mu \partial_\nu + m^2)\,\varphi = 0 \tag{1}$$

with the *free* dynamics - described now by a group $\mathcal{T}^0(t)$ - for the flat-space-time Klein Gordon equation

$$(\Box + m^2)\,\varphi = 0 \tag{13}$$

Correspondingly, we have to compare the interacting and free automorphisms $\alpha(t, s)$ and $\alpha^0(t)$ on the same equal-time algebra.

Our *goal* is to find conditions on g under which we can construct *in* and *out* fields - given in some formal sense by

$$\begin{pmatrix} \hat{\varphi}_{\substack{in\\out}} \\ \hat{\pi}_{\substack{in\\out}} \end{pmatrix} = \mathcal{T}^0(\pm\infty)\,\mathcal{T}(\mp\infty, 0)\begin{pmatrix} \hat{\varphi} \\ \hat{\pi} \end{pmatrix} \tag{14}$$

and the related scattering automorphism $\mathcal{S}$ given formally by

$$\mathcal{S} = \alpha^0(-\infty)\,\alpha(\infty, -\infty)\,\alpha^0(\infty) \tag{15}$$

In this section, we sketch the general strategy which is (by now this should come as no surprise!) to exploit the linearity of the problem to reduce it to an equivalent classical problem. In §§3, we shall discuss some of the specific classical problems which arise.

Before we begin, let us note that much of what we say here (as, indeed, much of what we say in the whole talk) would apply also to external scalar or electromagnetic or Yang-Mills etc. fields i.e. to equations such as

$$(\Box + m^2 + A^\mu \partial_\mu + V)\,\varphi = 0 \tag{16}$$

(of the large literature on such equations, [32] (for example) is close in spirit to our work. Cf. also [33]). All of our results could easily be adapted to such cases. And indeed some of our results would, as far as we know, be new even in those contexts. Let us stress however that the case of gravity presents important additional problems: Many of the perturbation-theoretic ideas which are effective for (16) fail completely for (1). - The reason being that, in gravity, one has, in addition to terms like $A^{\mu}\partial_{\mu} + V$, a perturbation of the form $(g^{\mu\nu} - \eta^{\mu\nu})\partial_{\mu}\partial_{\nu}$ which is just as "unsmoothing" as the leading term $\Box$ in the unperturbed equation.

We describe then a basic strategy for all external field scattering problems: splitting the problem into six stages:

Stage 1: Introduce the free energy-norm on time-zero Cauchy data:

$$\| (f, p)\|_{\mathcal{A}} = \frac{1}{2} \int (p^2 + (\nabla f)^2 + m^2 f^2)\, d^3 x \qquad (17)$$

and complete to form the (real) Hilbert space $\mathcal{A}$ of finite-energy Cauchy data. The free time-evolution $\mathcal{T}^{0}(t)$ preserves the energy-norm and thus extends to a unitary group $\mathcal{T}^{0'}(t)$ on $\mathcal{A}$ (later, we shall drop the prime). Furthermore, noting that the symplectic form σ is continuous in energy-norm:

$$\sigma(\phi, \psi) \leqslant c \, \|\phi\|_{\mathcal{A}} \|\psi\|_{\mathcal{A}} \qquad \phi, \psi \in C_0^{\infty}(\mathcal{C}) + C_0^{\infty}(\mathcal{C}) \qquad (18)$$

it also extends to a symplectic form σ' on $\mathcal{A}$. Also, σ' will be preserved by $\mathcal{T}^{0'}(t)$.

Stage 2: Now, assume that the interacting dynamics $\mathcal{T}(t_2, t_1)$ are uniformly (in time) bounded in the free energy norm. (Sufficient conditions for this are discussed in §3a.) Then the $\mathcal{T}(t_2, t_1)$ will extend to a uniformly bounded family of operators on $\mathcal{A}$. Unlike $\mathcal{T}^{0}(t)$, they will not preserve the norm. However, they remain symplectic (i.e. preserve σ') and, of course, invertible.

Stage 3: Define *classical wave-operators:*

$$\begin{aligned} \Omega^{\pm} &= \operatorname*{s-lim}_{t \to \mp\infty} \mathcal{T}(0, t)\, \mathcal{T}^{0}(t) \\ \tilde{\Omega}^{\pm} &= \operatorname*{s-lim}_{t \to \mp\infty} \mathcal{T}^{0}(-t)\, \mathcal{T}(t, 0) \end{aligned} \qquad (19)$$

as strong limits on $\mathcal{A}$ in analogy with the familiar quantum mechanical case. Here the domains $D(\Omega^{\pm})$, $D(\tilde{\Omega}^{\pm})$ are understood to consist of the set of vectors for which convergence holds (and may, a priori even be null). It follows from uniform boundedness (and invertibility) of the $\mathcal{T}(0, t)$'s that $\Omega^{\pm}$, $\tilde{\Omega}^{\pm}$ are bounded, $D(\Omega^{\pm})$ and $D(\tilde{\Omega}^{\pm})$ are closed subspaces of $\mathcal{A}$, and further that $\mathrm{Ran}(\tilde{\Omega}^{\pm}) = D(\Omega^{\pm})$; $\mathrm{Ran}(\tilde{\Omega}^{\pm}) = D(\Omega^{\pm})$ with

$$\tilde{\Omega}^{\pm}\Omega^{\pm} = \mathrm{id} \text{ on } D(\Omega^{\pm})$$

and

$$\Omega^{\pm}\tilde{\Omega}^{\pm} = \mathrm{id} \text{ on } D(\tilde{\Omega}^{\pm}) \tag{20}$$

Moreover, it follows from the symplecticness of $\mathcal{T}(t_2, t_1)$; $\mathcal{T}^0(t)$ that $(D(\Omega^{\pm}), \sigma')$; $(D(\tilde{\Omega}^{\pm}), \sigma')$ are symplectic spaces in their own right and that $\Omega^{\pm}$, $\tilde{\Omega}^{\pm}$ induce symplectic isomorphisms between them. The crucial question is then to find conditions on the metric ensuring the existence and uniqueness of scattering states - i.e. such that

$$D(\Omega^{+}) = D(\Omega^{-}) = \mathcal{A} \tag{21}$$

In the next section, we present such a result. In the remainder of this section, we assume this proved in which case, we define

$$\mathcal{A}_{\substack{\mathrm{in}\\ \mathrm{out}}} = \mathrm{Ran}(\Omega^{\pm}) = D(\tilde{\Omega}^{\pm}) \tag{22}$$

And we also assume that *weak asymptotic completeness* holds:

$$\mathcal{A}_{\mathrm{in}} = \mathcal{A}_{\mathrm{out}} \tag{23}$$

(There are several other important properties one might ask for - cf. [2] volume III). In these circumstances, one can define the *classical scattering operator* (S-matrix)

$$S = \tilde{\Omega}^{-}\Omega^{+} \tag{24}$$

which will then be a symplectic automorphism of $\mathcal{A}$.

<u>Stage 4</u>: One easily converts information about classical scattering to information about quantum scattering at the algebraic level. First, we extend the Weyl algebra $\mathcal{W}(C_0^\infty(\mathcal{C}) + C_0^\infty(\mathcal{C}), \sigma)$ to the Weyl algebra $\mathcal{W}(\mathcal{A}, \sigma')$ over the space of finite-energy solutions (Note also that $W(\mathcal{A}_{\mathrm{in}}, \sigma')$ constitutes a subalgebra). Then, we define *wave-isomorphisms:*

$$\omega^{\pm} : \mathcal{W}(\mathcal{A}, \sigma') \longrightarrow \mathcal{W}(\mathcal{A}_{\mathrm{in}}, \sigma') \qquad \tilde{\omega}^{\pm} : \mathcal{W}(\mathcal{A}_{\mathrm{in}}, \sigma') \longrightarrow \mathcal{W}(\mathcal{A}, \sigma')$$

via $$\omega^{\pm} W(\Phi) = W(\Omega^{\pm} \Phi)$$

$$\tilde{\omega}^{\pm} W(\Phi) = W(\tilde{\Omega}^{\pm} \Phi) \tag{25}$$

Clearly, they will inherit the obvious properties from the classical wave operators:

$$\tilde{\omega}^{\pm} \omega^{\pm} = \text{id on } \mathcal{W}(\mathcal{A}, \sigma') \qquad \omega^{\pm} \tilde{\omega}^{\pm} = \text{id on } \mathcal{W}(\mathcal{A}_{in}, \sigma') \tag{26}$$

In analogy with (6), we then define in and out (algebraic) fields:

$$W_{\substack{in \\ out}} (\Phi) = \omega^{\pm}(W(\Phi)) \tag{27}$$

which formally correspond to the in and out fields of (14) via

$$W_{\substack{in \\ out}} (\Phi) = \exp i \sigma (\hat{\varphi}_{\substack{in \\ out}}, \hat{\pi}_{\substack{in \\ out}} ; f, p) \qquad \mathcal{A} \ni \Phi = (f, p) \tag{28}$$

Finally, we define the *scattering automorphism* $\mathfrak{s}$ (which corresponds to the formal expression (15)) via

$$\mathfrak{s}(W(\Phi)) = W(S \Phi) \tag{29}$$

Stage 5: Now, choose an $\mathcal{A}$*-regular representation* ϱ of the time-zero algebra $W(\mathcal{A}, \sigma')$ over some Hilbert space $\mathcal{K}$. By an $\mathcal{A}$-regular representation, we mean one satisfying the property

$$\Phi_i \longrightarrow \Phi \text{ in } \mathcal{A} \implies W(\Phi_i) \xrightarrow{s.} W(\Phi) \text{ on } \mathcal{K} \tag{30}$$

Note that the standard (Fock) representation of the free field (which one will usually take in applications) is $\mathcal{A}$-regular *. It then immediately follows that the - thus concretely represented - in and out fields arise as strong limits on $\mathcal{K}$:

$$\varrho(W_{\substack{in \\ out}} (\Phi)) = \underset{t \to \mp\infty}{\text{s-lim}}\ \varrho\ (\alpha(0, t)\, \alpha^{0}(t)\, W(\Phi)) \qquad \Phi \in \mathcal{A} \tag{31}$$

In fact, the unbounded fields now make sense and e.g. in the standard free-field (Fock) representation, one recovers

---

* This is true for non-zero mass - where it follows from the usual continuity property over the "one-particle Hilbert space" $\mathcal{H}$ ($\mathcal{K} = \mathcal{F}(\mathcal{H})$: Fock space over $\mathcal{H}$) together with the fact that $\mathcal{A}$ is continuously embedded in $\mathcal{H}$.

$$\mathcal{S}(\hat{\varphi}_{\substack{in\\out}}, \hat{\pi}_{\substack{in\\out}}; f, p) = \underset{t \to \mp\infty}{\text{s-lim}} \mathcal{S}(\hat{\varphi}, \hat{\pi}; \mathcal{T}(0, t)\, \mathcal{T}^{0}(t)\, (f, p)) \tag{32}$$

on the domain in Fock-space of "finite-particle vectors". (Here $\hat{\varphi}, \hat{\pi}$ etc. are the time-zero field and field-momentum in the standard free-field (Fock) representation).

Stage 6: Finally, one wants to know if the $\mathcal{S}$-automorphism is unitarily implemented in the representation ϱ :

$$\varrho[\mathcal{S}(W(\Phi))] \overset{?}{=} \mathcal{J}\, \varrho[W(\Phi)]\, \mathcal{J}^{-1} \tag{33}$$

This is equivalent (as one can easily see) to asking if the in and out fields are unitarily equivalent in the in (or out) representation $\varrho' = \varrho \circ \tilde{\omega}^{+}$ (or $\varrho \circ \tilde{\omega}^{-}$):

$$\varrho'[W_{out}(\Phi)] \overset{?}{=} \mathcal{J}^{-1}\, \varrho'[W_{in}(\Phi)]\, \mathcal{J} \tag{34}$$

This final question reduces (at least for a class of representations known as Fock representations *) to yet another classical question which we do not have time to go into here. (Very briefly: (see the literature - especially [34] - for further details) the criterion is that S - when considered on the "one-particle Hilbert space" $\mathcal{H}$ (which is yet another completion of $C_0^{\infty}(\mathcal{C}) + C_0^{\infty}(\mathcal{C})$ - maybe related to but not the same as $\mathcal{A}$) satisfy $S^{+}S$ - id be Hilbert-Schmidt).

The answer is known in the special case where $g_{\mu\nu}$ differs from $\eta_{\mu\nu}$ only on a compact set (in which case, the classical and algebraic aspects of scattering are of course trivial):

Theorem [10 , 19]

Let $g_{\mu\nu}$ *satisfy condition* (A) *above and be equal to* $\eta_{\mu\nu}$ *outside of a compact set. Let* ϱ *(cf* (33)*) be the standard free-field representation. Then the scattering automorphism is unitarily implemented.*

§§3. PROVING SCATTERING THEORY RESULTS

§3a. An Existence Theorem for the Classical Wave Operators

As we showed in §2d, the questions of the existence of in and out fields and of a scattering automorphism may all be reduced to questions about the domains of the classical wave-operators.

---

* The standard free-field representation is a particular case.

We have recently obtained - in collaboration with Jonathan Dimock - a result for the existence of $\Omega^{\pm}$ on (all of) $\mathcal{A}$ which we now describe (full details will appear in [13]). First of all, one imposes conditions (cf §2d. Stage 2) guaranteeing that $\mathcal{T}(t, s)$ are uniformly bounded: Sufficient conditions are:

(B) * *The components* $g^{\mu\nu}$ *are bounded and* $\exists\ \varepsilon > 0$ s.t

$$\text{(i)} \quad g^{00} > \varepsilon > 0$$

$$\text{(ii)} \quad -\,{}^{4}g^{jk}\,\xi_j\,\xi_k > \varepsilon\,|\xi_j|^2$$

$$\text{(C)} \quad \int_{-\infty}^{\infty} \sup_{x} \left| \frac{\partial g_{\mu\nu}}{\partial t} \right| dt < \infty$$

(Here, $|\ |$ is any norm defined for multi-index objects.- A convenient choice is to take for $|\ |^2$ the sum of the squares of appropriate components: Greek indices ranging over 0,1,2,3; Latin indices over 1,2,3).

Condition (B) ensures that the interacting energy-norm is uniformly (in time) equivalent to the free-energy norm. Condition (C) ensures - via a standard-type energy-inequality - that the energy of the interacting field is bounded in time.

Condition (C) says that the space-time tends in a certain sense to be stationary at the future and past.

To guarantee convergence of $\Omega^{\pm}$, one extra condition is needed on the fall off of the gravitational field in space-like directions.

A sufficient such condition is

(D)** *There exist positive constants* C; ε *s.t. for some* $p \geq 2$

$$\| g_{\mu\nu} - \eta_{\mu\nu} \|_{1,p} < C\,|t|^{3/p - 1 - \varepsilon}, \quad |t| > 1$$

Here, $\|\ \|_{1,p}$ is the Sobolev norm given by ($p < \infty$)

$$\| f_{\alpha\beta \dots \gamma} \|_{1,p}^{p} = \int |f_{\alpha\beta\dots\gamma}|^{p}\, d^3x + \int |\partial_i f_{\alpha\beta\dots\gamma}|^{p}\, d^3x$$

and in the case $p = \infty$ by

$$\| f_{\alpha\beta\dots\gamma} \|_{1,\infty} = \sup_{\underline{x}} \left\{ |f_{\alpha\beta\dots\gamma}| \ , |\partial_i f_{\alpha\beta\dots\gamma}| \right\}$$

We then have the:

Theorem: When (A),(B),(C),(D) *are satisfied,* $\Omega^{\pm}$ *exist on* $\mathcal{A}$. ((A) appears in §2d.)

---

* Note that condition (B) has as a consequence also boundedness of $g_{\mu\nu}$; $g_{00} > \varepsilon > 0$; $g_{ij}\,\xi^i\,\xi^j > \varepsilon . |\xi^i|^2$ and $\det(-g_{\mu\nu}) > \varepsilon > 0$

** Note that, in the presence of conditions (B) and (C); this can be written in several equivalent ways e.g. one may replace $g_{\mu\nu} - \eta_{\mu\nu}$ by $g^{\mu\nu} - \eta^{\mu\nu}$ or $(-{}^4g)^{1/2} g^{\mu\nu} - \eta^{\mu\nu}$ etc.

We sketch the proof which is based on Cook's method (described below - see also [2] vol III) and the following Lemma on the decay of solutions of the Klein Gordon equation:

Decay Lemma: Let φ *be a solution of* $(\Box + m^2)\varphi = 0$ *with Cauchy data in* $C_0^\infty(\mathbb{R}^3) + C_0^\infty(\mathbb{R}^3)$; *then*

(a) $\exists\ C$ *s.t.* $\sup_{\underline{x}} |\varphi(t, \underline{x})| < C t^{-3/2}$, $|t| > 1$

(b) $\exists\ C$ *s.t. the* L^p *norm* $(1 \leq p \leq \infty)$

$$\|\varphi(t, \cdot)\|_p < C t^{-3/2 + 3/p}, \quad |t| > 1$$

Note (1) that (a) is a special case ($p = \infty$) of (b). However, once one knows (a); (b) follows trivially by finite propagation speed. (2) In §3b., we discuss this Lemma further and present a new geometrical proof (for part (a)).

Proof of Theorem:

Firstly, note that since the domains of $\Omega^{\pm}$ must be closed, it suffices to check convergence on the dense set $C_0^\infty(\mathbb{R}^3) + C_0^\infty(\mathbb{R}^3)$. Take Ω^-. We need to check, for a given pair of smooth time-zero Cauchy data (f, p) that $\mathcal{T}(0, t)\,\mathcal{T}^0(t)$ is Cauchy in $\mathcal{A}$-norm - i.e. that given $\varepsilon > 0$; there is a T s.t. $t, t' > T \implies$

$$\left\| \left[\mathcal{T}(0, t)\,\mathcal{T}^0(t) - \mathcal{T}(0, t')\,\mathcal{T}^0(t')\right] \begin{pmatrix} f \\ p \end{pmatrix} \right\|_{\mathcal{A}} < \varepsilon$$

Cook's method, familiar from the theory of Schrödinger operators consists in rewriting this as

$$\left\| \int_{t'}^{t} \frac{d}{ds} \left(\mathcal{T}(0, s)\,\mathcal{T}^0(s)\right) \begin{pmatrix} f \\ p \end{pmatrix} ds \right\|_{\mathcal{A}} < \varepsilon$$

Now, defining the first-order form Hamiltonians:

$$\mathbb{h}(t) = -\frac{\partial}{\partial t}\mathcal{T}(t', t)\Big|_{t'=t} \qquad \mathbb{h}^0 = -\frac{\partial}{\partial t}\mathcal{T}(t)\Big|_{t=0}$$

we may rewrite this as

$$\left\| \int_{t}^{t'} \mathcal{T}(0, s)\left[\mathbb{h}(s) - \mathbb{h}^0\right]\mathcal{T}^0(s) \begin{pmatrix} f \\ p \end{pmatrix} \right\|_{\mathcal{A}} < \varepsilon$$

$\mathbb{h}(s)$ and $\mathbb{h}^0$ are, of course, given by 2x2 matrices of first-order (elliptic) differential operators (the first-order form of the equations (1), (13) respectively, see [6], [13] for explicit expressions). Now, our expression is majorized if we interchange the integral and the norm. Then, using the fact that $\mathcal{T}(0, s)$ is uniformly bounded, and writing $\mathcal{T}^0(s)\begin{pmatrix} f \\ p \end{pmatrix}$ as $\begin{pmatrix} f_t^0 \\ p_t^0 \end{pmatrix}$ - the Cauchy-data for a free wave, it suffices to show:

$$\int_{-\infty}^{\infty} \| (\mathbb{h}(s) - \mathbb{h}^{o}) \begin{pmatrix} f_t^{\,o} \\ p_t^{\,o} \end{pmatrix} \|_{\mathcal{A}} < \infty$$

The integral here turns out to be just the L^2-norm of a sum of terms of the schematic form $[D(g - \eta)]\,[\varphi^{o}]$. The first piece, $[D(g - \eta)]$ involving essentially * the difference between some component of g and the corresponding component of η - or first derivatives thereof; the second piece $[\varphi^{o}]$ being a solution of the free equation. Using

$$\| [D(g - \eta)]\,[\varphi^{o}] \|_2 \leqslant \| [D(g - \eta)] \|_p \, \| [\varphi^{o}] \|_q$$

with $q^{-1} = \frac{1}{2} - p^{-1}$, condition (D) and our *decay lemma* the result follows.

<u>Note</u>: From a Newtonian point of view (where one writes $g_{oo} = 1 + A/r^{x}$ etc.), condition (D) allows gravitational potentials like $r^{-1-\varepsilon}$ (set $p = 3 - \varepsilon$) which are constant in time; or potentials like $1/r$ (set $p = 3$) which decay in time like $|t|^{-\varepsilon}$. Thus in analogy with the well-known situation for Schrödinger operators (see [2] vol. III), this simple approach to scattering just misses the stationary Coulomb (= Schwarzschild) case.

§3b. A Geometrical Approach to Proving Dispersion

As explained in §2d., the other half of the scattering problem - proving various versions of asymptotic completeness etc. - may be approached by studying the inverse wave operators $\tilde{\Omega}^{\pm}$. If one follows through the argument in §3a. replacing $\Omega^{\pm}$ by $\tilde{\Omega}^{\pm}$; one finds that everything would go through if one had a suitable replacement for the decay lemma (say for certain p) for solutions to the interacting equation (and certain of their derivatives). More generally, one would like to answer the

Question: What remains true about the decay of (massive) waves as one curves the space-time? (And under what conditions on decay of the metric?)

In certain cases, one can exploit energy arguments. For this, see e.g. the treatment in [13] of asymptotic completeness for certain transient cases (- corresponding essentially to $p = \infty$ in condition (D) i.e. to $p = 2$ in the decay lemma). We would like to conclude this talk however, by describing some recent ideas for another approach to the decay problem.

---

* There are also terms like $(-\,{}^{4}g)^{1/2}$, but they do not affect the argument.

We present these ideas in the form of a new proof * of the decay lemma in flat space-time. Note that standard proofs of the decay lemma (for textbook proofs, see [2] vol. III, [35]; see also [36]) are heavily dependent on translational invariance and Fourier-transform methods. Our proof, on the other hand, has a strongly differential-geometric flavour and thus it seems appropriate to reproduce it here. Work is in progress [9] on adapting the ideas of the proof to get information about decay in curved space-times and is briefly discussed at the end.

The idea of the proof is based on two observations:

(1) The corresponding *zero-mass* decay result in n space-time dimensions:

$$\sup_{\underline{x}} |\varphi(t, \underline{x})| < C\, t^{-\frac{n-2}{2}}, \quad |t| > 1, \quad n \geqslant 2$$

for ${}^{n}\Box\, \varphi = 0$; $\varphi(t\ \cdot), \dot{\varphi}(t, \cdot) \in C_0^{\infty}(\mathbb{R}^{n-1})$ may be proved by exploiting conformal invariance and Penrose's well-known conformal compactification of flat space-time (see [3] or [37]. See also Geroch [38] who uses related methods to prove related results on fall-off of massless waves).

(2) A solution of the massive Klein Gordon equation $({}^{n}\Box + m^2)\,\varphi = 0$ in n-dimensions may be considered as arising from a special solution of the *massless* Klein Gordon equation in (n + 1)-dimensions:
Simply set $\varphi(t, \underline{x}, \zeta) = \tilde{\varphi}(t, \underline{x}) \sin m\zeta$. Then we have

$${}^{(n+1)}\Box\, \varphi = 0 \iff ({}^{n}\Box + m^2)\, \tilde{\varphi} = 0$$

Thus "explaining" why massive fields decay like massless fields in one higher dimension i.e. like $t^{-\frac{n-1}{2}}$.

Actually, this remark as it stands is only suggestive since φ - being periodic in ζ - can never have compact support (or even decay at spacelike infinity). Nevertheless, we shall see that by using the more detailed information we actually obtain on decay in the massless case, and reformulating (2) in a slightly more roundabout way, it turns out to be essentially justified.

The proof is slightly tedious. But note that the only deep result used is the Leray theorem on the Cauchy problem - as quoted in §2a.! The rest is straightforward calculation. Specializing now (for simplicity) to the case of interest of (massive fields in) four and (massless fields in) five dimensions, we divide the proof into two

---

* At the time of the talk, only partial results had been obtained. The proof we give here was then still partially in conjecture form.

parts corresponding roughly to the above two points:

A "Generalizable" Proof of (Part (a) of) the Decay Lemma

Part I: The well-known Penrose construction identifies the whole of (5-dimensional) Minkowski space (i.e. $\mathbb{R}^5$ with metric $ds^2 = dt^2 - dx^{1\,2} - dx^{2\,2} - dx^{3\,2} - d\zeta^2$) with the region (of compact closure) $\diamond$ given by $-\pi < t' + r' < \pi$, $-\pi < t' - r' < \pi$; $r' > 0$ of the (5-dimensional) Einstein static universe (i.e. $\mathbb{R} \times S^4$ with metric

$$d\bar{s}^2 = dt'^2 - dr'^2 - \sin^2 r'^2 (d\theta^2 + \sin^2\theta \, d\chi^2 + \sin^2\theta \sin^2\chi \, d\varphi^2))$$

- The identification being given by the sequence of coordinate transformations:
(i) Introduce usual angular variables θ, χ, φ and $\xi = (\underline{x}^2 + \zeta^2)^{1/2}$
(giving $ds^2 = dt^2 - d\xi^2 - \xi^2(d\theta^2 + \sin^2\theta \, d\chi^2 + \sin^2\theta \, \sin^2\chi \, d\varphi^2)$)

(ii) $v = t + \xi$ $\qquad w = t - \xi$
(iii) $\tan p = v$ $\qquad \tan q = w$
(iv) $t' = p + q$ $\qquad r' = p - q$

Under this identification, the two metrics are conformally related:

$$ds^2 = \rho^{-2} \, d\bar{s}^2 \qquad (\alpha)$$

where $\rho = \cos p \cos q = \left([1 + (t - \xi)^2][1 + (t + \xi)^2] \right)^{-\frac{1}{2}}$
Note also that the Cauchy surface $t = 0$ in Minkowksi space-time maps into the Cauchy surface $t = 0$ in the Einstein universe. (In fact, on this surface, the above coordinate transformations reduce to the usual stereographic projection of $\mathbb{R}^4$ into S^4). Furthermore, it follows from (α) above by the well-known conformal properties of the Laplace-Beltrami operator [39] (or by direct calculation) that* :

$$\rho^{-\frac{7}{2}} \Box \varphi = (g^{\mu\nu} \nabla_\mu \partial_\nu + \tfrac{3}{16} R) \rho^{-\frac{3}{2}} \varphi$$

where g and R are the metric and curvature scalar for the Einstein universe. It follows (by the Leray theorem) that to any C^∞ solution φ with C_0^∞ Cauchy data on Minkowski space, the function $\rho^{-3/2} \varphi$ (expressed in t', r' coordinates, say) is the restriction to $\diamond$ of a C^∞ solution (arising from the corresponding Cauchy data) on the Einstein universe. One now invokes the fact that a continuous function on a compact set is bounded. (Our compact set is the closure of $\diamond$) to conclude that

---

* in n-dimensions: $\rho^{-\frac{n+2}{2}} \Box \varphi = \left[g^{\mu\nu} \nabla_\mu \partial_\nu + \frac{n-2}{4(n-1)} R \right] \rho^{-\frac{n-2}{2}} \varphi$

$\rho^{-3/2}\varphi$ and all its (t' and r' or p and q) derivatives * are bounded on $\diamond$.

Now, forgetting the Einstein universe and returning to our 5-dimensional Minkowski space, one recovers in particular the

Lemma: Let $\varphi(t, \underline{x}, \zeta)$ *be a* C^∞ *solution of* ${}^5\Box\varphi = 0$ *with compact support on Cauchy surfaces*
Then, defining $\xi = (\underline{x}^2 + \zeta^2)^{1/2}$, *and usual angular coordinates* θ, χ, φ, *(with* $\theta = \tan^{-1}\frac{|\underline{x}|}{\zeta}$ *), there is a constant* C *s.t.*

(i) $$|\varphi(t,\xi,\theta,\chi,\varphi)| \; ; \; \left|\frac{\partial}{\partial\theta}\varphi(t,\xi,\theta,\chi,\varphi)\right| < \frac{C}{[1+(t-\xi)^2]^{3/4}\,[1+(t+\xi)^2]^{3/4}}$$

(ii) $$\left|\frac{\partial\varphi}{\partial t}(t, \underline{x}, \zeta)\right| < \frac{C}{[1+(t-\xi)^2]^{5/4}\,[1+(t+\xi)^2]^{3/4}} + \frac{C}{[1+(t-\xi)^2]^{3/4}\,[1+(t+\xi)^2]^{5/4}}$$

(i) is just $|\rho^{-3/2}\varphi| < C$ and $\left|\frac{\partial}{\partial\theta}\rho^{-3/2}\varphi\right| < C$ rewritten in t, ξ coordinates.
(ii) follows from the boundedness of $\frac{\partial}{\partial q}(\rho^{-3/2}\varphi)$ and $\frac{\partial}{\partial p}(\rho^{-3/2}\varphi)$ after a little calculation using

$$\frac{\partial}{\partial\xi}\varphi = \left(\frac{\partial}{\partial\xi}\rho^{3/2}\right)\left(\rho^{-3/2}\varphi\right) + \rho^{3/2}\frac{\partial}{\partial\xi}(\rho^{-3/2}\varphi)$$

and

$$\frac{\partial}{\partial\xi} = \cos^2 p\,\frac{\partial}{\partial p} - \cos^2 q\,\frac{\partial}{\partial q}$$

<u>Part II</u>:

Now let $\tilde{\varphi}(t, \underline{x})$ be any C^∞ solution of $({}^4\Box + m^2)\varphi = 0$ having C_0^∞ Cauchy data $(\tilde{\varphi}(0, \underline{x}), \dot{\tilde{\varphi}}(0, \underline{x}))$. We now proceed to give a representation for $\tilde{\varphi}$ in terms of some corresponding (C^∞ with compact support on Cauchy surfaces) solution of ${}^5\Box\varphi = 0$:

Choose any (say odd) C_0^∞ function $f(\zeta)$ on $\mathbb{R}$ which satisfies:

$$\int_{-\infty}^{\infty} f(\zeta)\sin m\zeta \; d\zeta = 1$$

(There are clearly many such functions!) Then, construct the solution $\varphi(t, \underline{x}, \zeta)$

---

* We shall need the function and its first derivative only.

to $^5\Box\varphi = 0$ having Cauchy data $(\tilde{\varphi}(0, \underline{x})\, f(\zeta),\ \dot{\tilde{\varphi}}(0, \underline{x})\, f(\zeta))$. We then have:

$$\tilde{\varphi}(t, \underline{x}) = \int_{-\infty}^{\infty} \varphi(t, \underline{x}, \zeta)\sin m\zeta\, d\zeta$$

(*Reason:* It satisfies the same equation, and has the same Cauchy data!) We can now estimate this integral by applying the Lemma of Part I to φ: First, we change to $t, \xi, \theta, \chi, \varphi$ variables:

$$\tilde{\varphi}(t, \underline{x}) = 2\int_{|\underline{x}|}^{\infty} \varphi(t, \xi, \sin^{-1}\frac{|\underline{x}|}{\xi}, \chi, \varphi)\ \frac{\sin(m(\xi^2 - \underline{x}^2)^{1/2})}{(\xi^2 - \underline{x}^2)^{1/2}}\ \xi\, d\xi$$

Integrating by parts, this is:

$$\frac{2}{m}\varphi(t, |\underline{x}|, 0, \chi, \varphi) + \frac{2}{m}\int_{|\underline{x}|}^{\infty}\Big[\frac{\partial}{\partial\xi}(\sin^{-1}\frac{|\underline{x}|}{\xi})\frac{\partial\varphi}{\partial\theta}(t, \xi, \sin^{-1}\frac{|\underline{x}|}{\xi}, \chi, \varphi)$$

$$+ \frac{\partial\varphi}{\partial\xi}(t, \xi, \sin^{-1}\frac{|\underline{x}|}{\xi}, \chi, \varphi)\Big]\cos m(\xi^2 - \underline{x}^2)^{1/2}\, d\xi$$

$$\leqslant \frac{2}{m}\left|\varphi(t, |\underline{x}|, 0, \chi, \varphi)\right| + \frac{\pi C}{m}\sup_{\xi}\left|\frac{\partial\varphi}{\partial\theta}\right| + \frac{2}{m}\int_0^{\infty}\left|\frac{\partial\varphi}{\partial\xi}\right| d\xi$$

One now majorizes the first two terms using part (i) of the Lemma; and the integral by part (ii).

Then, taking the case $t > 0$ ($t < 0$ is similar) and noting $\xi + t \geqslant t$, we have

$$|\tilde{\varphi}(t, \underline{x})| \leqslant \frac{2 + \pi}{m}\ \frac{C}{(1 + t^2)^{3/4}}$$

$$+ \frac{C}{(1 + t^2)^{3/4}}\int_0^{\infty}\frac{d\xi}{(1 + (t - \xi)^2)^{3/4}}$$

$$+ \frac{C}{(1 + t^2)^{3/4}}\int_0^{\infty}\frac{d\xi}{(1 + (t - \xi)^2)^{5/4}}$$

Finally, bounding the integrals by

$$\int_{-\infty}^{\infty}\frac{d\xi}{(1 + \xi^2)^{3/4}} \qquad \int_{-\infty}^{\infty}\frac{d\xi}{(1 + \xi^2)^{5/4}}$$

we obtain the desired result - in the form

$$\sup_{\underline{x}}|\tilde{\varphi}(t, \underline{x})| \leqslant \frac{C'}{(1 + t^2)^{3/4}}$$

Note: In generalizing such ideas to curved space-times, Part 2 presents little difficulty: $({}^4g^{\mu\nu}\nabla_\mu\partial_\nu + m^2)\tilde{\varphi} = 0$ in 4 dimensions may be viewed as arising from $({}^5g^{\mu\nu}\nabla_\mu\partial_\nu)\varphi = 0$ in 5-dimensions where ${}^5g = \begin{pmatrix} g & 0 \\ 0 & -1 \end{pmatrix}$. I.e., one adds an extra flat dimension. Note that $\frac{\partial}{\partial\xi}$ is then a Killing vector, so we retain the desired translational symmetry in the 5th direction. This latter equation is of course not quite conformally invariant in curved space-time. But, it differs (cf. Part I of proof) from a conformally invariant equation by a "potential" term $\frac{3R}{16}$ which now vanishes at infinity at a rate determined by the vanishing of $g - \eta$.

In this way, one can hope to use conformal ideas (as in Part 1) to discuss the decay of massive fields in curved space-times. It is hoped to discuss this and related matters in [9].

ACKNOWLEDGMENTS

It is a pleasure to thank Jonathan Dimock for many stimulating conversations and for collaboration on part of this work.

I thank the British Science Research Council (under contract no. GR/A/61463) and the Schweizerische Nationalfonds for financial support. Thanks also go to S. Adler and the Institute for Advanced Study, Princeton, for support and hospitality.

REFERENCES

[1] Hawking, S.W., Commun. math. Phys. 43 (1975) 199.

[2] Reed, M. and Simon, B., Methods of Modern Mathematical Physics, Vols I-IV. Academic, New York-London. (1972, 75, 79, 78) and further volumes to appear.

[3] Hawking, S.W. and Ellis G.F.R., The large scale structure of space-time. Cambridge University Press, Cambridge 1973.

[4] Chernoff, P.R., J. Funct. Anal. 12 (1973) 401.

[5] Kay, B.S., Commun. math. Phys. 62 (1978) 55.

[6] Kay, B.S., Commun. math. Phys. 71 (1980) 29.

[7] Kay, B.S., J. Math. Phys. 20 (1979) 1712.

[8] Kay, B.S., in Mathematical problems in theoretical physics: Proceedings Lausanne (1979), ed. K. Osterwalder. Springer Lecture notes in physics No. 116. Springer-Verlag, Berlin-Heidelberg-N.Y. (1980).

[9] Kay, B.S., Decay of Massless and Massive Waves in Curved Space-Times (in prep.).

[10] Dimock, J., J. Math. Phys. 20 (1979) 2549.

[11] Dimock, J., Commun. math. Phys. 77 (1980) 219.

[12] Dimock, J., Dirac Quantum Fields on a Manifold (Princeton IAS Preprint 1980) (to appear).

[13] Dimock, J. and Kay, B.S., Classical Wave-Operators and Asymptotic Field Operators on Curved Space-Times (to appear).

[14] Ashtekar, A. and Magnon, A., Proc. Roy. Soc. Lond. A346 (1975) 375.

[15] Hajicek, P. in Differential geometric methods in mathematical physics II. Proceedings Bonn, 1977, ed. K. Bleuler, H.R. Petry, A. Reetz. Springer-Verlag, Berlin-Heidelberg-N.Y. (1978).

[16] Isham, C.J., in Bonn proceedings (as above).

[17] Moreno, C., J. Math. Phys. 18 (1977) 2153; 19 (1978) 92.

[18] Sewell, G.L., Phys. Lett. 79A (1980) 23.

[19] Wald, R.M., Ann. Phys. (N.Y.) 118 (1979) 490.

[20] Wald, R.M., J. Math. Phys. 21 (1980) 2802.

[21] DeWitt, B.S., Phys. Reports (Phys. Lett.) 19C (1975) 297.

[22] Isham, C.J., in 8th Texas symposium on relativistic astrophysics (Ann. N.Y. Acad. Sci. 302 (1977) 114.)

[23] Parker, L., in Asymptotic structure of space-times, eds. F.P. Esposito, L. Witten. Plenum N.Y.-London (1977).

[24] General relativity: an Einstein centenary survey, eds. S.W. Hawking and W. Israel. Cambridge University Press. Cambridge (1979).

[25] General relativity and gravitation, Vols. I, II, ed. A. Held. Plenum, N.Y.-London (1980).

[26] Recent developments in gravitation. Cargèse 1978, eds. M. Lévy and S. Deser. Plenum, N.Y.-London (1979).

[27] Segal, I.E., in Cargèse lectures on theoretical physics: Application of Mathematics to Problems in Theoretical Physics, ed. F. Lurçat. Gordon and Breach, N.Y. (1967).

[28] Wightman, A.S., in Invariant wave equations. Proceedings Erice 1977. Springer Lecture notes in physics No. 73. Springer-Verlag Berlin-Heidelberg-N.Y. (1978).

[29] Avis, S.J. and Isham, C.J., in Cargèse (1978), ([26] above).

[30] Haag, R. and Kastler, D., J. Math. Phys. 5 (1964) 848.

[31] Penrose, R., in [24] (above).

[32] Bongaarts, P., Ann. Phys. (N.Y.) 56 (1970) 108.

[33] Dimock, J., J. Math. Phys. 20 (1979) 1791.

[34] Shale, D., Trans. Am. Math. Soc. 103 (1962) 149.

[35] Bogolubov, N.N., Logunov, A.A. and Todorov, I.T., Introduction to axiomatic quantum field theory. Benjamin, Reading, Mass. (1975).

[36] Segal, I.E., in Conference on math. theory of elementary particles. Proceedings Dedham, Mass. 1965, eds. R. Goodman, I.E. Segal. M.I.T. Press, Cambridge, Mass. (1966).

[37] Penrose, R., Proc. Roy. Soc. Lond. A284 (1965) 159.

[38] Geroch, R., in Asymptotic structure of space-times, eds. F.P. Esposito, L. Witten. Plenum, N.Y.-London (1977).

[39] Friedlander, F.G., The wave equation on a curved space-time. Cambridge University Press, Cambridge (1975).

THE QUANTIZED MAXWELL FIELD AND ITS GAUGES; A GENERALIZATION OF WIGHTMAN THEORY.

P.J.M. Bongaarts

Instituut Lorentz, Nieuwsteeg 18
2311 SB Leiden, The Netherlands.

Abstract

A mathematically rigorous approach to the quantized Maxwell field is discussed, supporting the point of view that an extension of the standard Wightman formalism for quantum field theory is desirable and indicating a possible direction for such a generalization.

1. Wightman's axiomatic formulation of quantum field theory was developed in the fifties and early sixties; it gave a framework in which basic concepts of field theory as it then existed could be discussed in a non-heuristic, mathematically satisfactory way.

From the beginning the quantized electromagnetic field was seen as a special case having additional difficulties. As a result one will find that in any of the by now standard text-books on axiomatic field theory the Maxwell field is excluded, usually right on the first pages. See e.g. [1,2,3]. This is understandable as the mathematical problems of interacting scalar fields are already formidable enough, nevertheless it is a strange state of affairs in view of the fact that in elementary particle physics quantum electrodynamics was then, as it is now, by far the most successful field theory and the only one of which the physical validity could not be doubted. The situation has become even stranger due to the rise of non-abelian gauge theories in recent years. It seems no longer possible to see the Maxwell field as an isolated exceptional case; on the contrary, its typical features appear in more complicated form in the new field theories that at present dominate particle physics.

It is therefore important to obtain a better understanding of the quantized electromagnetic field and subsequently of quantum electrodynamics, not only for its own sake but also as a first step towards the study of the mathematical basis of general non-abelian gauge quantum fields.

The situation in field theory now is in some ways similar to that of the fifties. There has been a rapid expansion and again one is faced with a large body of heuristic ideas and methods, part of which does not seem to fit naturally in the framework given by the standard Wightman axioms. This suggests that a rethinking of the formulation of the foundations of quantum field theory has become desirable.

The discussion of the Maxwell field in the next sections will on one hand support the view that Wightman theory as it stands is in need of extension and will on the other hand give a first indication of a possible direction for such a modification.

2. The special problems of the quantized Maxwell field are connected with the following two well-known phenomena, both of which appear again, although in a more complicated form in non-abelian gauge theories.

a. The theory contains two field operators, the tensor field $F_{\mu\nu}(x)$ and the vector field $A_\mu(x)$. In classical electromagnetism $F_{\mu\nu}(x)$ is the physical field consisting of electric and magnetic field strengths; the potential $A_\mu(x)$ may be introduced purely as an auxiliary quantity. In the quantized theory the role of A_μ is an essential one; its use cannot really be avoided. Moreover the meaning of the relation $F_{\mu\nu} = \partial_\mu A_\nu - \partial_\nu A_\mu$ and of the non-uniqueness of A_μ is different and more complex.
b. There seems to be a general incompatibility between manifest Lorentz covariance and the use of a Hilbert space (i.e. with a positive definite inner product) as the space of quantum states. Since both these properties are required in the Wightman scheme, the absence of one of these places the quantized Maxwell field firmly outside standard Wightman theory.

A practical way of dealing with this situation is to do everything in a single non-covariant but positive-metric gauge such as the Coulomb gauge. For explicit calculations this is certainly sufficient as is demonstrated by text-books on quantum electrodynamics such as [4]. However for a general axiomatic formulation this seems to me a point of view that is too restricted.

An approach that has a wider theoretical scope is the one that accepts the indefinite metric state space of the covariant formulation but employs along with the "natural" Lorentz-invariant but indefinite inner product a second positive but non-invariant one by which one tries to keep the situation within the reach of familiar Hilbert space methods. This is the original approach of Gupta and Bleuler. It has more recently been developed further by Strocchi and others. See e.g. [5,6]. The formalism has awkward mathematical properties. As an example one may note that the operators that represent the Lorentz group in the Hilbert space of states are non-unitary and even unbounded as can be verified easily in the free field case. It should also be realised that an important property of normal Wightman theory is lost; the Wightman functions no longer determine the complete theory in a unique way. This is because the n-point functions give only the Lorentz invariant inner product. The second positive definite inner product has to be chosen separately. Different choices can be made, which lead to different associated Hilbert space structures and possibly to different physical theories. See [7].

I shall discuss an alternative approach, which also accepts the possibility of an indefinite metric as an unavoidable aspect of the quantized Maxwell field but which as a consequence abandons Hilbert space as the general state space in quantum field theory in favour of more general topological inner product spaces that appear when the Wightman formalism is generalized in a fairly natural way.

In a standard Wightman field theory there are two distinct mathematical structures. There is a locally convex space on which the field operators are defined; its topology comes from the distribution properties of the Wightman functions. Superimposed on this one has a weaker topology defined by the positive definite inner product. Completion under this topology gives the Hilbert space of states. A large part of Wightman theory, in particular the reconstruction theorem depends on the first structure. The Hilbert space structure is less important in this respect but has to do with the physical interpretation of the theory.

With this in mind one is led quite naturally to a generalization of Wightman theory which is explicitly based on the first structure, i.e. on the mathematical structure connected with the distribution properties of the n-point functions. Positivity of inner products can then be dropped as a general requirement and will appear only at places where this is appropriate for physical reasons. A quantum field is now described by a sequence of distributions, the n-point functions, with certain characteristic properties, not necessarily including full positivity. By using the reconstruction theorem in a slightly modified form such a theory can be represented as an operator theory in a topological vector space with a natural inner product, i.e. a non-degenerate, not necessarily positive definite hermitian form. The locally convex spaces that occur have been studied extensively and have attractive properties (e.g. nuclearity). It would of course be premature to claim that this will be sufficient for rigorously constructing general gauge theories or even full quantum electrodynamics, however it is encouraging that it provides a natural and effective language for the complexities of the quantized Maxwell field.

In the next sections it will be shown how by using this language the relation between A_μ and $F_{\mu\nu}$ and the occurrence of different gauges can be understood and how for the free field a complete and rigorous treatment can be given in which the known gauges, covariant and non-covariant, emerge as consequences of a few basic and simple properties.

A transparent way of handling sets on n-point functions is to see such sets as continuous linear functionals on a topological tensor algebra over a basic testfunction space. This picture due to Borchers, see [8,9], will not be used in this brief discussion. I shall also suppress testfunctions and use non-smeared "generalized function" language for distributions. For the proper formulations, precise definitions, proofs and in general for a more detailed and systematic exposition, see [10,11].

Finally one will notice that part of what follows could have been formulated conveniently in terms of differential forms. Further developments in the direction of non-abelian gauge fields will probably benefit greatly by a more explicit use of such differential geometric concepts.

3. The analysis of the relation between the A_μ and $F_{\mu\nu}$ fields can be carried out in the following steps.

a) Suppose a field theory for $F_{\mu\nu}(x)$ is given in the sense of ordinary Wightman theory, i.e. there are field operators $F_{\mu\nu}(x)$ in a Hilbert space $\mathcal{H}$ with vacuum vector Ω^F, defined on a suitably chosen dense subspace $\mathcal{H}^F$. In addition to the usual Wightman axioms we suppose that $F_{\mu\nu}$ satisfies

$$\partial_\mu F_{\nu\rho} + \partial_\nu F_{\rho\mu} + \partial_\rho F_{\mu\nu} = 0 \tag{1}$$

as an *operator equation.*

At this point one might try to find the possible operators $A_\mu(x)$ in $\mathcal{H}$ such that

$$F_{\mu\nu} = \partial_\mu A_\nu - \partial_\nu A_\mu \tag{2}$$

This would be a problem with very unpleasant technical aspects because the operators A_μ and $F_{\mu\nu}$ must be expected to be unbounded (even after smearing with testfunctions). No satisfactory general solution could be expected; in any case the best known covariant realisations of A_μ would not be found in this way. In the spirit of our approach we shall follow a different route which does lead to a simple and complete solution.

b) The theory as given in a) has n-point functions

$$\omega^F_{\mu_1\nu_1\ldots\mu_n\nu_n}(x_1\ldots x_n) = (\Omega^F, F_{\mu_1\nu_1}(x_1)\ldots F_{\mu_n\nu_n}(x_n)\Omega^F) \tag{3}$$

The properties of the theory are reflected in properties of these distributions. There is Lorentz invariance and because $\mathcal{H}$ is a Hilbert space a positivity property. The important fact here is however that the $\omega^F_{\mu_1\nu_1\ldots\mu_n\nu_n}(x_1\ldots x_n)$ satisfy equation (1), in the sense of distributions, in *each variable* x_j *separately*.

Classically equation (1) for a function $F_{\mu\nu}(x)$ is equivalent to the existence of a function $A_\mu(x)$ such that (2) holds. This remains true for distributions, furthermore one has the following generalization:

Theorem 1. Tempered distribution $\omega^F_{\mu_1\nu_1\ldots\mu_n\nu_n}(x_1\ldots x_n)$ satisfy equation (1) in every variable x_j separately if and only if there exist tempered distributions $\omega^A_{\mu_1\nu_1\ldots\mu_n\nu_n}(x_1\ldots x_n)$ such that

$$\omega^F_{\mu_1\nu_1\ldots\mu_n\nu_n} = \sum_{\substack{\sigma_j\in S_2\\ j=1,\ldots n}} \varepsilon(\sigma_1)\ldots\varepsilon(\sigma_n)\, \partial_{\sigma_1(\mu_1)}\cdots\partial_{\sigma_n(\mu_n)}\omega^A_{\sigma_1(\nu_1)\ldots\sigma_n(\nu_n)} \tag{4}$$

in which for $j=1,\ldots n$, σ_j is one of the two possible permutations of the pair of indices (μ_j, ν_j) ; $\varepsilon(\sigma_j)$ is the sign of the permutation σ_j and $\partial_{\sigma_j(\mu_j)}$ means $\frac{\partial}{\partial x_j^{\sigma_j(\mu_j)}}$.

The proof of this uses tensor product and duality properties of continuous open linear mappings between spaces of testfunctions. It can be found in [10] .

Because of this theorem we know that there exist sets of distributions $\omega^A_{\mu_1\ldots\mu_n}(x_1,\ldots,x_n)$ from which the n-point functions of the given $F_{\mu\nu}$ theory can be obtained using formula (4). We interpret the $\omega^A_{\mu_1\ldots\mu_n}$ as n-point functions for an A_μ field theory. It is a Wightman theory in the generalized sense. Theorem (1) does not allow us to conclude that positivity and Lorentz invariance of the $\omega^F_{\mu_1\nu_1\ldots\mu_n\nu_n}$ implies that the $\omega^A_{\mu_1\ldots\mu_n}$ also possess these properties.

c) The reconstruction theorem in a slightly modified form permits us to construct from the $\omega^A_{\mu_1\ldots\mu_n}$ an A_μ operator field theory, i.e. field operators $A_\mu(x)$ in a locally convex topological vector space $\mathcal{H}^A$, which has a non-degenerate hermitian form as inner product and a unit vector Ω^A as vacuum vector. A set ω^A of n-point functions $\omega^A_{\mu_1\cdots\mu_n}$ or equivalently the corresponding A_μ operator field theory will be called a *gauge* for the original $F_{\mu\nu}$ theory.

d) Just as in the classical situation there are different gauges for the same $F_{\mu\nu}$ theory. Only in special cases two such gauges are connected by $A'_\mu = A_\mu + \partial_\mu \Phi$ as an operator relation. In general there is only a more complicated relation between the n-point functions.

Theorem 2. ω^A and ω'^A are gauges for the same $F_{\mu\nu}$ theory if and only if for each n=1,2,... there exist tempered distributions $\Phi^{(j)}_{\mu_1\cdots\mu_{n-1}}(x_1,\ldots x_n)$, j=1,2,...n, such that

$$\omega^A_{\mu_1\cdots\mu_n} - \omega'^A_{\mu_1\cdots\mu_n} = \sum_{j=1}^{n} \partial_{\mu_j} \Phi^{(j)}_{\mu_1\cdots\mu_{j-1}\mu_{j+1}\cdots\mu_n} \tag{5}$$

The proof is similar in spirit to that of theorem (1) and can be found in ref. [11].

Theorem (1) and (2) together determine the mathematical formalism for the Maxwell quantum field of which the outlines are given in the present discussion and of which it is hoped that it may also have some significance for more general gauge theories. Its characteristic feature is the way it contains with each $F_{\mu\nu}$ theory all the associated A_μ theories as gauges. The operator $F_{\mu\nu}(x)$ and the different $A_\mu(x)$ field operators act in spaces that are a priori distinct. That for the $F_{\mu\nu}$ field is a Hilbert space, those for the A_μ fields more generally locally convex inner product spaces. The gauges are connected with the $F_{\mu\nu}$ theory and with each other through relations between the n-point functions, as given by formulae (4) and (5). There are additional relations connecting the spaces $\mathcal{H}^A$ with $\mathcal{H}$. Each set $\omega^A_{\mu_1\cdots\mu_n}$ inherits from the positivity of the $\omega^F_{\mu_1\nu_1\cdots\mu_n\nu_n}$ a *partial* positivity property which determines a subspace $\mathcal{H}^A_{ph}$ of $\mathcal{H}^A$ on which the inner product is non-negative definite together with a linear map W^{FA} from $\mathcal{H}^A_{ph}$ onto $\mathcal{H}^F$ which is isometric with respect to the inner products on $\mathcal{H}^A$ and $\mathcal{H}$ and with consistency properties such as $W^{FA}\Omega^A = \Omega^F$.

4. All the different gauges for a fixed $F_{\mu\nu}$ theory are physically completely equivalent, choosing a gauge is a matter of mathematical convenience. The variety of gauges allowed by theorem (2) is very large but in practice only a small number characterized by simple general properties are used.

A convenient property for gauges is Lorentz covariance. If the distributions $\omega^A_{\mu_1\ldots\mu_n}$ are Lorentz invariant the corresponding A_μ field theory is Lorentz covariant, i.e. there exist in $\mathcal{H}^A$ a representation of the Poincaré group by operators $U(a,\Lambda)$, such that $U(a,\Lambda)A_\mu(x)U(a,\Lambda)^{-1} = \Lambda^{\mu'}_{\ \mu} A_{\mu'}(\Lambda x+a)$ and $U(a,\Lambda)\,\Omega^A = \Omega^A$. The operators $U(a,\Lambda)$ are isometric and have moreover strong continuity properties as can be seen from the details of the reconstruction theorem.

It is in general impossible to combine Lorentz invariance with a second obvious requirement, (full) positivity of the $\omega^A_{\mu_1\ldots\mu_n}$. For a positive gauge there occurs a considerable simplification of the relation between the A_μ and $F_{\mu\nu}$ fields. $\mathcal{H}^A$ is then a pre-Hilbert space. The map W^{FA} becomes injective, its inverse injects $\mathcal{H}^F$ onto $\mathcal{H}^A_{ph}$. This can be used to identify $\mathcal{H}^F$ with $\mathcal{H}^A_{ph}$. The completion $\mathcal{H}$ of $\mathcal{H}^F$ is a subspace of the Hilbert space completion of $\mathcal{H}^A$ and may even be equal to it. In that case we are back in the situation of a single state space $\mathcal{H}$, a Hilbert space in which both operators $A_\mu(x)$ and $F_{\mu\nu}(x)$ act. One also has now the relation $F_{\mu\nu} = \partial_\mu A_\nu - \partial_\nu A_\mu$ as an operator relation.

In classical electromagnetism the current J_μ is defined as $J_\mu = \partial^\nu F_{\mu\nu} = \Box A_\mu - \partial_\mu(\partial^\nu A_\nu)$ and is clearly a gauge invariant quantity. The Lorentz gauge condition $\partial^\mu A_\mu = 0$ for A reduces the relation between J_μ and A_μ to the equation $J_\mu = \Box A_\mu$.

For the quantized field the situation is more complicated. The current operator $J_\mu = \partial^\nu F_{\mu\nu}$ is defined as an operator in $\mathcal{H}$ and is therefore by definition gauge independent but it is in general not equal to $\Box A_\mu - \partial_\mu(\partial^\nu A_\nu)$ which is an operator in $\mathcal{H}^A$. Instead of this one has a relation between the $\omega^A_{\mu_1\ldots\mu_n}$ and the vacuum expectation values $\omega^J_{\mu_1\ldots\mu_n}(x_1,\ldots x_n)$ in $\mathcal{H}$ of products of the current operator $J_\mu(x)$

$$\omega^J_{\mu_1\ldots\mu_n} = \partial^{\nu_1}\ldots\,\partial^{\nu_n}\,\omega^F_{\nu_1\mu_1\cdot\,\nu_n\mu_n}$$

$$= \partial^{\nu_1}\ldots\,\partial^{\nu_n}\sum_{\substack{\sigma_j\in S_2\\ j=1,\ldots n}} \varepsilon(\sigma_1)\ldots\varepsilon(\sigma_n)\,\partial_{\sigma_1(\nu_1)}\cdots\partial_{\sigma_n(\nu_n)}\,\omega^A_{\sigma_1(\mu_1)\ldots\,\sigma_n(\mu_n)} \tag{6}$$

with $\partial^{\nu_j} = g^{\nu_j\alpha}\dfrac{\partial}{\partial x_j^\alpha}$ and the usual summation convention.

The Lorentz condition $\partial^\mu A_\mu = 0$ is a natural and convenient gauge condition when used as an operator identity. It is then equivalent to the equation

$$\partial^{\mu_j} \omega^A_{\mu_1 \dots \mu_n} = 0 \tag{7}$$

for every x_j separately ($j=1,\dots n$). It simplifies relation (6) to

$$\omega^J_{\mu_1 \dots \mu_n} = \Box_1 \dots \Box_n \, \omega^A_{\mu_1 \dots \mu_n} \tag{8}$$

with $\Box_j = g^{\alpha\beta} \frac{\partial}{\partial x_j^\alpha} \frac{\partial}{\partial x_j^\beta}$. However the effect of this is not quite the same as in the classical situation. For instance for the free field where $\omega^j_{\mu_1 \dots \mu_n} = 0$, it does not imply that the wave equation $\Box A_\mu = 0$ holds for the field operator A_μ. In fact this last condition gives rise, as will be seen in the next section, to a different important set of Lorentz invariant free field gauges.

5. The discussion in the preceding sections can be made more explicit by considering the free Maxwell field. One starts from the unique free $F_{\mu\nu}$ field theory which is known. It satisfies the ordinary Wightman axioms, equation (1) and in addition to this the free field equation $\partial^\mu F_{\mu\nu} = 0$, again as an operator equation for $F_{\mu\nu}(x)$. Its n-point functions $\omega^F_{\mu_1\nu_1 \dots \mu_n\nu_n}$ are in the usual free-field manner made up from a 2-point function which is

$$\omega^F_{\mu_1\nu_1\mu_2\nu_2}(x_1,x_2) = -i \Big\{ g_{\mu_1\mu_2} \overset{1}{\partial}_{\nu_1} \overset{2}{\partial}_{\nu_2} - g_{\mu_1\nu_2} \overset{1}{\partial}_{\nu_1} \overset{2}{\partial}_{\mu_2} + $$

$$- g_{\nu_1\mu_2} \overset{1}{\partial}_{\mu_1} \overset{2}{\partial}_{\nu_2} + g_{\nu_1\nu_2} \overset{1}{\partial}_{\mu_1} \overset{2}{\partial}_{\mu_2} \Big\} D^{(+)}(x_1 - x_2) \tag{9}$$

with $D^{(+)}(x_1-x_2) = \Delta^{(+)}(x_1-x_2, m=0) = \frac{-i}{(2\pi)^3} \int \frac{d\vec{k}}{2|\vec{k}|} e^{-ik(x_1-x_2)}$

$$= -\frac{i}{(2\pi)^3} \int \delta_+(k^2) e^{-ik(x_1-x_2)} d^4k \quad \text{and}$$

$$\overset{j}{\partial}_{\mu_j} = \frac{\partial}{\partial x_j^{\mu_j}}, \text{ etc.}$$

All the possible realizations of the free photon field in terms of A_μ appear as gauges ω^A or sets of n-point functions $\omega^A_{\mu_1 \dots \mu_n}$ associated with these given $\omega^F_{\mu_1\nu_1 \dots \mu_n\nu_n}$

by means of (4). If one has found one particular gauge all the others may be obtained by adding arbitrary generalized gradient functions in the sense of theorem (2) to the n-point functions of the known gauge. The simplest gauge to serve this purpose is the set ω^A having the free-field form and based on the 2-point function

$$\omega^A_{\mu_1\mu_2}(x_1,x_2) = -ig_{\mu_1\mu_2} D^{(+)}(x_1-x_2).$$

The full collection of gauges is too large for practical purposes and therefore we restrict it immediately by two reasonable requirements. The first is invariance with respect to space-time translations, the second is that the n-point functions have what we shall call Gaussian form, i.e. are again just like the $\omega^F_{\mu_1\nu_1 \ldots \mu_n\nu_n}$ based on a 2-point function in the manner which is customary for free (boson) fields. This reduces the discussion of gauges to a discussion of translation invariant 2-point functions. In ref. [11] one may find a proof of the following theorem:

<u>Theorem 3.</u> The translation invariant 2-point functions for the free $A_\mu(x)$ field have the form

$$\omega^A_{\mu_1\mu_2}(x_1,x_2) = -ig_{\mu_1\mu_2} D^{(+)}(x_1-x_2) + (\partial_{\mu_1}\Phi_{\mu_2})(x_1-x_2)$$

$$+ \overline{(\partial_{\mu_2}\Phi_{\mu_1})(x_2-x_1)} + C_{\mu_1\mu_2\varrho}(x_1^\varrho - x_2^\varrho) \qquad (10)$$

with $\Phi_\mu(y)$ arbitrary tempered distributions, the bar meaning complex conjugation, and $C_{\mu_1\mu_2\varrho}$ arbitrary real constants, totally antisymmetric in the indices $\mu_1\mu_2\varrho$

The operator field theories that we consider in this way as gauges for the free photon field have essentially the same simple mathematical structure, due to the restriction to sets $\omega^A_{\mu_1\ldots\mu_n}$ with "Gaussian" form. Reconstruction from such $\omega^A_{\mu_1\ldots\mu_n}$ gives field operators $A_\mu(x)$ that are linear expressions in creation and annihilation operators generating the state space $\mathcal{H}^A$ from the vacuum vector Ω^A. As a "many-particle space" $\mathcal{H}^A$ can be thought of as constructed from a "one particle space" $\mathcal{H}^{(1)A}$ which is the quotient space of the space of testfunctions $f^\mu(x)$ used for smearing the field $A_\mu(x)$, obtained by eliminating the degeneracy of the hermitian form

$$(f,g)^{(1)} = \int \omega^A_{\mu_1\mu_2}(x_1,x_2)\, \overline{f^{\mu_1}(x_1)}\, g^{\mu_2}(x_2)\, d^4x_1\, d^4x_2 \qquad (11)$$

If $(\cdot,\cdot)^{(1)}$ is non-negative definite $\mathcal{H}^{(1)A}$ and therefore also $\mathcal{H}^A$ are pre-Hilbert spaces. Completion gives a Fock space structure in the standard Hilbert space sense. For more details on the general locally convex version of Fock space see [11].

Information on aspects of the operator field theory such as possible Lorentz covariance, definiteness or indefiniteness of the metric, special subspaces, etc. can be read off from the properties of the "one-particle subspace" $\mathcal{H}^{(1)A}$ which in turn depend only on the 2-point function $\omega^A_{\mu_1\mu_2}(x_1,x_2)$. The remaining part of this section is therefore devoted to a discussion of typical 2-point functions, with brief indications of the consequences of their properties for the corresponding $A_\mu(x)$ operator field theories. Vectors in $\mathcal{H}^{(1)A}$ are obtained as equivalence classes. For practical purposes these are represented by suitably chosen multi-component momentum "wave functions". This may introduce complications that tend to obscure the underlying simple general structure.

Because of translation invariance it is convenient to use fourier transformed 2-point functions defined by

$$\omega^A_{\mu_1\mu_2}(x_1,x_2) = \frac{1}{(2\pi)^4}\int \hat{\omega}^A_{\mu_1\mu_2}(k)e^{-ik(x_1-x_2)}d^4k$$

$$f^\mu(x) = \frac{1}{(2\pi)^2}\int \hat{f}^\mu(k)e^{-ikx}d^4k \tag{12}$$

The result of theorem (3) then becomes

$$\hat{\omega}^A_{\mu_1\mu_2}(k) = -2\pi g_{\mu_1\mu_2}\delta_+(k^2) - 2\pi(k_{\mu_1}\hat{\Phi}_{\mu_2}(k) + k_{\mu_2}\overline{\hat{\Phi}_{\mu_1}}(k))$$

$$+ iC'_{\mu_1\mu_2\varrho}\,\partial^\varrho\delta(k) \tag{13}$$

The hermitian form $(\cdot,\cdot)^{(1)}$ is now

$$(f,g)^{(1)} = \int \hat{\omega}^A_{\mu_1\mu_2}(k)\,\overline{f^{\mu_1}}(k)\,g^{\mu_2}(k)\,d^4k \tag{14}$$

The Lorentz covariant gauges are obtained by imposing Lorentz invariance on the $\hat{\omega}^A_{\mu_1\mu_2}(k)$ in (13).

<u>Theorem 4.</u> The Lorentz invariant 2-point functions for the free $A_\mu(x)$ field are given by expressions of the form

$$\hat{\omega}^A_{\mu_1\mu_2}(k) = -2\pi(g_{\mu_1\mu_2}\delta_+(k^2) + k_{\mu_1}k_{\mu_2}\Phi(k)) \tag{15}$$

with $\Phi(k)$ an arbitrary real tempered Lorentz invariant distribution.

If one adds to Lorentz invariance the *Lorentz gauge* condition from section 4, $\partial^{\mu}A_{\mu}(x) = o$ as an operator identity for the field operator $A_{\mu}(x)$, which is for a translation invariant "Gaussian" theory equivalent to $k^{\mu_1}\hat{\omega}^A_{\mu_1\mu_2}(k) = o$, then one obtains a well-known class of gauges which might be called *generalized Landau gauges* and which are conveniently given by

$$\hat{\omega}^A_{\mu_1\mu_2}(k) = -2\pi(g_{\mu_1\mu_2} + \lambda k_{\mu_1\mu_2} + \tfrac{1}{4}(k_{\mu_1}\partial_{\mu_2} + k_{\mu_2}\partial_{\mu_1}))\,\delta_+(k^2) \tag{16}$$

with λ an arbitrary real number and ∂_μ here $\partial_\mu = \frac{\partial}{\partial k^\mu}$, etc. For $\lambda = o$ one has the standard Landau gauge. Because $k^2(\frac{1}{4}(k_{\mu_1}\partial_{\mu_2} + k_{\mu_2}\partial_{\mu_1})\delta_+(k^2)) = -k_{\mu_1}k_{\mu_2}\delta_+(k^2)$ it is often given by the rather ambiguous expression

$$\hat{\omega}^A_{\mu_1\mu_2}(k) = -2\pi(g_{\mu_1\mu_2} - \frac{k_{\mu_1}k_{\mu_2}}{k^2})\,\delta_+(k^2) \tag{17}$$

The form $(\cdot,\cdot)^{(1)}$ is indefinite, for all real λ, so the Landau gauges have indefinite metric state spaces $\mathcal{H}^A$. These spaces when realized in terms of momentum "wave functions" look quite complicated, even for $\lambda = o$. See e.g. [12,13].

A second natural gauge condition for the free field is $\Box A_\mu = o$, as an equation for the field operator $A_\mu(x)$. It is equivalent to $k^2\,\hat{\omega}^A_{\mu_1\mu_2}(k) = o$, again for translation invariant "Gaussian" theories. It does not hold for the Landau gauges but defines another important and well-known class of Lorentz covariant gauges, which may be called the *generalized Gupta-Bleuler gauges*. Imposing $k^2\,\hat{\omega}^A_{\mu_1\mu_2} = o$ on (15) gives $k^2 k_{\mu_1}k_{\mu_2}\,\Phi(k) = o$ with general solution $\Phi(k) = b_+\delta_+(k^2) + b_-\delta_-(k^2) + b_o\delta(k) + c\,\partial_\mu\,\delta\,(k)$; $b^{\pm}$, b_o, c real constants. Omitting terms that lead to negative frequencies in time translation and irrelevant behaviour at $k=o$ one is left with

$$\hat{\omega}^A_{\mu_1\mu_2}(k) = -2\pi(g_{\mu_1\mu_2} + \lambda k_{\mu_1}k_{\mu_2})\,\delta_+(k^2) \tag{18}$$

in which λ is an arbitrary real number. The case $\lambda = o$ is known as *Feynman gauge*. The corresponding operator field theory is the rigorous form of the standard indefinite metric Gupta-Bleuler formalism. It is the simplest gauge in terms of realization by momentum functions. Because

$$(f,g)^{(1)} = -2\pi \int \overline{\hat{f}^{\mu}(k)}\, \hat{g}_{\mu}(k)\, \delta_{+}(k^2)\, d^4k$$

$$= -2\pi \int \overline{\hat{f}^{\mu}(k)}\, \hat{g}_{\mu}(k)\, \frac{d\vec{k}}{2\,|k^0|} \qquad (19)$$

$$k^2=o, \quad k^o \geq o$$

the vectors in $\mathcal{H}^{(1)A}$ given as equivalence classes can be represented by functions on the forward light cone, that are restrictions of fourier transforms $\hat{f}^{\mu}(k)$ of testfunctions $f^{\mu}(x)$. The state spaces $\mathcal{H}^{A}_{(\lambda)}$ for the Gupta-Bleuler gauges for $\lambda \neq o$ can be identified with that for $\lambda = o$ by defining $\Omega^{A}_{(\lambda)} = \Omega^{A}_{(o)}$ and $A^{(\lambda)}_{\mu}(x) = A^{(o)}_{\mu}(x) - \frac{1}{2}\lambda\, \partial_\mu \partial^\nu A^{(o)}_{\nu}(x)$. We are here in the special situation where different gauges are connected by the classical relation $A'_{\mu}(x) = A_{\mu}(x) + \partial_\mu \phi(x)$ in operator form.

Finally one has as a typical example of a gauge that is not Lorentz covariant but has positive definite metric, the *Coulomb gauge*. It can be obtained from the general formula (13) by requiring rotational invariance together with for instance the condition $\hat{\omega}^{A}_{oo} = \hat{\omega}^{A}_{oj} = \hat{\omega}^{A}_{jo} = o$ (j=1,2,3) equivalent to the vanishing of the component $A_o(x)$ of the field operator. It is then given by

$$\hat{\omega}^{A}_{\mu_1\mu_2}(k) = -2\pi\left(g_{\mu_1\mu_2}\,\delta_{+}(k^2) + k_{\mu_1}\hat{\phi}_{\mu_2}(k) + k_{\mu_2}\hat{\phi}_{\mu_1}(k)\right) \qquad (20)$$

with

$$\hat{\phi}_o(k) = -\frac{\delta_+(k^2)}{2k^o}, \qquad \hat{\phi}_j(k) = -\frac{k_j}{2k^o}\,\hat{\phi}_o(k) \qquad (21)$$

(j=1,2,3)

Both distributions are well-defined as integrals over $\vec{k}$. Formula (20) can be rewritten in the more convenient standard form

$$\hat{\omega}^{A}_{jl}(k) = 2\pi\left(\delta_{lj} - \frac{k_j k_l}{|\vec{k}|^2}\right)\delta_{+}(k^2) \qquad (22)$$

for j,l=1,2,3, otherwise = o. One verifies easily that $(\,\cdot\,,\,\cdot\,)^{(1)}$ is positive definite, translation and rotation but not Lorentz invariant. $\mathcal{H}^{A}$ can be completed to a Hilbert space. The realization of $\mathcal{H}^{(1)A}$ by momentum "wave-functions" involves the choice of two transverse polarization directions for every $\vec{k}$. The field operator $A_{\mu}(x)$ satisfies the free wave equation $\Box A_{\mu} = o$ and also the Lorentz condition

$\partial^\mu A_\mu = o$, which because $A_o(x)=o$ implies $\vec{\nabla}\cdot\vec{A}=o$, the transversality of the field.

The isometric map W^{FA} from $\mathcal{H}^A_{ph} \subset \mathcal{H}^A$ onto $\mathcal{H}^F$, the state space of the $F_{\mu\nu}(x)$ field, is injective as discussed in sections 3 and 4. $\mathcal{H}^F$ can be identified with $\mathcal{H}^A_{ph}$ which is strictly smaller than $\mathcal{H}^A$ but has the same Hilbert space completion. The result is a single Hilbert space $\mathcal{H}$ as state space, with a single vacuum vector $\Omega^F = \Omega^A$. In $\mathcal{H}$ one has the vector field operator $A_\mu(x)$, defined on the dense domain $\mathcal{H}^A$ and the tensor field operator $F_{\mu\nu}(x)$ which is in this special case connected with $A_\mu(x)$ by the relation $F_{\mu\nu} = \partial_\mu A_\nu - \partial_\nu A_\mu$ and which is defined on the smaller dense domain $\mathcal{H}^F = \mathcal{H}^A_{ph}$. There is also in $\mathcal{H}$ a representation of the Poincaré group by unitary operators $U(a,\Lambda)$ that leave the vacuum vector and the dense domain $\mathcal{H}^F = \mathcal{H}^A_{ph}$ invariant and transform the field operator $F_{\mu\nu}(x)$ in the ordinary tensorial way. The domain $\mathcal{H}^A$ is however not invariant and neither does the operator $A_\mu(x)$ transform properly, this is the way the non-covariance of the $A_\mu(x)$ theory shows up.

This ends our brief review of the free photon field and its gauges. Although still other gauges are known and used, the discussion of these typical cases should be sufficient to make it clear how the often heuristic and unrelated results on the subject that can be found scattered in the literature can be derived in a unified and rigorous manner from the simple general theory of the quantized Maxwell Field of which an outline was given in section 3.

References

[1] Jost, R.: The general theory of quantized fields. Providence: American Mathematical Society 1965.

[2] Streater, R.G., Wightman, A.S.: PCT, spin and statistics and all that. New York: W.A. Benjamin, Inc. 1964

[3] Bogolubov, N.V., Logunov, A.A., Todorov, I.T.: Introduction to axiomatic quantum field theory. Reading: W.A. Benjamin, Inc. 1975.

[4] Bjorken, J.D., Drell, S.D.: Relativistic quantum fields. New York: Mc Graw-Hill, 1965.

[5] Strocchi, F., Wightman, A.S.: J. Math. Phys. 15, 2198 (1974).

[6] Fröhlich, J., Morchio, G., Strocchi, F.: Ann. Phys. 119, 241 (1979).

[7] Yngvason, J.: Rep. Math. Phys. 12, 57 (1977).

[8] Borchers, H.J.: Nuovo Cimento 24, 214 (1962).

[9] Borchers, H.J.: Lectures given at the Europhysics Conference on Statistical Mechanics and Field Theory, Haifa 1971. Jerusalem: Israel Universities Press 1972.

[10] Bongaarts, P.J.M.: J. Math. Phys. 18, 1510 (1977).

[11] Bongaarts, P.J.M.: Maxwell's Equations in Axiomatic Quantum Field Theory. II: Covariant and non-covariant gauges. Pre-print, Leiden 1981.

[12] Rideau, G.: Lett. in Math. Phys. 1, 17 (1975).

[13] Carey, A.L., Hurst, C.A.: Lett. in Math. Phys. 2, 227 (1978).